LINEAR CAUSAL MODELING WITH STRUCTURAL EQUATIONS

Aims and scope

Large and complex datasets are becoming prevalent in the social and behavioral sciences and statistical methods are crucial for the analysis and interpretation of such data. This series aims to capture new developments in statistical methodology with particular relevance to applications in the social and behavioral sciences. It seeks to promote appropriate use of statistical, econometric and psychometric methods in these applied sciences by publishing a broad range of reference works, textbooks and handbooks.

The scope of the series is wide, including applications of statistical methodology in sociology, psychology, economics, education, marketing research, political science, criminology, public policy, demography, survey methodology and official statistics. The titles included in the series are designed to appeal to applied statisticians, as well as students, researchers and practitioners from the above disciplines. The inclusion of real examples and case studies is therefore essential.

Published Titles

Analysis of Multivariate Social Science Data, Second Edition
David J. Bartholomew, Fiona Steele, Irini Moustaki, and Jane I. Galbraith

Bayesian Methods: A Social and Behavioral Sciences Approach, Second Edition
Jeff Gill

Foundations of Factor Analysis, Second Edition
Stanley A. Mulaik

Linear Causal Modeling with Structural Equations
Stanley A. Mulaik

Multiple Correspondence Analysis and Related Methods
Michael Greenacre and Jorg Blasius

Multivariable Modeling and Multivariate Analysis for the Behavioral Sciences
Brian S. Everitt

Statistical Test Theory for the Behavioral Sciences
Dato N. M. de Gruijter and Leo J. Th. van der Kamp

Chapman & Hall/CRC
Statistics in the Social and Behavioral Sciences Series

LINEAR CAUSAL MODELING WITH STRUCTURAL EQUATIONS

STANLEY A. MULAIK

CRC Press
Taylor & Francis Group
Boca Raton London New York

CRC Press is an imprint of the
Taylor & Francis Group, an **informa** business

A CHAPMAN & HALL BOOK

Chapman & Hall/CRC
Taylor & Francis Group
6000 Broken Sound Parkway NW, Suite 300
Boca Raton, FL 33487-2742

© 2009 by Taylor and Francis Group, LLC
Chapman & Hall/CRC is an imprint of Taylor & Francis Group, an Informa business

No claim to original U.S. Government works

Printed in the United States of America on acid-free paper
10 9 8 7 6 5 4 3 2 1

International Standard Book Number: 978-1-4398-0038-6 (Hardback)

Library of Congress Cataloging-in-Publication Data

Mulaik, Stanley A., 1935-
Linear causal modeling with structural equations / Stanley A. Mulaik.
p. cm. -- (Chapman & Hall/CRC statistics in the social and behavioral sciences series)
Includes bibliographical references and index.
ISBN 978-1-4398-0038-6 (hard back : alk. paper)
1. Social sciences--Mathematics. I. Title. II. Series.

H61.25.M84 2009
515'.63--dc22   2009015830

Visit the Taylor & Francis Web site at
http://www.taylorandfrancis.com

and the CRC Press Web site at
http://www.crcpress.com

Contents

Preface

This book is intended for quantitative methodologists and graduate students in methodology programs, but the book will be useful for researchers and graduate students in the behavioral and social sciences seeking a deeper understanding of causation, linear causal modeling, and structural equation modeling than is offered in other texts on the topic. Some sections, such as those found in the chapter on parameter estimation and polychoric correlation, will be mathematically challenging requiring a knowledge of calculus, but even there I try to accompany mathematical developments with verbal explanations and graphs showing what is occurring. These sections concern the kind of knowledge one needs to understand the inner workings of structural equation modeling programs, and to write these kinds of programs, of which quantitative methodologists should have some general knowledge.

Compared with common factor analysis, structural equation modeling is less intensive in the use of matrix algebra. There are no eigenvectors and eigenvalues. In fact, many commercial programs for structural equation modeling do not use matrices to input the models, but rather use individual equations, which are easy to comprehend by anyone with a background in multiple regression. Still, matrix algebra is a compact notation, and general model equations are still best expressed with matrices, which I use when appropriate. Perhaps more essential for the reader is a knowledge of college algebra. To be a scientist in almost any field requires some mastery of mathematics. In Chapter 2, I briefly lay out the basic ideas of college algebra, linear algebra, and calculus that the student might need later in the book.

Still this book is not all mathematical. Since it is about causation, and causation has been a topic of philosophers for over 2000 years, I attempt to sketch the history of the idea of causation in Chapter 3. (Although my PhD in 1963 was in clinical psychology, I had a postdoctoral fellowship in quantitative psychology at the University of North Carolina and made my career in that field, teaching factor analysis, multivariate analysis, psychometric theory, and structural equation modeling. But in dealing with empiricist critics of causal applications for structural equation modeling in the 1980s, I was led into the philosophy of causality and from there to seeking to master Immanuel Kant and Ludwig Wittgenstein. In the process I spent 10 years studying them and most modern philosophy intensively, especially the philosophy of science. I am a member of the Philosophy of Science Association and have published four articles in the Association's journal, *Philosophy of Science*, since 1985.) I seek to present a theory of causation involving variables that is central for structural equation modeling. I argue that causation is best understood as a functional relation between variables. I also introduce a conception of probabilistic causal modeling with functional relations between variables inspired by the remarks of Herbert Simon: Causes determine probability distributions. The constraints on the functional relation will arise in the links between causes and effects, found, say, in exchanges of conserved quantities in the case of physical causation. I also touch on recent developments in experimental psychology on studies of the perception of causation.

However, my current philosophy of science is naturalistic and cognitive in approach, which means I draw heavily on the cognitive sciences for understanding how the mind works to form concepts and knowledge. I call my philosophy of science *objective realism*. I am heavily indebted to the innovative ideas of the cognitive linguist George Lakoff and his philosopher co-author Mark Johnson on metaphor and the role it plays in abstract thought and concept formation. So, I recognize that many causal connections (along with most concepts) in both the physical and nonphysical sciences are metaphoric, which is not to say that they are fanciful or merely entertaining, but rather metaphors are the framework of concepts. This has led me to recognize that causality concerns objects, as a determining relation between their attributes. And a cognitive science view of objects (given, say, by J. J. Gibson, 1950, 1966) is that they are invariants in the perceptual field formed from the stimulus information. Objects furthermore are independent of the actions, perspectives, and changes in the perceptual field due to the actions and motions of the observer, information about which is also given to the observer in perception.

But Lakoff and Johnson (1999) argue that most metaphors underlying our concepts are taken from our embodied perceptual and motor experience as schemas that are mapped onto unfamiliar and abstract domains of experience to give us a structure for understanding those domains. From this it is understandable why many causal concepts are metaphors, for the causal schema of cause and effect is so minimal that it is easily used metaphorically to understand the effects of forces, substances, purposes, settings, and of intentions and goal-seeking behavior. It also suggests that causal relations are invariants whose invariance must be tested to establish their "objectivity."

This led me then to a realization that science itself functions with a basic metaphor, that "science is a knowledge of objects." Not all scientific "objects" are directly and literally perceived by the human observer. Often they are regarded as unobservables or latent entities or variables. Their status is often conceptual, representing metaphoric invariants that unify or synthesize diverse perceptual experiences recalled from memory or taken from graphically or even electronically recorded information. Their objectivity comes from successfully passing tests of hypotheses asserted of them as invariants. This explains why scientists demand replication in different laboratories, with different instruments, and different observers. It also explains why they seek to test hypotheses asserted as invariants with data not used in their formulation to make the results of the test not dependent on the researcher's data. To begin with, the researcher would have adjusted the hypothesis until it optimally fit the data. So comparing the hypothesis to the same data is then not a test of the hypothesis, since it cannot logically fail to fit optimally, and a test must have two possible outcomes, fail or pass. Second, hypotheses that may be adjusted to fit that data are not unique, so the hypothesis may be an idea or concept uniquely linked to the researcher—a subjective concept about the original data and not one with objective credentials obtained by passing a real test. The data with which to test the hypothesis must be logically independent of any data used to formulate the hypothesis. In other words, the hypothesis must add something not in the original data alone that allows the researcher to deduce something else potentially in the world that could not be deduced from the original data alone without the hypothesis. It is also the reason that scientists prefer more parsimonious theories and models, because they are more testable by asserting more invariants against the data.

Recent developments in computer science and among philosophers of science have led to graph theoretic analyses of causal relations. I refer to the works of Judea Pearl (2000) and Spirtes, Glymour, and Scheines (1993, 2000). In Chapter 4, I survey the basic concepts of graph theory that will be useful in the formulation of structural models: acyclic and cyclic graphs, the Markov condition, d-separation, the minimality condition, and faithfulness.

Chapter 5 concerns structural equation modeling itself. Here I introduce the student to path diagrams, structural coefficients, disturbances, exogenous variables, and endogenous variables. I then show how one can write a set of structural equations corresponding to the path diagram. Then I show two ways of computing variances and covariances of variables in a structural equation model: (1) the algebraic method and (2) the use of path tracing rules. The latter becomes very convenient. Finally, I introduce matrix equations for the general structural equation model, from which I derive the formulas for the variances and covariances among the observed variables as functions of the model parameters.

In Chapter 6, I consider the problem of identification of a model, which occurs in models that are incompletely specified, requiring the use of free parameters that must be estimated. The problem is whether unique solutions exist for the free parameters.

Chapter 7 concerns parameter estimation. I show how parameter estimates in identified models depend on the discrepancy functions to minimize. Then I get into the problem of finding solutions for the free parameters that minimize a discrepancy function. The mathematically challenged may want to skip this section, because it involves obtaining partial derivatives of the discrepancy function with respect to the respective free parameters and then setting the derivatives to zero and solving the resulting equations using nonlinear optimization algorithms. Since the illustration of these algorithms would be too complex with a real structural equation model, I instead use contour graphs for the solution of a simple nonlinear system of equations involving two variables to show how the algorithms work. I discuss the method of steepest descent and the quasi-Newton methods which are universally used in structural equation modeling programs. This material shows how structural equation programs work and how to develop the algorithms on which such programs are based.

Chapter 8 discusses issues involved in designing structural equation models. I consider the importance of thinking in terms of variables rather than labels for constructs when formulating models, to force the researcher to think concretely rather than abstractly. Next, I consider the values of using multiple indicators to establish the objectivity of latent constructs. But I also point out circumstances under which, even using multiple indicators, may be misleading in establishing correct inferences of causal relations among latent variables. Then I consider a four-step procedure for isolating, where lack of fit arises in a model using a nested sequence of models implicit in the structural equation model of interest. I also consider how one would test invariance of models across groups of subjects sampled from different populations. Finally, I end with a discussion of models with mean structures, followed by a discussion of multigroup comparisons of mean structures.

In Chapter 9 I introduce the application of confirmatory factor analysis, illustrating it with a first-order factor model. Next, I examine a multirater–multioccasion study that involves first- and second-order common factors and a correlation between second-order factors that represents an objective assessment of trait stability over a yearlong interval. I also consider multitrait–multirater models and their limitations arising from having only single indicators for a single trait–rater combination.

Chapter 10 introduces students to equivalent models: Models with the same constraints and free parameters in the same positions, but with reversals in causal direction, or replacement of causal paths with correlations among disturbances. The existence of equivalent models arises with respect to incompletely specified models with free parameters that allow for numerous distinct models with the same constraints to fit the covariance matrix equally well. The existence of equivalent models must be considered when seeking to infer the validity of a given model.

Chapter 11 introduces the use of instrumental variables to resolve issues of causal direction and mediated causation, while Chapter 12 discusses multi-level models in which subjects may be nested within a hierarchy of categories or classes. Variation will be due not simply to the causal variables explicit in the study, but due to variation in influences from the higher levels within which the subject is classed. I illustrate with a multilevel factor analysis model and with a multilevel path analysis model.

Chapter 13 concerns longitudinal modeling. I begin with a consideration of a simplex model and a bi-simplex model. Then I turn to latent curve models.

Chapter 14 takes up the case of nonrecursive models with loops. I introduce basic concepts and then do a brief survey of flow graph analysis to show how models may be reduced or paths condensed to produce equivalent models. I consider the effect nonrecursive models have on the variances and covariances among variables, and on the total effect of causes in such models, which must be analyzed somewhat differently from the way in which we do it with recursive, acyclic models. I introduce Mason's rule for finding the total effect of a cause on another variable in the model. One consequence is that correlations between variables in nonrecursive models can be quite misleading in suggesting the importance and influence of causes in such models. I also consider the problem of identifying nonrecursive models and the need for instrumental variables to accomplish this and consider questions of parameter estimation with nonrecursive models. I end with a discussion of applications for nonrecursive models.

Chapter 15 is the longest chapter in the book and concerns model evaluation. Models are evaluated on several dimensions: (1) What is the degree of fit of the reproduced model covariance matrix to the observed covariance matrix? Does the model have small discrepancy function values, nonsignificant chi-squares indicating fit to within sampling error, or high approximate fit indices or near-zero lack of fit indices of approximation? (2) To what degree has the model been specified with fixed parameters and equality constraints, leaving the model free to differ from the data, as indicated by the degrees of freedom of the model, if the model is incorrect? (3) To what extent do the degrees of freedom of the model approach the number of distinct data elements to be fit by the model, which indicates the proportion of distinct ways in which the model could fail to fit the data if it were wrong in some way? Each of these concerns is discussed as I take up the noncentrality parameter, the chi-square statistic, the Satorra–Bentler corrected chi-square, the goodness-of-fit indices of approximation such as LISREL$^©$'s GFI, Steiger's GFI, Bentler's CFI, the Tucker–Lewis Index, McDonald's μ, Steiger and Lind's RMSEA index of lack of fit, Mulaik's conversion of the RMSEA to a goodness-of-fit index by an exponential transformation named by Paul Dudgeon the ER index. Parsimony in model formulation is discussed as the fewness of estimated parameters relative to the number of data points to be fit by the model, expressed as a ratio. Mulaik's parsimony ratio is discussed as a way of evaluating the disconfirmability of a model, and how it may be combined with a goodness-of-fit index value. Information theoretic indices of fit such as the AIC, the BIC and Bozdogan's ICOMP, used for comparing models, are also discussed. I also discuss my reservation that these indices are not appropriate for a hypothesis-testing approach to model evaluation. Furthermore, I derive results that show that information theoretic indices in samples beyond some sample size will tend to prefer a less constrained model to a model that has a small discrepancy but is more constrained and in theory would prefer a saturated model. Information theoretic indices also seem not to take into account how free parameters make possible the existence of equivalent models and so they really do not consider parsimony or the importance of degrees of freedom at extremely large sample sizes to reducing the possibility of equivalent models. I show similar limitations for cross-validation indices involving two random samples from the same population. I then turn to discussing the Lagrange multiplier, Wald, and likelihood ratio tests and their derivation. (The derivation is mathematically challenging in parts.) I consider a related index used in LISREL, Sorbom's modification index of expected parameter change. Given an understanding of the Lagrange multiplier test in EQS and the modification index in LISREL, I consider the problems with post hoc modifications of the model under guidance of these tests and indices. Then I turn to developments reviewed by Ke-Hai Yuan (2005) concerning the problems of violations of distributional assumptions in using chi-square tests and indices of approximation and how they may be remedied. A final issue is criticism given of indices of approximation by some researchers. I seek a middle ground between them and the proponents of the indices of approximation; both kinds of indices can and should be used. Much of the heat in these discussions involves the proponents' use of different metaphors for assessing fit: distance in parameter space versus improbability as degree of difference and discounting their antagonists' metaphor.

Chapter 16 is the final chapter and concerns the polychoric and polyserial correlation coefficients and their derivation. Polychoric correlation concerns correlation between two variables that are polytomous, with multiple categories that nevertheless are regarded as ordered in some way. The assumption made is that there is an underlying bivariate normal distribution for these variables but the continuum of each variable has been divided into a finite number (usually small) of intervals and category scores assigned. The aim is to first estimate the interval boundaries, and then infer what the correlation coefficient would be for the underlying bivariate normal distribution under the constraint that the variances of the underlying continuous variables are fixed to unity. The polyserial correlation concerns the case when a continuous variable is paired with a polytomous variable. Again, the underlying correlation between corresponding continuous variables is sought. Since many variables in the social and behavioral sciences are polytomous, the use of these indices is now recommended for those cases.

This should be only the first book on linear causal modeling with structural equations for the advanced student in quantitative methods, as well as the researcher who seeks deeper understanding of these methods. There are now numerous books on special applications of structural equation modeling.

Acknowledgments

Numerous sources and persons have contributed to my acquisition of knowledge of causality and structural equation modeling. But those that specifically stand out in my mind are Karl Jöreskog's and Peter Bentler's papers in *Psychometrika*. I am ever grateful and indebted to my former student and now colleague Larry James for inviting me in 1981 to join him and Jeanne Brett in presenting a workshop on *Structural Equation Modeling* sponsored by Division 14 of the American Psychological Association and held at the Center for Creative Leadership. From that workshop we wrote a small book *Causal Analysis: Assumptions, Models, and Data* (James, Mulaik, and Brett, 1982), and that started me down the road of studying causality, the philosophy of science, and structural equation modeling. While preparing that book I was strongly influenced by the writings of Herbert Simon on causality and that influence continues in me today. I also wish to acknowledge my debt to Jack McArdle who, over hamburgers in a New York City café during an American Psychological Association Convention, explained path tracing rules to me on a paper napkin. There are also a number of professors and colleagues, influenced by the logical empiricists, whom I will not name, who stimulated me to find arguments to counter their rejection of the idea of causality in modern science.

None of this would have occurred had I not had the support of, first, Henry Kaiser in 1966 in getting me a postdoctoral fellowship at the L. L. Thurstone Psychometric Laboratory, at the University of North Carolina, and, second, of its Director, Lyle Jones, during three additional years as a member of the laboratory's faculty. That exposed me to leading statisticians, and gave me the opportunity to complete my first book *The Foundations of Factor Analysis*.

Writing that book led me to learn the mathematics needed to understand the algorithms of structural equation modeling and their derivation.

I also wish to acknowledge the loving support of my wife, Jane, who for 45 years has stood behind my writing efforts, not the least that of writing this book.

Author

Stanley A. Mulaik is a professor emeritus of the School of Psychology, Georgia Institute of Technology in Atlanta, Georgia. He received his PhD in 1963 in clinical psychology from the University of Utah, had a year postdoctoral fellowship in quantitative psychology at the University of North Carolina at Chapel Hill, and was on the faculty of the Psychometric Laboratory of the Department of Psychology at the University of North Carolina at Chapel Hill from 1967 to 1970. He joined the faculty of the School of Psychology at the Georgia Institute of Technology in 1970, and remained there until his retirement in 2000. He was president of the Society for Multivariate Experimental Psychology from 1991 to 1992, editor of *Multivariate Behavioral Research* from 1988 to 1995, and a fellow of Division 5 of the American Psychological Association. His special areas of interest are factor analysis, structural equation modeling, the philosophy of science and of causality, and individual differences. He has published books and numerous articles on factor analysis and linear causal modeling.

1

Introduction

The Rise of Structural Equation Modeling

During the first 60 years of the twentieth century, factor analysis was the dominant method in the behavioral and social sciences for representing causal relations between latent variables and observed variables. But in this regard the method is overly restrictive for representing the great variety of possible ways in which variables may be related to one another causally by linear functions and thereby account for the correlations among them.

Given any two variables A and B that have a nonzero correlation between them, A may be the cause of B, B may be the cause of A, or A and B may have other variables that are common causes of both. And even if they have a zero correlation, this may be because causes between them cancel. So, between pairs of variables, causality is indeterminate from the nature of their correlation. Nevertheless, common factor analysis arbitrarily assumes that if pairs of variables are correlated, it is due to the presence of latent common causes of them. Consequently, exploratory factor analysis goes off in search of these common causes. Sometimes, maybe even often, this is misleading.

Meanwhile, in 1921, a geneticist by the name of Sewell Wright (1921) developed a different approach to using correlations in connection with the idea of linear causal relations. Whereas up to that time correlations had been used in an exploratory fashion to discover which variables were related to other variables, Wright sought to infer what the correlations or pattern of correlations should be among a set of variables *if* they had certain specified linear causal relations between them. He then compared the inferred correlations among these variables with those observed among them as a way of testing his inference. His technique came implemented with *path diagrams* consisting of arrows between variables representing *causal paths*, and the method was dubbed "path analysis." As for statistical analysis, the method depended heavily on partial correlations.

In the 1930s, the economist John Maynard Keynes (1936) developed models of the economy using systems of simultaneous linear equations relating one set of variables to another set of variables. Some of the variables were inputs into the system and were not caused by any other variable in the system. These were *exogenous variables*. Other variables in the system were dependent on even other variables in the system. These were *endogenous variables*. Econometricians saw their problem as a logical extension of regression analysis, and used matrix algebra to represent the equations. They then sought various methods for estimating the parameters of the structural equations. But this led them to become aware of the identification problem. When does the system of equations have unique solutions for the parameters of the system? Sometimes it does not. Can we specify conditions in which the parameters are identified? So, whereas some worked on the identification problem and others worked on estimation methods such as two-stage and three-stage least squares, the method came to the attention of a few sociologists, who saw the connections between simultaneous linear equations and Wright's path analysis. They began formulating sociological theories using path analysis and simultaneous linear equations.

In the meantime, after Bock and Bargmann (1966) proposed a method for testing models of linear structural equations between variables with covariation among variables, known as "analysis of covariance structures," Karl Jöreskog (1969) proposed the general confirmatory factor analysis model and described the mathematical solution for the maximum-likelihood estimates of parameters of the model. He additionally helped develop a computer program to perform these estimates using the algorithm of Fletcher and Powell (1963), which he so successfully used earlier in developing an efficient program for performing maximum-likelihood exploratory factor analysis. Then Jöreskog (1970) saw the similarities of Bock and Bargman's (1966) model to the confirmatory factor analysis model and provided a full-information maximum-likelihood estimation algorithm for estimating the parameters of the analysis of covariance structures model. Shortly afterward Jöreskog collaborated with the well-known econometrician Arthur S. Goldberger and developed his own general model (Jöreskog, 1973) of simultaneous linear equations with latent variables (a new twist). Furthermore, he provided it with an algorithm and subsequent computer program for obtaining estimates of the parameters and chi-square tests for goodness of fit. The program was named LISREL for "linear structural relations." With this program he united the latent variable idea of common factor analysis with simultaneous structural equations. In the process, he initiated a methodological revolution in the behavioral sciences.

Whereas correlational methodology had been principally exploratory and hypothesis-developing, the LISREL program led researchers to think in terms of causal hypotheses and testing these hypotheses. Numerous articles began to appear using the methodology and program. A new journal, *Structural Equation Modeling*, devoted to the method was established in 1994.

Others have followed Jöreskog with programs of their own. Jöreskog's structural equation model was formulated in matrix algebra, and this carried over into the LISREL program, with the user being required to input his/her model in the form of matrices. The model also required the user to segregate variables according to whether they were dependent on exogenous or endogenous latent variables. It became difficult to formulate models in which a variable could depend on both exogenous and endogenous latent variables. Still tricks were developed, cumbersome though they were to implement, by which it could be done.

Bentler and Weeks (1982) proposed a different but equivalent way of modeling the variables in which it was not necessary to segregate the observed variables according to whether they were dependent on exogenous or endogenous latent variables. Any observed variable could be dependent on both exogenous and endogenous variables. In fact, the latents were not explicitly distinguished according to whether they were *exogenous* or *endogenous*. Instead, they were distinguished by being functions or not of other variables, that is, dependent or independent variables.

One then specified the model by specifying an equation for each dependent variable in which one indicated those variables that were direct causes of it. Both latent variables and observed variables could have an equation that indicated it was dependent on other latent variables and even on an observed variable. So whether variables were dependent variables or not depended on whether there was an equation indicating it was a function of other variables in the system. If there was no equation listing a variable otherwise found among the equations as a function of other variables, it was an independent variable.

Bentler then developed a program, EQS (Bentler, 1989, 1992, 1993, 1995), to implement his approach of specifying a model by specifying linear equations for each dependent variable in the system. Eventually, his programmers made this easy to do with a spreadsheet-like window in which one simply clicked on variables that the variable was dependent on. Subsequently, he and his programmers worked out a way in which one could specify the model by simply drawing the model as a path diagram in a program window in the computer monitor. Programmers of other commercial structural equation modeling programs adapted some of his ideas to their programs. So, today, besides LISREL and EQS there are the Amos program (Arbuckle and Wothke, 1999), Steiger's EZPath and SEPath programs, McDonald's COSAN, Neale's Mx, and Fox's SEM in R.

At the end of the twentieth century, computer scientists and philosophers of science made contributions to the theory of linear causal modeling.

Philosophers of science, Glymour et al. (1987), developed algorithms to implement Spearman's tetrad-differences criteria in looking for causal connections between variables. Pearl (2000) and Spirtes, Glymour, and Scheines (1993, 2000) developed causal theory in the context of graph theory, introducing new concepts into the field such as *directed acyclic graphs, Markov condition, collider, minimality condition, faithfulness,* and *d-separation,* among others. So, the field of structural equation modeling for linear causal relations is both widening and deepening conceptually, computationally, and statistically.

As of this writing interest has also shifted to specific applications of structural equation modeling, such as longitudinal and multilevel modeling.

In many applications of structural equation modeling with latent variables, a common factor model is embedded within the model, and we will begin studying examples of the structural equation model by considering the confirmatory factor analysis model as a special case.

An Example of Structural Equation Modeling

The following example of a structural equation model is taken from a study (Carlson and Mulaik, 1993) of how person descriptions may drive personality trait judgments and these in turn drive the ratings judges make of these persons. The subjects were 280 college students enrolled in undergraduate psychology courses at a major university in the southern United States. There were 177 male subjects and 103 female subjects. Each subject was presented with a written description of a person, and his or her task was then to rate the described person on 15 trait-rating scales. Unbeknownst to the raters, the trait-rating scales were deliberately chosen to represent three personality trait factors found in other personality trait-rating studies: *friendliness, capability,* and *outgoingness.* Each rating scale was a 7-point Likert-type scale anchored by opposite trait word pairs. The scales were similar to the semantic differential scales used by Osgood, Suci, and Tannenbaum (1957). The scales were assumed to be equal-interval scales with a true zero point midway between the anchors of the scales (Osgood et al., 1957). The order of the trait-word anchors was varied to eliminate any bias in favor of one end of the scale.

Previous studies of other subjects' ratings with these scales had led to a theory of how they were related to the *friendliness, capability,* and *outgoingness* factors. There were four indicator scales for *friendliness* (friendly–unfriendly, sympathetic–unsympathetic, kind–unkind, and affectionate–unaffectionate), four indicator scales were indicators for *capability* (intelligent–unintelligent, capable–incapable, competent–incompetent, and smart–stupid), and five indicator scales were indicators for *outgoingness* (sociable–unsociable, talkative–untalkative, outgoing–withdrawn, gregarious–ungregarious, and extraverted–introverted). Two other scales were hypothesized to be dependent on both *friendliness and capable* (helpful–unhelpful and cooperative–uncooperative).

Personality Descriptions as Variable Stimuli

Each rater was given a sheet on which were three paragraphs describing a male person in a work situation. Each paragraph represented the degree to which the described person acted in ways suggestive of one of the judgment dimensions. For example, the first paragraph described typical behaviors of the individual relating to his friendliness, kindness, sympathy, and affectionateness. The words chosen to describe these behaviors were selected from a set of words similar in meaning to the rating-scale words according to Roget's *Pocket Thesaurus* (Mawson, 1963). The second and third paragraphs were similarly constructed with respect to the indicators of *capability* and *outgoingness*.

To allow for a controlled variation of the stimulus paragraphs presented to the subjects, four versions of each paragraph were written before constructing the individual's description. The first version of the first paragraph described a person who was the friendliest, kindest, most sympathetic, and most affectionate, and this paragraph was assigned the value of 4. These behaviors deteriorated in degree from version to version, until the fourth version described a person who was hostile, cruel, unsympathetic, and unaffectionate. The fourth version was assigned the value of 1. The two intermediate versions were assigned values of 3 and 2, respectively. In this way, a stimulus variable was created and its values were assigned for the stimulus variable of *friendliness*. Four versions were similarly created for each of the other two judgment dimensions, thus creating stimulus variables for *capability* and *outgoingness*.

With three sets of four paragraphs from which to choose a paragraph to enter into a person's description, there was a possibility of 64 different personality descriptions. With only 280 subjects, there was concern that selecting the paragraphs totally at random would not yield a sufficiently full range of each stimulus dimension. So, 38 versions of the stimulus person were constructed from the pools of stimulus paragraphs. The four versions of each paragraph were approximately equally represented across the 38 versions. In approximately one-half of the versions, the levels were similar for the three paragraphs. In the other half, dissimilar levels of the three paragraphs were combined, yielding an inconsistent behavior pattern.

Each rater then received a randomly selected version of a stimulus person from the appropriate set of 38 stimulus-person versions. Within the set of 38 stimulus-person versions, each version was given one of the two digit numbers between 01 and 38. Then each rater was given one of the 38 descriptions chosen at random according to a random number generator designed to uniformly generate numbers at random between 01 and 38.

Then the rater read the person description and proceeded to make his or her ratings of the person on the 15 trait-rating scales. Each rater was then assigned a vector of scores representing the values of the three stimulus person variables of the person rated along with the 15 rated values of the 15 rating scale variables produced by the rater. So, each rater had 18 scores on 18 variables representing the values of the stimulus person rated and the ratings on the 15 rating scales. From these scores on each rater, a covariance matrix for the 18 variables was obtained, and this was the basis for estimating parameters and testing the fit of a reproduced covariance matrix based on the model to the data.

A path diagram summarizing the model of causal relations between the stimulus variables, the mediating judgment variables, and the resulting rating variables is shown in Figure 1.1. Observed variables are indicated by squares. Latent variables are indicated by circles. Single-headed arrows indicate causal pathways with causes proceeding in the direction of the arrows. Each arrow has corresponding to it a structural parameter that represents how much a unit change in the causal variable produces a change in the effect variable. Arrows in bold were given fixed values for their corresponding structural parameters, meaning they were prespecified by hypothesis and not changed during the iterations for estimating the remaining "free" parameters. The free parameters were estimated using maximum-likelihood estimation assuming that the dependent variables had approximately multivariate normal distributions.

Parameter estimates were determined conditional on any fixed or constrained parameter values in the model in such a way as to minimize any discrepancy between the reproduced variance–covariance matrix for the 18 variables and the observed sample covariance matrix for the same 18 variables.

Variances of independent variables are indicated by short two-headed arrow loops pointing to the same variable. Covariances between different independent variables are shown by curves with two-headed arrows connecting the variables in question. The covariance values are shown on the curves.

On dependent variables there will be a short arrow pointing to it representing a "disturbance variable," which is like a unique factor of factor analysis. The structural parameter of a disturbance variable is fixed at unity, but the variance of the disturbance is allowed to be free to be estimated.

What may be initially unusual to those already familiar with structural equation modeling are the three latent variables on the left with arrows from them pointing to the three squares of the observed variables to their left. Note that the arrows are bold and the loadings are also bold 1.00's. None of the three observed variables on the left has a disturbance arrow, or, if one wishes, each has a zero disturbance variance. This means that the three observed variables on the left and the three latent variables to their immediate right are the same variable. This was just a device for tricking LISREL into allowing an observed exogenous cause to be the cause of a latent endogenous variable.

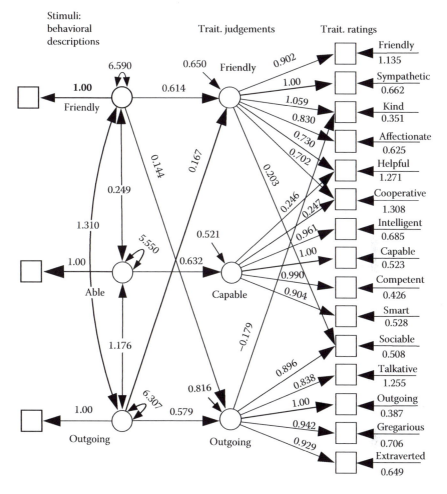

FIGURE 1.1 Path diagram of the hypothesized causal connections between the variables of the personality rating study of Carlson and Mulaik (1993). (Adapted from Carlson, M. and Mulaik, S. A. (1993). *Multivariate Behavioral Research, 28*, 111–159.)

Because of LISREL's origins in factor analysis, LISREL allows latent variables to be causes of other latent variables, but observed variables cannot be causes of latent variables unless you trick the program by creating the dummy latent variables standing for the observed stimulus variables on the left.

After obtaining estimates of the free parameters of the model, a reproduced covariance matrix based on this model was obtained and compared with the original sample covariance matrix. A chi-square goodness-of-fit index with 126 degrees of freedom had a value of 322.36, which was significant, indicating statistical lack of fit. An index of the degree of approximation of the reproduced covariance matrix to the observed sample covariance matrix was the comparative fit index (CFI) of Bentler (1990), and this had a value of 0.964. Researchers deemed that a model with a CFI greater than or equal to 0.95 was a "good fit." So, while there was a significant statistical lack of fit, meaning that somewhere in the model there was something wrong with it, the degree of approximation was high enough to consider tentatively that the model as specified was worth pursuing further in future research. Although there are diagnostics that give clues as to the possible sources of lack of fit, any hypotheses as to what in the world corresponds to these sources would have to be tested by further studies. One obvious source of potential minor lack of fit is that the polytomous variables only have approximate multivariate normal distributions and the chi-square statistic assumes that the variables have multivariate normal distributions.

Moreover, there was good reason to believe that the model showed that the person description variables actually drove the ratings via three latent judgment variables serving as mediators. In other words, stimulus variables did not directly influence the ratings; rather the effects of the stimulus variables on the ratings were mediated through the judgment variables. This conclusion is especially strengthened because the stimulus variables were manipulated by the experimenters and preceded in time the judgments and then the ratings.

Furthermore, the friendly stimulus variable principally influenced (0.614) the friendly judgment variable, which in turn mostly influenced its four indicators of friendly, sympathetic, kind, and affectionate. However, as expected the friendly judgment variable also influenced the helpful and cooperative ratings.

The friendly stimulus variable did not influence the capability judgment variable, as expected, but did influence slightly (0.144) the outgoing judgment variable, which otherwise was principally influenced by the outgoing stimulus variable. Furthermore, the friendly judgment variable had a modest (0.203) influence on one of the outgoing rating scales, sociable–unsociable, which otherwise was influenced principally (0.896) by the outgoing judgment variable.

On the other hand, the influence of the ability stimulus variable (0.632) was confined to the capability judgment variable, which in turn influenced only its indicators, the four ability rating scales, and, as expected, to a modest degree, the helpful and cooperative ratings. The outgoing stimulus variable principally influenced (0.579) the outgoing judgment variable, and to a modest degree (0.167) the friendly judgment variable, but to no degree the capable judgment variable. In turn the outgoing judgment variable had an unusual negative relationship (−0.179) to the kind–unkind rating scale, which suggests that kind persons are friendly and loving but not bubbling overextroverts. Otherwise, the outgoing judgment variable only influenced its own indicator rating scales.

It is important to understand that what is tested in a structural equation model are constraints placed upon the parameters. The estimated parameters are not tested by the chi-square goodness-of-fit statistic. It may be possible to formulate numerous different but equivalent models that reproduce the data in the same way, but all have the same constraints while varying directions of arrows in paths with free parameters. Furthermore, in early studies of a research program, you may not have values for structural parameters to indicate how causes produce their effects. You may simply be able to indicate that some variables are not causes of other variables by omitting arrows/paths from the former to the latter in the path diagram. But by specifying effectively that a path coefficient is zero, meaning there is no linear causal effect of a variable on another, you introduce a constraint in the model. So, it is important early on to recognize that what you may not see in a path diagram, that is, an omitted path, is what is being tested by the model. In later studies, after you have obtained values for structural parameters by estimating them, you may then use these values as fixed, constrained parameters of the model in later studies with new data, and then you will be testing these values in the new contexts.

One of the most important features of structural equation modeling is that it allows us to formulate complex models that are more realistic, involving numerous variables. Unlike factor analysis that always assumes that all variables are measured simultaneously, structural equation modeling may also model relations among variables that are not measured simultaneously while being ordered in time or in space. This occurs in studies of growth and development, or even the current example, where stimulus variables preceded the rating variables in time. And structural equation models can also represent all forms of linear causation; hence causes are not always latent common causes but may be direct effects of one variable on another. Next, structural equation modeling encourages the researcher to make explicit his/her ideas about what causes what and to what degree in the worldly phenomenon studied and to put these ideas to the test.

If there is a limitation, and there are several for structural equation modeling, it will be that researchers will tend not to attend to experimental controls when conducting their studies, which may be done in natural settings rather than in the laboratory. Much thought must be paid to controlling for possible extraneous causes influencing the outcomes of a study. Finally, another limitation is that structural equation modeling involves linear functional relations between variables. The physical world, as we know it, behaves according to nonlinear relations in many important cases. But science often models nonlinear monotonic relations with linear relations as "first approximations," and some causal relations are indeed linear relations. So, there is a place, an important place at this time, in the behavioral and social sciences for structural equation modeling.

2

Mathematical Foundations for Structural Equation Modeling

Introduction

Ideally, one begins a study of structural equations modeling with a mathematical background of up to a year of calculus. This is not to say that structural equations modeling requires an extensive knowledge of calculus, because calculus is used in only a few instances, such as in finding equations for estimates of parameters using a quasi-Newton algorithm or in finding formulas for polychoric correlation. These are very advanced topics, and since this text may also be used in courses for quantitative psychologists, they must be included. Those topics may be skipped over by the usual reader. But having calculus in one's background provides sufficient exposure to working with mathematical concepts so that one will have overcome reacting to a mathematical subject such as structural equation modeling as though it were an esoteric subject comprehensible only to select initiates to its mysteries. One will have learned those subjects such as trigonometry, college algebra, matrix algebra, and analytic geometry upon which structural equation modeling draws heavily.

In practice, however, the author recognizes that many students who now undertake a study of structural equation modeling come from the behavioral, social, and biological sciences, where mathematics is not greatly stressed. Consequently, in this chapter the author will attempt to provide a brief introduction to those mathematical topics from modern algebra, trigonometry, analytic geometry, and calculus, that will be necessary in the study of structural equation modeling. The author will also provide a background in operations with vectors and matrix algebra, which some students with just first-year calculus will find new to them. In this regard, most students will find themselves on even ground if they have a knowledge of college-level algebra, because operations with vectors and matrix algebra are extensions of algebra using a new notation.

Scalar Algebra

By the term scalar algebra we refer to the ordinary algebra applicable to real numbers, which the reader should already be familiar with. The use of this term distinguishes ordinary algebra from operations with vectors and matrix algebra, which we will take up shortly.

Fundamental laws of scalar algebra. The following laws govern the basic operations of scalar algebra such as addition, subtraction, multiplication, and division:

  (i) Closure law for addition: $a + b$ is a unique real number.

  (ii) Commutative law for addition: $a + b = b + a$.

  (iii) Associative law for addition: $(a + b) + c = a + (b + c)$.

  (iv) Closure law for multiplication: ab is a unique real number.

  (v) Commutative law for multiplication: $ab = ba$.

  (vi) Associative law for multiplication: $a(bc) = (ab)c$.

  (vii) Identity law for addition: There exists a number 0 such that $a + 0 = 0 + a = a$.

  (viii) Inverse law for addition: $a + (-a) = (-a) + a = 0$.

  (ix) Identity law for multiplication: There exists a number 1 such that $a1 = 1a = a$.

  (x) Inverse law for multiplication: $a(1/a) = (1/a)a = 1$.

  (xi) Distributive law: $a(b + c) = ab + ac$.

The above laws are sufficient for dealing with the real-number system. However, the special properties of zero should be pointed out:

$$\frac{0}{a} = 0.$$

$a/0$ is undefined.
$0/0$ is indeterminate.

Rules of signs. The rule of signs for multiplication is given as

$$a(-b) = -(ab) \quad \text{and} \quad (-a)(-b) = +(ab).$$

The rule of signs for division is given as

$$\frac{-a}{b} = \frac{a}{-b} \quad \text{and} \quad \frac{-a}{-b} = \frac{a}{b}.$$

The rule of signs for removing parentheses is given as

$$-(a - b) = -a + b \quad \text{and} \quad -(a + b) = -a - b.$$

Rules for exponents. If n is a positive integer, then x^n will stand for

$$xx \cdots x \text{ with } n \text{ terms.}$$

If $x^n = a$, then a is known as the nth root of a. The following rules govern the use of exponents:

(i) $x^a x^b = x^{a+b}$.

(ii) $(x^a)^b = x^{ab}$.

(iii) $(xy)^a = x^a y^a$.

(iv) $\left(\dfrac{x}{y}\right)^a = \dfrac{x^a}{y^a}$.

(v) $\dfrac{x^a}{x^b} = x^{a-b}$.

Solving simple equations. Let x stand for an unknown quantity, and let a, b, c, and d stand for known quantities. Then, given the following equation,

$$ax + b = cx + d,$$

the unknown quantity x can be found by applying operations to both sides of the equation until only an x remains on one side of the equation and the known quantities on the otherside. That is,

$$ax - cx + b = d \text{ (subtract } cx \text{ from both sides),}$$

$$ax - cx = d - b \text{ (subtract } b \text{ from both sides),}$$

$$(a - c)x = d - b \text{ (by reversing the distributive law),}$$

$$x = \frac{d - b}{a - c} \text{ (by dividing both sides by } a - c\text{).}$$

Vectors

It may be of some help to those who have not had much exposure to the concepts of modern algebra to learn that one of the essential aims of modern algebra is to classify mathematical systems—of which scalar algebra is an example—according to the abstract rules that govern these systems. To illustrate, a very simple mathematical system is a group. A group is a set of elements and a single operation for combining them (which in some cases is addition and in some others multiplication), behaving according to the following properties:

(i) If a and b are elements of the group, then $a + b$ and $b + a$ are also elements of the group, although not necessarily the same element (closure).

(ii) If a, b, and c are elements of the group, then $(a + b) + c = a + (b + c)$ (associative law).

(iii) There is an element 0 in the group such that for every a in the group $a + 0 = 0 + a = a$ (identity law).

(iv) For each a in the group there is an element $(-a)$ in the group such that $a + (-a) = (-a) + a = 0$ (inverse law).

One should realize that the nature of the elements in the group has no bearing upon the fact that the elements constitute a group. These elements may be integers, real numbers, vectors, matrices, or positions of an equilateral triangle; it makes no difference what they are as long as under the operator (+) they behave according to the properties of a group. Obviously, a group is a far simpler system than the system that scalar algebra exemplifies. Only four laws govern the group, whereas 11 laws are needed to govern the system exemplified by scalar algebra, which, by the way, is known to mathematicians as a field.

Among the various abstract mathematical systems, the system known as a vector space is the most important for factor analysis. (*Note:* One should not let the term space unduly influence one's concept of a vector space. A vector space need have no geometric connotations but may be treated entirely as an abstract system. Geometric representations of vectors are only particular examples of a vector space.) A vector space consists of two sets of mathematical objects—a set V of vectors and a set R of elements of a field (such as the scalars of scalar algebra)—together with two operations for combining them. The first operation is known as addition of vectors and has the following properties such that for every **u**, **v**, **w** in the set V of vectors:

(i) $\mathbf{u} + \mathbf{v}$ is also a uniquely defined vector in V.

(ii) $\mathbf{u} + (\mathbf{v} + \mathbf{w}) = (\mathbf{u} + \mathbf{v}) + \mathbf{w}$.

(iii) $\mathbf{u} + \mathbf{v} = (\mathbf{v} + \mathbf{w})$.

(iv) There exists a vector $\mathbf{0}$ in V such that $\mathbf{u} + \mathbf{0} = \mathbf{0} + \mathbf{u} = \mathbf{u}$.

(v) For each vector in V there exists a unique vector $-\mathbf{u}$ such that $\mathbf{u} + (-\mathbf{u}) = (-\mathbf{u}) + \mathbf{u} = \mathbf{0}$.

(One may observe that under addition of vectors the set V of vectors is a group.)

The second operation governs the combination of the elements of the field R of scalars with the elements of the set V of vectors and is known as scalar multiplication. Scalar multiplication has the following properties such that for all elements a, b from the field R of scalars and all vectors $\mathbf{u}$, $\mathbf{v}$ from the set V of vectors:

(vi) $a\mathbf{u}$ is a vector in V.

(vii) $a(\mathbf{u} + \mathbf{v}) = a\mathbf{u} + a\mathbf{v}$.

(viii) $(a + b)\mathbf{u} = a\mathbf{u} + b\mathbf{u}$.

(ix) $a(b\mathbf{u}) = ab(\mathbf{u})$.

(x) $1\mathbf{u} = \mathbf{u}$; $0\mathbf{u} = \mathbf{0}$.

In introducing the idea of a vector space as an abstract mathematical system, we have so far deliberately avoided considering what the objects known as vectors might be. Our purpose in doing so has been to have the reader realize at the outset that a vector space is an abstract system that may be found in connection with various kinds of mathematical objects. For example, the vectors of a vector space may be identified with the elements of any field such as the set of real numbers. In such a case the addition of vectors corresponds to the addition of elements in the field. In another example, the vectors of a vector space may be identified with n-tuples, which are ordered sets of real numbers. (In the upcoming discussion we will develop the properties of vectors more fully in connection with vectors represented by n-tuples.) Vectors may also be identified with the unidimensional random variables of mathematical statistics. This fact has important implications for multivariate statistics, and factor analysis and structural equation modeling in particular, because it means one may use the vector concept to unify the treatment of variables in both finite and infinite populations. The reader should take note at this point that the key idea in this book is that, in any linear analysis of variables of the behavioral, social, or biological sciences, the variables may be treated as if they are vectors in a linear vector space. The concrete representation of these variables may differ in various contexts (i.e., may be n-tuples or random variables), but they may always be considered vectors.

N-tuples as Vectors

In a vector space of n-tuples, by a vector we shall mean a point in n-dimensional space designated by an ordered set of numbers known as an

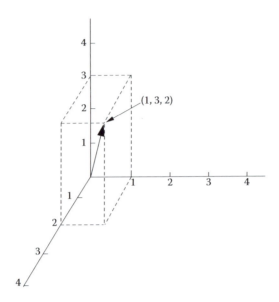

FIGURE 2.1   Graphical representation of vector (1, 3, 2) in three-dimensional space.

n-tuple, which are the coordinates of the point. For example, (1, 3, 2) is a 3-tuple that represents a vector in three-dimensional space that is 1 unit from the origin (of the coordinate system) in the direction along the x axis, 3 units from the origin in the direction of the y axis, and 2 units from the origin in the direction of the z axis. Note that in a set of coordinates the numbers appearing in special positions indicate the distance one must go along each of the reference axes to arrive at the point. In graphically portraying a vector, we shall use the convention of drawing an arrow from the origin of the coordinate system to the point. For example, the vector (1, 3, 2) is illustrated in Figure 2.1.

In a vector space of n-tuples, we will be concerned with certain operations to be applied to the n-tuples that define the vectors. These operations will define addition, subtraction, and multiplication in the vector space of n-tuples. As a notational shorthand to allow us to forgo writing the coordinate numbers in full when we wish to express the equations in vector notation, we will designate individual vectors by lowercase, bold letters. For example, let **a** stand for the vector (1, 3, 2).

Equality of Vectors

Two vectors are equal if they have the same coordinates.

For example, if **a** = (1, 2, 4) and **b** = (1, 2, 4), then **a** = **b**. A necessary condition that two vectors are equal is that they have the same number of coordinates. For example, if

$$\mathbf{a} = (1,2,3,4) \quad \text{and} \quad \mathbf{b} = (1,2,3),$$

then $\mathbf{a} \neq \mathbf{b}$. In fact, when two vectors have different numbers of coordinates, they refer to a different order of n-tuples and cannot be compared or added.

Scalars and Vectors

Vectors compose a system complete in themselves. But they are of a different order from the algebra we normally deal with when using real numbers. Sometimes we introduce real numbers into the system of vectors. When we do this, we call the real numbers scalars. In our notational scheme, we distinguish scalars from vectors by writing the scalars in italics and the vectors in lowercase bold characters. Thus $a \neq \mathbf{a}$.

In vector notation, we will most often consider vectors in the abstract. Then, rather than using actual numbers to stand for the coordinates of the vectors, we will use scalar quantities expressed algebraically. For example, let a general vector $\mathbf{a}$ in five-dimensional space be written as

$$\mathbf{a} = (a_1, a_2, a_3, a_4, a_5).$$

In this example, we use the character a to stand for the coordinates by the use of different subscripts. Whenever possible, algebraic expressions for the coordinates of vectors should take the same character as the character standing for the vector itself. This will not always be done however.

Multiplying a Vector by a Scalar

Let $\mathbf{a}$ be a vector such that

$$\mathbf{a} = (a_1, a_2, \ldots, a_n)$$

and λ a scalar; then the operation

$$\lambda \mathbf{a} = \mathbf{c}$$

produces another vector $\mathbf{c}$ such that

$$\mathbf{c} = (\lambda a_1, \lambda a_2, \ldots, \lambda a_n).$$

In other words, multiplying a vector by a scalar produces another vector that has for components the components of the first vector each multiplied by the scalar. To cite a numerical example, let $\mathbf{a} = (1,3,4,5)$; then

$$2\mathbf{a} = (2 \times 1, \ 2 \times 3, \ 2 \times 4, \ 2 \times 5) = (2,6,8,10) = \mathbf{c}.$$

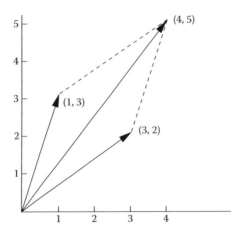

FIGURE 2.2 Graphical representation of sum of vectors (1, 3) and (3, 2) by vector (4, 3) in two-dimensional space, illustrating the parallelogram law for addition of vectors.

Addition of Vectors

If $\mathbf{a} = (1,3,2)$ and $\mathbf{b} = (2,1,5)$, then their sum, denoted as $\mathbf{a} + \mathbf{b}$, is another vector $\mathbf{c}$ such that

$$\mathbf{c} = \mathbf{a} + \mathbf{b} = ((1+2),(3+1),(2+5)) = (3,4,7).$$

When we add two vectors together, we add their corresponding coordinates together to obtain a new vector.

The addition of vectors is found in physics in connection with the analysis of forces acting upon a body where vector addition leads to the resultant by the well-known parallelogram law. This law states that if two vectors are added together, then lines drawn from the points of these vectors to the point of the vector produced by the addition will make a parallelogram with the original vectors. In Figure 2.2 we show the result of adding two vectors (1, 3) and (3, 2) together.

If more than two vectors are added together, the result is still another vector. In some factor-analytic procedures this vector is known as a centroid, because it tends to be at the center of the group of vectors added together.

Scalar Product of Vectors

In some vector spaces, in addition to the two operations of addition of vectors and scalar multiplication, a third operation known as the scalar product of two vectors (written as $\mathbf{xy}$ for each pair of vectors $\mathbf{x}$ and $\mathbf{y}$) is defined, which associates a scalar with each pair of vectors. This operation has the following abstract properties, given that **x**, **y**, and **z** are arbitrary vectors and a is an arbitrary scalar:

(xi) $\mathbf{xy} = \mathbf{yx} = a$ scalar.

(xii) $\mathbf{x(y + z)} = \mathbf{xy} + \mathbf{xz}$.

(xiii) $\mathbf{x}(a\mathbf{y}) = a(\mathbf{xy})$.

(xiv) $\mathbf{xx} \geq 0$; $\mathbf{xx} = 0$ implies $\mathbf{x} = \mathbf{0}$.

When the scalar product is defined in a vector space, the vector space is known as a unitary vector space. In a unitary vector space it is possible to establish the length of a vector as well as the cosine of the angle between pairs of vectors. Factor analysis as well as other multivariate linear analyses is concerned exclusively with unitary vector spaces.

We will now consider the definition of the scalar product for a vector space of n-tuples. Let **a** be the vector $(a_1, a_2, \ldots, a_n)$ and **b** the vector $(b_1, b_2, \ldots, b_n)$; then the vector product of **a** and **b**, written **ab**, is the sum of the products of the corresponding components of the vectors, that is,

$$\mathbf{ab} = a_1 b_1 + a_2 b_2 + \cdots + a_n b_n. \tag{2.1}$$

To use a simple numerical example, let $\mathbf{a} = (1, 2, 5)$ and $\mathbf{b} = (3, 3, 4)$; then $\mathbf{ab} = 29$.

As a further note on notation, consider that an expression such as Equation 2.1, containing a series of terms to be added together that differ only in their subscripts, can be shortened by using the summational notation. For example, Equation 2.1 can be rewritten as

$$\mathbf{ab} = \sum_{i=1}^{n} a_i b_i. \tag{2.2}$$

As an explanation of this notation, the expression $a_i b_i$ on the right of the sigma sign stands for a general term in the series of terms, as in Equation 2.1, to be added. The subscript i in the term $a_i b_i$ stands for a general subscript. The expression $\sum_{i=1}^{n}$ indicates that one must add together a series of subscripted terms. The expression $i = 1$ underneath the sigma sign indicates which subscript is pertinent to the summation governed by this summation sign—in the present example the subscript i is pertinent—as well as the first value in the series of terms that the subscript will take. The expression n above the sigma sign indicates the highest value that the subscript will take. Thus we are to add together all terms in the series, with the subscript i ranging in value from 1 to n.

Distance between Vectors

In high school geometry we learn from the Pythagorean theorem that the square on the hypotenuse of a right triangle is equal to the sum of the squares on the other two sides. This theorem forms the basis for finding the distance between two vector points. We shall define the distance between two vectors **a** and **b**, written as

$$|a - b| = \left[\sum_{i=1}^{n}(a_i - b_i)^2\right]^{1/2},$$ (2.3)

where **a** and **b** are both vectors with n components. This means that we find the sum of squared differences between the corresponding components of the two vectors and then take the square root of that. In Figure 2.3 we have diagrammed the geometric equivalent of this formula.

$$|a - b| = \sqrt{(a_1 - b_1)^2 + (a_2 - b_2)^2}$$

Length of a Vector

Using Equation (2.3), we can find an equation for the length of a vector, that is, the distance of its point from the origin of the coordinate system. If we define the zero vector as

$$0 = (0, 0, \ldots, 0),$$

then

$$|a| = \|a - 0\| = \left[\sum_{i-1}^{n}(a_i - 0)^2\right]^{1/2} = \left[\sum_{i=1}^{n}a_i^2\right]^{1/2}.$$ (2.4)

The length of a vector **a**, denoted $|a|$, is the square root of the sum of the squares of its components. (*Note*: Do not confuse $|a|$ with $|A|$, which is a determinant.)

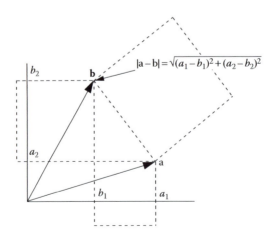

FIGURE 2.3 Graphical illustration of application of Pythagorean theorem for determination of the distance between two-dimensional vectors **a** and **b**.

Another Definition for Scalar Multiplication

Another way of expressing scalar multiplication of vectors is given by the formula

$$\mathbf{ab} = |\mathbf{a}|\,|\mathbf{b}|\cos\theta, \tag{2.5}$$

where θ is the angle between the vectors. In other words, the scalar product of one vector times another vector is equivalent to the product of the lengths of the two vectors times the cosine of the angle between them.

Cosine of the Angle between Vectors

Since we have raised the concept of the cosine of the angle between vectors, we should consider the meaning of the cosine function as well as other important trigonometric functions.

In Figure 2.4 there is a right triangle where θ is the value of the angle between the base of the triangle and the hypotenuse. If we designate the length of the side opposite the angle θ as a, the length of the base as b, and the length of the hypotenuse as c, the ratios of these sides to one another give the following trigonometric functions:

$$\tan\theta = \frac{a}{b},$$

$$\sin\theta = \frac{a}{c},$$

$$\cos\theta = \frac{b}{c}.$$

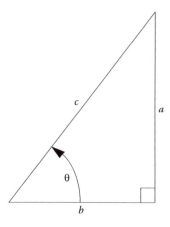

FIGURE 2.4   A right triangle with angle θ between base and hypotenuse.

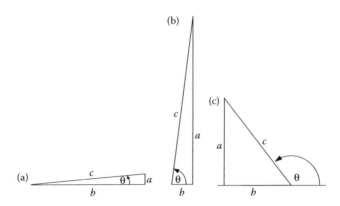

FIGURE 2.5   Illustrations of different right triangles.

Reflection on the matter will show that these ratios are completely determined by the value of the angle θ in the right triangle and are independent of the size of the triangle.

Of particular interest to us here is what happens to the value of cos θ when θ is close to 0° and when θ is close to 90°. In Figure 2.5a, we illustrate a triangle in which the angle θ is close to 0°. Hence the value of cos θ as the ratio of $b : c$ is close to 1.00.

In Figure 2.5b, we illustrate a triangle in which the angle θ is close to 90°. In this case the length b is quite small relative to the length of c. Hence the value of cos θ is close to 0.

In Figure 2.5c, we illustrate a triangle in which the angle is greater than 90°. In this case b is given a negative value, because the length of the base is measured from the origin of the angle in a direction opposite from the direction in which we normally measure the length of the base. By convention, the length of the hypotenuse is always positive. Thus the cosine of an angle between 90° and 180° is a negative value. Moreover, as the angle θ approaches 180°, cos θ approaches −1.

The cosine function thus serves as a useful index of relationship between vectors. When two vectors have a very small angle between them, the cosine of the angle between them is nearly 1. When the two vectors are 90° apart, nothing projects from one vector onto the other, and the cosine of the angle between them is 0. When the angle between them is between 90° and 180°, the cosine of the angle is negative, indicating that one vector points in the opposite direction, to some extent from the other.

We can use Equation 2.5 to find the cosine of the angle between two vectors in terms of their components. If we divide both sides of Equation 2.5 by the expression $|\mathbf{a}||\mathbf{b}|$, we have

$$\cos \theta = \frac{\mathbf{ab}}{|\mathbf{a}||\mathbf{b}|}. \tag{2.6}$$

But if we use Equations 2.2 and 2.4, Equation 2.6 can be rewritten as

$$\cos\theta = \frac{\sum_{i=1}^{n} a_i b_i}{\left[\left(\sum_{i=1}^{n} a_i^2\right)\left(\sum_{i=1}^{n} b_i^2\right)\right]^{1/2}}. \tag{2.7}$$

Equation (2.7) has a direct bearing on the formula for the correlation coefficient. To illustrate, let us construct an N-component vector $\mathbf{x}$ such that the components of this vector are the respective deviation scores of N individuals on a variable denoted by X. (A deviation score, by the way, is defined as the difference between an actual score and the mean of the scores of a variable. In other words, if X_i is the ith subject's actual score on variable X, then the ith subject's deviation score x_i equals $X_i - \bar{X}$, where $\bar{X}$ is the mean of scores on X.) Similarly, let us construct an N-component vector $\mathbf{y}$ such that the components of this vector are the respective deviation scores of the same N individuals on a variable Y. Then the correlation between variables X and Y is equivalent to the cosine of the angle between the vectors $\mathbf{x}$ and $\mathbf{y}$, that is,

$$r_{XY} = \frac{\sum_{i=1}^{n} x_i y_i}{\left[\left(\sum_{i=1}^{n} x_i^2\right)\left(\sum_{i=1}^{n} y_i^2\right)\right]^{1/2}}. \tag{2.8}$$

This formula is equivalent to one of the forms in which the sample correlation coefficient is given in most statistics texts.

Projection of a Vector onto Another Vector

In Figure 2.6, $\mathbf{x}$ and $\mathbf{y}$ are two vectors with an angle θ between them. The projection of $\mathbf{x}$ onto $\mathbf{y}$ is the vector $\mathbf{p}$ obtained by dropping a perpendicular from $\mathbf{x}$ to a line collinear with $\mathbf{y}$ and drawing a vector to that point. c and b

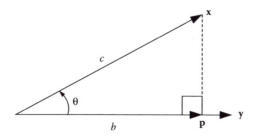

FIGURE 2.6 The vector **p** is the projection of the vector **x** onto the vector **y**. A line perpendicular to **y** is drawn from **x** to a line through **y**, and **p** is the vector collinear with **y** up to that point.

are the lengths of $\mathbf{x}$ and $\mathbf{p}$, respectively. Because $\cos\theta = b/c$, $b = c\cos\theta$ is the length of $\mathbf{p}$. Moreover,

$$c = \sqrt{\mathbf{x'x}} \quad \text{and} \quad \cos\theta = \frac{\mathbf{x'y}}{\sqrt{\mathbf{x'x}}\sqrt{\mathbf{y'y}}};$$

hence $b = \mathbf{x'y}/\sqrt{\mathbf{y'y}}$. The proportion of length of $\mathbf{y}$ due to length of $\mathbf{p}$ is $b/\sqrt{\mathbf{y'y}} = \mathbf{x'y}/\sqrt{\mathbf{y'y}}$. Projection of $\mathbf{x}$ on $\mathbf{y}$ is thus $\mathbf{p} = (\mathbf{x'y}/\sqrt{\mathbf{y'y}})\mathbf{y}$. The concept of projection is central to least-squares regression of one variable onto another. Here $\mathbf{p}$ is the predictable part of $\mathbf{y}$ due to $\mathbf{x}$.

Types of Special Vectors

The null vector $\mathbf{0}$ is the origin of the coordinate system in a space and is defined by

$$\mathbf{0} = (0, 0, \ldots, 0).$$

The sum vector $\mathbf{1} = (1, 1, \ldots, 1)$ is used when we wish to sum the components of another vector $\mathbf{a}$ as

$$\sum_{i=1}^{n} a_i = \mathbf{1a}.$$

Note that $\mathbf{1a} \neq \mathbf{1a}$.

We shall say that two vectors are orthogonal if their scalar product is zero. That is, if $\mathbf{a}$ and $\mathbf{b}$ are n-component vectors (neither of which is the null vector), then $\mathbf{a}$ is orthogonal to $\mathbf{b}$ if

$$\mathbf{ab} = \sum_{i=1}^{n} a_i b_i = 0.$$

It follows then from Equation 2.7 that any two vectors having for their scalar product the value zero must have the cosine of the angle between them equal to zero. Consequently, the angle between them must be $90°$.

Example   If $\mathbf{a} = (1, 3, 5)$ and $\mathbf{b} = (1, 3, -2)$, then

$$\mathbf{ab} = 1 + 9 - 10 = 0.$$

A normalized vector is a vector with a length of 1. Any given vector (other than the null vector) can be transformed into a normalized vector by dividing each of its components by the value of its length. In other words, if $\mathbf{v}$ is a vector, $\mathbf{v}$ can be transformed into a corresponding normalized vector $\mathbf{u}$ such that

$$\mathbf{u} = \frac{1}{|\mathbf{v}|}\mathbf{v}, \quad \text{where } |\mathbf{u}| = 1.$$

The unit vector is frequently used as a basis vector in a set of basis vectors from which all other vectors occupying the space can be derived as linear combinations. The ith unit vector in a space of n dimensions is denoted by e_i, where

$$e_1 = (1,0,0,\ldots,0),$$
$$e_2 = (0,1,0,\ldots,0),$$
$$e_3 = (0,0,1,\ldots,0),$$
$$\cdots\cdots\cdots\cdots$$
$$e_n = (0,0,0,\ldots,1).$$

Note that any two different unit vectors are orthogonal to one another, that is,

$$e_j e_k = 0, \quad j \neq k.$$

Linear Combinations

We shall say that a vector is a linear combination of some other vectors if it can be obtained as the sum of scalar multiples of the other vectors. That is to say, let $v_1, v_2, v_3, \ldots, v_n$ be vectors. Then the vector a is a linear combination of these vectors if

$$a = w_1 v_1 + w_2 v_2 + w_3 v_3 + \cdots + w_n v_n, \tag{2.9}$$

where $w_1, w_2, w_3, \ldots, w_n$ are scalars (numbers, not vectors). For example, if

$$a = (1,1), \quad v_1 = (2,3), \quad \text{and} \quad v_2 = (3,5),$$

it can be shown that

$$a = 2v_1 + (-1)v_2 = (4,6) - (3,5) = (1,1).$$

We can show that any vector in an n-dimensional space is a linear combination of the n unit vectors in that space. To illustrate, suppose we have a vector

$$a = (a_1, a_2)$$

and two other vectors $a_1 = (a_1, 0)$ and $a_2 = (0, a_2)$; then

$$a = a_1 + a_2 = (a_1, 0) + (0, a_2).$$

But $a_1 = a_1(1,0)$ and $a_2 = a_2(0,1)$.
Hence

$$a = (a_1, a_2) = a_1(1,0) + a_2(0,1) = a_1 e_1 + a_2 e_2.$$

Linear Independence

We shall say that a set of n vectors is linearly independent if no vector in that set can be derived as a linear combination from the rest of the vectors in the set. Conversely, if any vector in a set of n vectors is a linear combination of some of the other vectors in the set, then the set is linearly dependent.

The mathematical definition of linear dependence states that a set of vectors $\mathbf{v}_1, \mathbf{v}_2, \mathbf{v}_3, \ldots, \mathbf{v}_n$ is linearly dependent if some scalars $w_1, w_2, w_3, \ldots, w_n$ can be found (not all zero) such that

$$w_1\mathbf{v}_1 + w_2\mathbf{v}_2 + \cdots + w_n\mathbf{v}_n = 0. \tag{2.10}$$

If Equation 2.10 is true, we can derive any vector $\mathbf{v}_k$ in the set of vectors $\mathbf{v}_1, \mathbf{v}_2, \mathbf{v}_3, \ldots, \mathbf{v}_n$ as a linear combination of the rest. For proof, subtract the kth term $w_k\mathbf{v}_k$ from both sides of Equation 2.10. Then

$$w_1\mathbf{v}_1 + w_2\mathbf{v}_2 + \cdots + w_{k-1}\mathbf{v}_{k-1} + w_{k+1}\mathbf{v}_{k+1} + \cdots + w_n\mathbf{v}_n = -w_k\mathbf{v}_k$$

or

$$-\frac{w_1}{w_k}\mathbf{v}_1 - \frac{w_3}{w_k}\mathbf{v}_2 - \cdots - \frac{w_n}{w_k}\mathbf{v}_n = \mathbf{v}_k.$$

The significance of a linearly independent set is that it is a collection of vectors that contains no redundant information.

Basis Vectors

In a space of n dimensions, a linearly independent set of vectors occupying that space can contain at most n vectors. Such linearly independent sets of vectors, from which all other vectors in the space can be derived as linear combinations, are known as sets of basis vectors. There are an unlimited number of different sets of basis vectors that may be obtained for any n-dimensional space.

Matrix Algebra

Matrix algebra is in some ways a logical extension of n-tuple vector algebra. Whereas n-tuple vector algebra deals with the operations for manipulating individual vectors, matrix algebra deals with the operations for manipulating whole collections of n-tuple vectors simultaneously.

Definition of a Matrix

A matrix is defined as a rectangular array of numbers arranged into rows and columns. The following is the manner in which a matrix expressed algebraically is written in full:

$$\begin{bmatrix} a_{11} & a_{12} & a_{13} & a_{14} & a_{15} \\ a_{21} & a_{22} & a_{23} & a_{24} & a_{25} \\ a_{31} & a_{32} & a_{33} & a_{34} & a_{35} \\ a_{41} & a_{42} & a_{43} & a_{44} & a_{45} \end{bmatrix}.$$

This array represents a 4×5 matrix (4 rows by 5 columns). As a rule of notation, brackets [ ], parentheses ( ), or double lines $\| \ \|$ are used to enclose the rectangular array of numbers to designate it as a matrix. Remember that a matrix has no numerical value. Rather, it can be thought of as a collection of either row vectors or column vectors of numbers.

Because matrices are two-dimensional arrays of numbers, a system of double subscripts must be used to identify the elements in the matrix. According to convention among mathematicians, the first subscript (reading from left to right) designates the number of the row in which the element appears. The second subscript designates the column in which the element appears. Hence a_{23} designates an element in the second row and third column; a_{ij} refers to a general element appearing in the ith row and jth column.

The following arrays of numbers are matrices:

$$\begin{bmatrix} 1 & 2 \\ 3 & 1 \end{bmatrix} \quad \begin{bmatrix} 3 & 4 & 5 & 4 & 3 \\ 3 & 6 & 8 & 4 & 2 \\ 7 & 12 & 8 & 4 & 2 \end{bmatrix} \quad (4,5,6) \quad \begin{bmatrix} 2 \\ 5 \\ 7 \end{bmatrix}.$$

Other conventions that we will follow are the following:

Matrices will be denoted by uppercase, bold letters ($\mathbf{A}$, $\mathbf{B}$, etc.) and elements by lowercase, italicized letters (a_{ij}, b_{kj}, etc.). Usually the elements of a matrix take the same lowercase, italicized alphabetical letter as the uppercase letter standing for the matrix, but not always.

If we refer to a matrix as $\mathbf{A}$, this does not tell us how many rows or columns the matrix has. We must learn this from sources other than the notation before we start performing operations on the matrix. Usually this is indicated at the outset of a discussion of a matrix in the following manner: $\mathbf{A}$ is an $m \times n$ matrix . . . .

Another short way of denoting a matrix is with double lines and a single typical element, or with brackets and a single typical element:

$$\mathbf{A} = \|a_{ij}\|, \quad \mathbf{A} = [a_{ij}].$$

Matrix Operations

Just as with vectors, matrices also have operations of addition, subtraction, equality, and multiplication by a scalar and matrix multiplication.

Equality. Two matrices **A** and **B** are said to be equal, written **A** = **B**, if they have identical corresponding elements. That is, **A** = **B** only if $a_{ij} = b_{ij}$ for every i and j. Both **A** and **B** must have the same dimensions. If **A** is not equal to **B**, we write **A** ≠ **B**.

Multiplication by a scalar. Given a scalar v and a matrix **A**, the product v**A** is defined as

$$v\mathbf{A} = \begin{bmatrix} va_{11} & \cdots & va_{1n} \\ va_{21} & \cdots & va_{2n} \\ \cdots & \cdots & \cdots \\ va_{m1} & \cdots & va_{mn} \end{bmatrix}.$$

As with vectors, multiplying a scalar by a matrix implies multiplying the scalar by every element of the matrix. The order of multiplication does not matter so that

$$v\mathbf{A} = \mathbf{A}v.$$

Examples   If

$$v = 3 \quad \text{and} \quad \mathbf{A} = \begin{bmatrix} 2 & 3 \\ 1 & 0 \end{bmatrix}, \quad \text{then } v\mathbf{A} = \begin{bmatrix} 6 & 9 \\ 3 & 0 \end{bmatrix}.$$

If

$$v = -2 \quad \text{and} \quad \mathbf{A} = \begin{bmatrix} 1 & 3 & 2 \\ -2 & 1 & 3 \end{bmatrix}, \quad \text{then } v\mathbf{A} = \begin{bmatrix} -2 & -6 & -4 \\ 4 & -2 & -6 \end{bmatrix}.$$

Addition. The sum of two matrices **A** and **B**, both having the same number of rows and columns, is a matrix **C** whose elements are the sums of the corresponding elements in **A** and **B**. The sum is written as

$$\mathbf{A} + \mathbf{B} = \mathbf{C},$$

and for any element c_{ij} of **C**

$$c_{ij} = a_{ij} + b_{ij}.$$

(*Note:* If the two matrices **A** and **B** do not have the same number of rows and columns, then addition is undefined and cannot be carried out for them.)

Example   Let

$$\mathbf{A} = \begin{bmatrix} 3 & 2 & 1 \\ -1 & 3 & 4 \end{bmatrix}, \quad \mathbf{B} = \begin{bmatrix} 4 & -1 & 2 \\ 7 & 1 & 3 \end{bmatrix},$$

$$\mathbf{A} + \mathbf{B} = \begin{bmatrix} 7 & 1 & 3 \\ 6 & 4 & 7 \end{bmatrix}.$$

Subtraction. Subtraction is analogous to addition of matrices. In the case of the matrices **A** and **B** just defined,

$$\mathbf{A} - \mathbf{B} = \begin{bmatrix} -1 & 3 & -1 \\ -8 & 2 & 1 \end{bmatrix}.$$

Matrix multiplication. If **A** is an $n \times m$ matrix and **B** an $m \times p$ matrix, then the matrix multiplication of **A** times **B** produces another matrix **C** with n rows and p columns. This is written as

$$\mathbf{AB} = \mathbf{C}. \tag{2.11}$$

Perhaps the simplest way to understand how matrix multiplication works is to assume that the premultiplying matrix **A** consists of n row vectors and the postmultiplying matrix **B** consists of p column vectors. Then the matrix **C** is a rectangular array containing the scalar products of each row vector in **A** multiplied by each column vector in **B**. Each resulting scalar product is placed in the corresponding row of **A** and the corresponding column of **B** in **C**.

To illustrate, let us write Equation 2.11 in full, arranging the elements of **A** to represent n row vectors and the elements of **B** to represent p column vectors:

$$\begin{bmatrix} (a_{11} & a_{12} & \cdots & a_{1m}) \\ (a_{21} & a_{22} & \cdots & a_{2m}) \\ \cdots & \cdots & \cdots & \cdots \\ (a_{n1} & a_{n2} & \cdots & a_{nm}) \end{bmatrix} \begin{bmatrix} b_{11} \\ b_{21} \\ \vdots \\ b_{m1} \end{bmatrix} \begin{bmatrix} b_{12} \\ b_{22} \\ \vdots \\ b_{m2} \end{bmatrix} \cdots \begin{bmatrix} b_{1p} \\ b_{2p} \\ \vdots \\ b_{mp} \end{bmatrix}$$

$$= \begin{bmatrix} c_{11} & c_{12} & \cdots & c_{1p} \\ c_{21} & c_{22} & \cdots & c_{2p} \\ \cdots & \cdots & \cdots & \cdots \\ c_{n1} & c_{n2} & \cdots & c_{np} \end{bmatrix}.$$

To find the matrix **C**, the number of elements in the rows of **A** must equal the number of elements in the columns of **B**, to allow us to obtain the scalar products of the row vectors of **A** times the column vectors of **B**. In other words, the number of columns of **A** must equal the number of rows of **B**. **C** will have as many columns as **B**.

We begin with the first row vector of **A** and find the scalar product of this row vector times each of the column vectors of **B**. The resulting scalar products are placed in the first row and corresponding columns of **C**. Next, we take the second row of **A** and multiply it with each of the column vectors of **B**. Again the scalar products of these multiplications are placed in the second row of **C** in the position corresponding to the respective column of **B**. The process continues by multiplying a successive row vector of **A** times each of the column vectors of **B** and placing the resulting scalar product values in the respective row and columns of **C**. In general, if $\mathbf{a}'_i$ is the *i*th row vector in **A** and $\mathbf{b}_j$ is the *j*th column vector of **B**, then

$$c_{ij} = \mathbf{a}'_i \mathbf{b}_j = \sum_{k=1}^{m} a_{ik} b_{kj},$$

where c_{ij} is the element in the *i*th row and *j*th column of **C**. Defined in terms of the operation of the scalar product of vectors, matrix multiplication requires that the number of elements in the rows of the premultiplying matrix equals the number of elements in the columns of the postmultiplying matrix. This is necessary so that the components of the row and column vectors, respectively, will match in performing the scalar product of vectors.

Examples

(1) Let

$$\mathbf{A} = \begin{bmatrix} 1 & 3 & 2 & 1 \\ -1 & 2 & 3 & 5 \end{bmatrix} \quad \text{and} \quad \mathbf{B} = \begin{bmatrix} 1 & 3 & 4 \\ 2 & 1 & 2 \\ 3 & 4 & 5 \\ -2 & 1 & 2 \end{bmatrix};$$

then

$$\mathbf{AB} = \mathbf{C} = \begin{bmatrix} 11 & 15 & 22 \\ 2 & 16 & 25 \end{bmatrix}.$$

(2) Let

$$\mathbf{G} = \begin{bmatrix} 1 & 3 & 5 \\ 2 & 1 & -3 \end{bmatrix} \quad \text{and} \quad \mathbf{H} = \begin{bmatrix} 1 & 2 \\ 3 & 1 \\ -2 & 5 \end{bmatrix};$$

then

$$\mathbf{GH} = \begin{bmatrix} 0 & 30 \\ 11 & -10 \end{bmatrix}.$$

(3) Let

$$\mathbf{Q} = \begin{bmatrix} 1 \\ 5 \\ 7 \end{bmatrix} \quad \text{and} \quad \mathbf{QR} = \begin{bmatrix} 3 & 4 & 2 \\ 15 & 20 & 10 \\ 21 & 28 & 14 \end{bmatrix}.$$

Unlike in ordinary algebra, it does not always follow (in fact, it rarely follows) that $\mathbf{AB} = \mathbf{BA}$ in matrix algebra. This is because by reversing which matrix is the first matrix on the left, one may upset the match of row elements to column elements, in which case the necessary scalar product of vectors could not be found, leaving matrix multiplication undefined for the pair of matrices. This is seen in Example 1, where **BA** is not defined although **AB** is. Or the resulting matrix may be different in size if one reverses the order of multiplication (and multiplication is defined), which is seen in Example 2, where **GH** is a 2×2 matrix but **HG** is a 3×3 matrix, and in Example 3, where **QR** is a 3×3 matrix but **RQ** is a 1×1 matrix (or a single element).

However, the following expressions do hold:

$$(\mathbf{AB})\mathbf{C} = \mathbf{A}(\mathbf{BC}) = \mathbf{ABC} \text{ (associative law).}$$

$$\mathbf{A}(\mathbf{B} + \mathbf{C}) = \mathbf{AB} + \mathbf{AC} \text{ (distributive law).}$$

(*Note*: In manipulating matrix equations, students sometimes forget that the order of multiplication is important when they apply the distributive law.) For example,

$$\mathbf{A}(\mathbf{B} + \mathbf{C}) \neq \mathbf{BA} + \mathbf{CA}$$

or

$$(\mathbf{B} + \mathbf{C})\mathbf{A} \neq \mathbf{AB} + \mathbf{AC}.$$

Identity Matrix

The identity matrix is a special matrix that plays the role in matrix algebra that the number 1 plays in ordinary scalar algebra. The identity matrix is denoted by the symbol **I** or $\mathbf{I}_n$, where n is the order (size) of the identity matrix in question. It is a square matrix having as many rows as columns and with 1's running down its main diagonal from the upper left-hand corner to the lower right-hand corner. Zeros are found in every other position off the diagonal. That is,

$$\mathbf{I} = \begin{bmatrix} 1 & 0 & 0 & 0 & \cdots & 0 \\ 0 & 1 & 0 & 0 & \cdots & 0 \\ 0 & 0 & 1 & 0 & \cdots & 0 \\ 0 & 0 & 0 & 1 & \cdots & 0 \\ \vdots & \vdots & \vdots & \vdots & \ddots & \vdots \\ 0 & 0 & 0 & 0 & \cdots & 1 \end{bmatrix}.$$

Identity matrices come in different sizes: The second-order identity matrix is

$$\mathbf{I} = \mathbf{I}_2 = \begin{bmatrix} 1 & 0 \\ 0 & 1 \end{bmatrix}.$$

The identity matrix of order 3 is

$$\mathbf{I} = \mathbf{I}_3 = \begin{bmatrix} 1 & 0 & 0 \\ 0 & 1 & 0 \\ 0 & 0 & 1 \end{bmatrix}.$$

If $\mathbf{A}$ is a square matrix of order n and $\mathbf{I}_n$ the identity matrix of order n, then

$$\mathbf{I}_n \mathbf{A} = \mathbf{A}\mathbf{I}_n.$$

However, if $\mathbf{A}$ is an $n \times m$ matrix, then, although $\mathbf{A}\mathbf{I}_m$ is defined, $\mathbf{I}_m \mathbf{A}$ is not defined. However, $\mathbf{I}_n \mathbf{A}$ is defined. The result of these multiplications of $\mathbf{A}$ by the identity matrix yields

$$\mathbf{I}_n \mathbf{A} = \mathbf{A}\mathbf{I}_m = \mathbf{A}.$$

Scalar Matrix

Sometimes we wish to multiply every element in a matrix by a scalar number v. This can be accomplished by a square scalar matrix $\mathbf{S}$ such that

$$\mathbf{S} = v\mathbf{I}.$$

Then if $\mathbf{A}$ is a square matrix that commutes with $\mathbf{S}$,

$$\mathbf{S}\mathbf{A} = v\mathbf{A}.$$

Diagonal Matrix

A matrix with different values for its main diagonal elements but with zeros as the values of the off-diagonal elements is known as a diagonal matrix. $\mathbf{D}$ is often (but not always) used to designate a diagonal matrix.

Examples

$$\mathbf{D} = \begin{bmatrix} 3 & 0 & 0 \\ 0 & 2 & 0 \\ 0 & 0 & 1 \end{bmatrix}, \quad \mathbf{U} = \begin{bmatrix} 0.3 & 0 \\ 0 & 0.1 \end{bmatrix}.$$

Diagonal matrices have some interesting and useful properties. Suppose $\mathbf{D}$ is a diagonal matrix as just shown above and

$$\mathbf{A} = \begin{bmatrix} 4 & 1 & 3 & 2 \\ 5 & 2 & 6 & 7 \\ 1 & 3 & 8 & 4 \end{bmatrix}, \quad \mathbf{B} = \begin{bmatrix} -2 & 5 & 1 \\ 3 & 7 & 3 \\ 4 & 1 & 6 \\ 5 & -1 & 4 \end{bmatrix};$$

then

$$\mathbf{DA} = \begin{bmatrix} 3 & 0 & 0 \\ 0 & 2 & 0 \\ 0 & 0 & 1 \end{bmatrix} \begin{bmatrix} 4 & 1 & 3 & 2 \\ 5 & 2 & 6 & 7 \\ 1 & 3 & 8 & 4 \end{bmatrix} = \begin{bmatrix} 12 & 3 & 9 & 6 \\ 10 & 4 & 12 & 14 \\ 1 & 3 & 8 & 4 \end{bmatrix}$$

whereas

$$
\mathbf{BD} =
\begin{bmatrix}
-2 & 5 & 1 \\
3 & 7 & 3 \\
4 & 1 & 6 \\
5 & -1 & 4
\end{bmatrix}
\begin{bmatrix}
3 & 0 & 0 \\
0 & 2 & 0 \\
0 & 0 & 1
\end{bmatrix}
=
\begin{bmatrix}
-6 & 10 & 1 \\
9 & 14 & 3 \\
12 & 2 & 6 \\
15 & -2 & 4
\end{bmatrix}.
$$

Note that when $\mathbf{D}$ is the premultiplying matrix, as in $\mathbf{DA}$, its diagonal elements multiply respectively each of the elements in the corresponding row of postmultiplying $\mathbf{A}$. When $\mathbf{D}$ is the postmultiplying matrix, as in $\mathbf{BD}$, the diagonal elements of $\mathbf{D}$ are multiplied respectively with the elements of the corresponding columns of the premultiplying $\mathbf{B}$. Knowing these properties can save you from multiplying by a diagonal matrix in the usual, more complex way. By multiplying the above expressions in the usual way, you will obtain the same results.

Upper and Lower Triangular Matrices

An $n \times n$ square matrix some of whose elements on the principal diagonal and above the principal diagonal are nonzero, while all elements below the principal diagonal are zero is known as an upper triangular matrix.

An $n \times n$ square matrix whose nonzero elements are on the principal diagonal and below the principal diagonal, while those above are zero is a lower triangular matrix.

$$
\begin{bmatrix}
a_{11} & a_{12} & a_{13} & \cdots & a_{1n} \\
0 & a_{22} & a_{23} & \cdots & a_{2n} \\
0 & 0 & a_{33} & \cdots & a_{3n} \\
0 & 0 & 0 & \ddots & \vdots \\
0 & 0 & 0 & 0 & a_{nn}
\end{bmatrix}
\text{ is an upper triangular matrix.}
$$

$$
\begin{bmatrix}
a_{11} & 0 & 0 & 0 & 0 \\
a_{21} & a_{22} & 0 & 0 & 0 \\
a_{31} & a_{32} & a_{33} & 0 & 0 \\
\vdots & \vdots & \vdots & \ddots & 0 \\
a_{n1} & a_{n2} & a_{n3} & \cdots & a_{nn}
\end{bmatrix}
\text{ is a lower triangular matrix.}
$$

Null Matrix

A matrix of whatever dimensions having only zeros in its cells is known as a null matrix. It is denoted as $\mathbf{0}$. A null matrix has the following algebraic properties, provided that it commutes with the matrices in question:

$$
\mathbf{A} + \mathbf{0} = \mathbf{A}, \quad \mathbf{A} - \mathbf{A} = \mathbf{0}, \quad \mathbf{A0} = \mathbf{0}.
$$

However, if $\mathbf{AB} = \mathbf{0}$, this does not necessarily imply that either $\mathbf{A}$ or $\mathbf{B}$ is a null matrix. It is easy to find nonnull matrices $\mathbf{A}$, $\mathbf{B}$ whose product $\mathbf{AB} = \mathbf{0}$. This is an instance where matrix algebra is not like ordinary algebra.

Transpose Matrix

The transpose of a matrix $\mathbf{A}$ is another matrix designated $\mathbf{A}'$, which is formed from $\mathbf{A}$ by interchanging rows and columns of $\mathbf{A}$ so that row i becomes column i and column j becomes row j of $\mathbf{A}'$. That is, if

$$\mathbf{A} = \|a_{ij}\|, \quad \text{then } \mathbf{A}' = \|a_{ji}\|.$$

Examples

$$\mathbf{A} = \begin{bmatrix} 1 & 3 & 2 \\ 2 & 0 & 1 \end{bmatrix}, \quad \mathbf{A}' = \begin{bmatrix} 1 & 2 \\ 3 & 0 \\ 2 & 1 \end{bmatrix}.$$

$$\mathbf{B} = \begin{bmatrix} 1 & 3 \\ 2 & 4 \end{bmatrix}, \quad \mathbf{B}' = \begin{bmatrix} 1 & 2 \\ 3 & 4 \end{bmatrix}.$$

A way of visualizing transposing a matrix is to imagine that it is flipped over, using as a hinge the first diagonal elements of the original matrix going from left to right and down. For example, the elements 1 in the first row and 0 in the second row of $\mathbf{A}$ constitute the first diagonal of $\mathbf{A}$. The elements 1 and 4 in the first and second row are the principal diagonal of $\mathbf{B}$.

The transpose matrix $(\mathbf{AB})'$ equals $\mathbf{B}'\mathbf{A}'$, reversing the order of multiplication and then transposing the individual matrices. The pattern holds for any number of multiplied matrices

$$(\mathbf{ABCDE} \cdots \mathbf{XYZ}) = (\mathbf{Z}'\mathbf{Y}'\mathbf{X}' \cdots \mathbf{E}'\mathbf{D}'\mathbf{C}'\mathbf{B}'\mathbf{A}').$$

But there is no effect on summing matrices, except to transpose them individually:

$$(\mathbf{A} + \mathbf{B} + \mathbf{C})' = \mathbf{A}' + \mathbf{B}' + \mathbf{C}'.$$

Note that

$$\mathbf{I}' = \mathbf{I} \quad \text{and} \quad (\mathbf{A}')' = \mathbf{A}.$$

Symmetric Matrices

A symmetric matrix is a square matrix for which its transpose is equal to itself. It is a matrix in which any element a_{ij} equals the corresponding element a_{ji}. That is, let $\mathbf{R}$ be a square symmetric matrix; then

$$\mathbf{R}' = \mathbf{R}.$$

All covariance and intercorrelation matrices for a single set of variables are symmetric matrices.

Example

$$\mathbf{R} = \begin{bmatrix} 1.00 & -0.21 & 0.32 \\ -0.21 & 1.00 & 0.43 \\ 0.32 & 0.43 & 1.00 \end{bmatrix}.$$

Matrix Inverse

In scalar algebra, if $ab = 1$, then b is a special element with respect to a, that is, $1/a$, such that multiplying a by b yields 1. Is there an analogous relationship among matrices? If such a matrix exists for a given matrix $\mathbf{A}$, we shall call it the matrix inverse of $\mathbf{A}$. It will be denoted by $\mathbf{A}^{-1}$.

The following necessary (but not sufficient) condition must be met for the matrix $\mathbf{A}$ to have an inverse: the matrix must be square to have an inverse. However, not every square matrix has an inverse.

When the inverse matrix exists for a matrix, the following algebraic relationships hold:

$$(\mathbf{AB})^{-1} = \mathbf{B}^{-1}\mathbf{A}^{-1},$$

$$(\mathbf{A}^{-1})^{-1} = \mathbf{A},$$

$$(\mathbf{A}')^{-1} = (\mathbf{A}^{-1})',$$

$$(\mathbf{A}^{-1})'\mathbf{A}' = \mathbf{A}'(\mathbf{A}^{-1})' = \mathbf{I}.$$

Methods for finding the inverse of any given square matrix are too complicated to be given here. However, for matrices up to order 3, a method based on determinants can be used to find the matrix inverse, if it exists. Before going into this method, we should therefore take up the topic of determinants.

Orthogonal Matrices

A square matrix $\mathbf{P}$ is said to be orthogonal if

$$\mathbf{PP}' = \mathbf{P}'\mathbf{P} = \mathbf{I}.$$

Looking at this on the level of vector multiplication, we see that different column (or row) vectors of the matrix are orthogonal to one another, and the length of each column (row) vector is 1. It also follows that the inverse of an orthogonal matrix is its transpose.

Example

$$\mathbf{P} = \begin{bmatrix} 1/\sqrt{3} & 1/\sqrt{6} & -1/\sqrt{2} \\ 1/\sqrt{3} & 1/\sqrt{6} & 1/\sqrt{2} \\ 1/\sqrt{3} & -2/\sqrt{6} & 0 \end{bmatrix}, \quad \mathbf{P}' = \begin{bmatrix} 1/\sqrt{3} & 1/\sqrt{3} & 1/\sqrt{3} \\ 1/\sqrt{6} & 1/\sqrt{6} & -2/\sqrt{6} \\ -1/\sqrt{2} & 1/\sqrt{2} & 0 \end{bmatrix},$$

$$\mathbf{P}\mathbf{P}' = \begin{bmatrix} 1 & 0 & 0 \\ 0 & 1 & 0 \\ 0 & 0 & 1 \end{bmatrix}.$$

Orthogonal matrices will be important to exploratory factor analysis because maximum likelihood and other solutions obtain eigenvector matrices that are orthogonal matrices.

Sometimes an orthogonal matrix is called an orthonormal matrix because its column or row vectors have unit lengths and are mutually orthogonal.

Trace of a Matrix

Let

$$\mathbf{A} = \begin{bmatrix} a_{11} & a_{12} & a_{13} & a_{14} & a_{15} \\ a_{21} & a_{22} & a_{23} & a_{24} & a_{25} \\ a_{31} & a_{32} & a_{33} & a_{34} & a_{35} \\ a_{41} & a_{42} & a_{43} & a_{44} & a_{45} \\ a_{51} & a_{52} & a_{53} & a_{54} & a_{55} \end{bmatrix};$$

then

$$\text{tr}\mathbf{A} = a_{11} + a_{22} + a_{33} + a_{44} + a_{55}$$

is known as the trace of $\mathbf{A}$. The trace of any square matrix is the sum of its diagonal elements. The result is a scalar. In general,

$$\text{tr}\mathbf{A} = \sum_{i=1}^{n} a_{ii}.$$

Invariance of traces under cyclic permutations. Suppose the $n \times n$ square matrix $\mathbf{A}$ is itself the product of other matrices that are not all square: Given

$$\mathbf{A}_{n\times n} = \mathbf{F}_{n\times r}\mathbf{C}_{r\times r}\mathbf{F}'_{r\times n},$$

then

$$\text{tr}\mathbf{A}_{n\times n} = \text{tr}\mathbf{F}_{n\times r}\mathbf{C}_{r\times r}\mathbf{F}'_{r\times n} = \text{tr}\mathbf{F}'_{r\times n}\mathbf{F}_{n\times r}\mathbf{C}_{r\times r} = \text{tr}\mathbf{C}_{r\times r}\mathbf{F}'_{r\times n}\mathbf{F}_{n\times r}.$$

Although the products of the matrices in the equality above are all square matrices, they are not of the same dimension. The first product produces an $n \times n$ matrix, and the second and third produce $r \times r$ matrices. Nevertheless, each of their traces will be the same. Note that the square matrices are produced by cycling the rightmost matrix of a matrix product around to the leftmost position. This produces a different ordering or permutation of the component matrices, and hence cyclic permutations. In general, if **ABCDE** is square,

$$\text{tr}(\textbf{ABCDE}) = \text{tr}(\textbf{EABCD}) = \text{tr}(\textbf{DEABC}) = \text{tr}(\textbf{CDEAB}) = \text{tr}(\textbf{BCDEA}).$$

Traces are often used in defining in matrix notation a least-squares criterion to be optimized in least-squares estimation. For example, given **S** we seek a matrix $\Sigma = \Lambda\Phi\Lambda' + \Psi^2$ by varying the elements of the matrices Λ, Φ, Ψ^2 on the right, such that $\textbf{E} = \textbf{S} - \Sigma = \textbf{S} - (\Lambda\Phi\Lambda' + \Psi^2)$ and $\text{tr}\textbf{E}'\textbf{E}$ is a minimum. **E** represents the element-by-element difference between the matrix **S** and the matrix Σ. The diagonal elements of $\textbf{E}'\textbf{E}$ contain the column sum of squares of the elements of **E**, which are corresponding differences between elements of **S** and Σ. Then by summing the column sum of squares, $\text{tr}\textbf{E}'\textbf{E}$ contains the sum of all squared differences between elements of **S** and Σ. And the values for elements of Λ, Φ, Ψ^2, which make $\text{tr}\textbf{E}'\textbf{E}$ a minimum, are least-squares estimates of them.

Determinants

For any square matrix **A**, there exists a number uniquely determined by the elements of **A** known as the determinant of **A**. This number is designated $|\textbf{A}|$ but is also written as

$$|\textbf{A}| = \begin{vmatrix} a_{11} & a_{12} & \cdots & a_{1n} \\ a_{21} & a_{22} & \cdots & a_{2n} \\ \vdots & \vdots & \ddots & \vdots \\ a_{n1} & a_{n2} & \cdots & a_{nn} \end{vmatrix},$$

that is, by enclosing the matrix elements within vertical straight lines.

For a 1×1 matrix $\textbf{A} = a$, the determinant of **A** is simply a. For a 2×2 matrix **A**, the determinant of **A** is

$$|\textbf{A}| = \begin{vmatrix} a_{11} & a_{12} \\ a_{21} & a_{22} \end{vmatrix} = a_{11}a_{22} - a_{12}a_{21}.$$

For a 3×3 matrix $\mathbf{A}$, the determinant of $\mathbf{A}$ is written as

$$|\mathbf{A}| = \begin{vmatrix} a_{11} & a_{12} & a_{13} \\ a_{21} & a_{22} & a_{23} \\ a_{31} & a_{32} & a_{33} \end{vmatrix} = a_{11}a_{22}a_{33} + a_{12}a_{23}a_{31} + a_{13}a_{21}a_{32}$$

$$- a_{13}a_{22}a_{31} - a_{11}a_{23}a_{32} - a_{12}a_{21}a_{33}.$$

One computational device for finding determinants of 2×2 and 3×3 matrices is to copy all but the last column of the determinant to the right side of the determinant. Then draw arrows through the elements as shown. Then for each

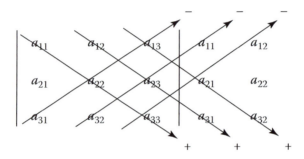

arrow, multiply together the terms through which the arrow passes. Multiply the product by $+1$ if the arrow points down, and multiply the product by -1 if the arrow points up. Add the products together, each multiplied by 1 or -1, respectively. (*Note:* This does not work with 4×4 matrices or larger.)

Unfortunately in factor analysis and structural equation modeling, the determinants are of matrices that are much larger. Usually they are determinants of the covariance or correlation matrix among a large number of variables subjected to analysis. So, we need a more general definition of a determinant to include large matrices.

The determinant of an nth-order $n \times n$ matrix $\mathbf{A}$ is defined by mathematicians as

$$|\mathbf{A}| = \sum (-1)^t a_{1i} a_{2h} a_{3r} \cdots a_{ns},$$

where the sum is taken across all $n!$ permutations of the right-hand subscripts, and t is the number of inversions or interchanges of pairs of adjacent elements in the product needed to bring the right-hand subscripts into ascending numerical order. In effect, the different product terms in the sum are obtained by selecting each element to enter the term from a different row and column of the matrix. (One should recognize that this does not denote the absolute value of a number because the symbol between the vertical lines is a matrix and not a scalar).

An interchange means exchanging the position of two adjacent numbers in a permutation of the integers $1, 2, 3, \ldots, n$, so that if one number is smaller than the other, it is placed before the larger number on the left in the resulting new permutation. For example, suppose $n = 6$, and we have the permutation

$$3\,1\,4\,2\,6\,5$$

of the first six integers; then an interchange would be

$$\mathbf{3\,1}\,4\,2\,6\,5 \longrightarrow \mathbf{1\,3}\,4\,2\,6\,5,$$

where the pair 3 1 is interchanged to become 1 3 because 1 comes before 3. Another interchange might then be to interchange 4 2 to 2 4, so we would have

$$1\,3\,\mathbf{4\,2}\,6\,5 \longrightarrow 1\,3\,\mathbf{2\,4}\,6\,5.$$

And even further, we would interchange 3 2 to become 2 3:

$$1\,\mathbf{3\,2}\,4\,6\,5 \longrightarrow 1\,\mathbf{2\,3}\,4\,6\,5.$$

And finally we could interchange 6 5 to 5 6:

$$1\,2\,3\,4\,\mathbf{6\,5} \longrightarrow 1\,2\,3\,4\,\mathbf{5\,6}.$$

Now, in the present example, we made four interchanges to bring the original permutation into ascending order. The quantity t in the formula for the determinant is the number of interchanges of the elements of the permutation needed to bring its right-hand subscripts into ascending order. If t is odd then $(-1)t$ is equal to -1, and if t is even then $(-1)t$ is equal to 1.

Although considerable mathematical research was conducted on determinants in the nineteenth and early twentieth centuries, their importance today has somewhat lessened. Many of their uses, in the theory of inverses, areas of parallelopipeds, in the concept of rank of a matrix, in the theory of multiple correlation, have been replaced by other concepts that allow one to achieve much the same thing as with determinants. However, they do play an important role in multivariate statistics, insofar as the determinant of a covariance matrix is found in the formula for the multivariate normal distribution, and the concept of a generalized variance is based on the determinant of the variance–covariance matrix of a set of variables. The concept of the rank of a matrix, of nonsingular and singular matrices, and tests for linear independence of variables also derive from the theory of determinants, and these are important concepts in factor analysis and structural equation modeling.

Nevertheless, in some cases the use of determinants leads to the simplest way of deriving some mathematical results. Computationally, the formula in the definition of a determinant is not a very efficient way of obtaining the value of a determinant. Alternative methods for finding the value of determinants usually involve transforming the matrix into an upper or lower triangular matrix or into a diagonal matrix, and then multiplying the diagonal elements of the resulting matrix together to yield the determinant. In factor analysis the correlation matrix is transformed into a diagonal matrix of eigenvalues, and then the eigenvalues are multiplied together to obtain the determinant of the correlation matrix.

Examples

$$\begin{vmatrix} 1 & 2 \\ 3 & 4 \end{vmatrix} = -2$$

$$\begin{vmatrix} 3 & 2 & 1 \\ 1 & 2 & 2 \\ 3 & 1 & 3 \end{vmatrix} = 13.$$

Minors of a Matrix

The determinant of an rth-order square submatrix of a square matrix $\mathbf{A}$ is known as a minor of the matrix $\mathbf{A}$. In the above examples, the individual elements of the two matrices are first-order minors, whereas in the case of the 3×3 matrix

$$\begin{vmatrix} 3 & 2 \\ 1 & 2 \end{vmatrix} \quad \begin{vmatrix} 1 & 2 \\ 3 & 3 \end{vmatrix} \quad \begin{vmatrix} 2 & 2 \\ 1 & 3 \end{vmatrix}$$

are second-order minors.

Rank of a Matrix

The rank of a matrix is the order of the highest order nonzero determinant obtainable from square submatrices of elements of the matrix obtained by deleting rows and/or columns of the matrix. For example, the matrix

$$\begin{bmatrix} 2 & 3 & 5 \\ 1 & 1 & 6 \\ 3 & 5 & 6 \\ 3 & 1 & 7 \end{bmatrix}$$

has at most a rank of 3 because the largest square submatrix that can be formed from this matrix by deleting rows and/or columns is a 3×3 matrix, and no determinants of larger order may be formed. It will have a rank of 3 if there exists a 3×3 submatrix of this matrix with a determinant not equal to zero.

If no 3×3 submatrix has a determinant not equal to zero, then the rank may be 2, and will be determined to be 2 if at least one 2×2 nonzero determinant can be formed by deleting rows and columns of the original matrix. And it will have a rank of 1 if at least one nonzero element exists in the matrix. The matrix will have rank zero if all elements in the matrix are zero, for then all 1×1 determinants will be zero.

The rank of a matrix indicates the maximum number of linearly independent rows (columns) of the matrix.

Thurstone (1947) used the concept of minimum rank of a correlation matrix whose diagonal elements had been replaced with communalities as the number of common factors to retain.

Cofactors of a Matrix

In a matrix $\mathbf{A}$, we will designate A_{ij} to be the cofactor of the element a_{ij}, found by deleting the ith row and jth column from the matrix $\mathbf{A}$, taking the determinant of the remaining submatrix of elements, and giving it the sign of $(-1)^{i+j}$.

Example

$$\mathbf{A} = \begin{bmatrix} a_{11} & a_{12} & a_{13} \\ a_{21} & a_{22} & a_{23} \\ a_{31} & a_{32} & a_{33} \end{bmatrix}, \quad A_{12} = - \begin{vmatrix} a_{21} & a_{23} \\ a_{31} & a_{33} \end{vmatrix}.$$

Expanding a Determinant by Cofactors

Let $\mathbf{A}$ be an $n \times n$ matrix. The determinant of $\mathbf{A}$ can be found by taking any row or column of the matrix and summing the products of the elements of that row or column, each multiplied by its corresponding cofactor. For example, if we choose the second row of the matrix $\mathbf{A}$ to find the determinant of $\mathbf{A}$, we have

$$|\mathbf{A}| = a_{21}A_{21} + a_{22}A_{22} + \cdots + a_{2n}A_{2n}.$$

Example

$$\mathbf{A} = \begin{bmatrix} 1 & 3 & 5 \\ 4 & 2 & 3 \\ 1 & 4 & 2 \end{bmatrix},$$

$$|\mathbf{A}| = -4 \begin{vmatrix} 3 & 5 \\ 4 & 2 \end{vmatrix} + 2 \begin{vmatrix} 1 & 5 \\ 1 & 2 \end{vmatrix} - 3 \begin{vmatrix} 1 & 3 \\ 1 & 4 \end{vmatrix}$$

$$= -4(6 - 20) + 2(2 - 5) - 3(4 - 3) = 47.$$

Adjoint Matrix

If $\mathbf{A}$ is an $n \times n$ matrix, the transpose of a matrix obtained from $\mathbf{A}$ by replacing its elements by their corresponding cofactors is known as the adjoint matrix for $\mathbf{A}$ and is designated by $\mathbf{A}^+$. For example, if

$$\mathbf{A} = \begin{bmatrix} a_{11} & a_{12} & a_{13} \\ a_{21} & a_{22} & a_{23} \\ a_{31} & a_{32} & a_{33} \end{bmatrix}, \quad \mathbf{A}^+ = \begin{bmatrix} A_{11} & A_{21} & A_{31} \\ A_{12} & A_{22} & A_{32} \\ A_{13} & A_{23} & A_{33} \end{bmatrix}.$$

The inverse of a matrix $\mathbf{A}$ can be obtained by multiplying its adjoint matrix $\mathbf{A}^+$ by the reciprocal of the determinant of $\mathbf{A}$, that is,

$$\mathbf{A}^{-1} = \frac{1}{|\mathbf{A}|} \mathbf{A}^+.$$

Example

$$\mathbf{A} = \begin{bmatrix} 1 & 3 & 5 \\ 4 & 2 & 3 \\ 1 & 4 & 2 \end{bmatrix}, \quad \mathbf{A}^{-1} = \frac{1}{47} \begin{bmatrix} -8 & 14 & 1 \\ -5 & -3 & 17 \\ 14 & -1 & -10 \end{bmatrix}.$$

For proof that the right-hand matrix is the inverse of the left-hand matrix, multiply the matrices together to see if the result is an identity matrix.

Important Properties of Determinants

1. The determinant of the transpose matrix equals the determinant of the original matrix, that is, $|\mathbf{A}'| = |\mathbf{A}|$.

2. Any theorem about $|\mathbf{A}|$ that is true for rows (columns) of a matrix $\mathbf{A}$ is also true for columns (rows).

3. If any two rows (columns) of the matrix $\mathbf{A}$ are interchanged, the determinant of the resulting matrix equals $-|\mathbf{A}|$.

4. Let $\mathbf{B}$ be a matrix formed from the matrix $\mathbf{A}$ by multiplying one of its rows (columns) by k; then $|\mathbf{B}| = k|\mathbf{A}|$.

5. Adding k times one row (column) of $\mathbf{A}$ to another row (column) of $\mathbf{A}$ produces a matrix whose determinant is still $|\mathbf{A}|$.

6. If two rows (columns) of a matrix $\mathbf{A}$ are identical, then $|\mathbf{A}| = 0$.

7. If any column of a matrix $\mathbf{A}$ is a linear combination of other columns of $\mathbf{A}$, then $|\mathbf{A}| = 0$.

8. If any row (column) of $\mathbf{A}$ consists of only zero's, then $|\mathbf{A}| = 0$.

9. If the determinant of a square matrix $\mathbf{A}$ is zero, we say the matrix is singular; otherwise we say that the matrix is nonsingular. Singular matrices have no inverses.

10. If **A** and **B** are $n \times n$ square matrices, then $|\mathbf{AB}| = |\mathbf{A}||\mathbf{B}|$. This result extends to the determinant of the product of any number of $n \times n$ matrices.

11. The determinant of a diagonal matrix equals the product of its diagonal elements, that is, $|\mathbf{D}| = \prod_{i=1}^{n} d_{ii}$.

12. The determinant of an upper (lower) triangular matrix equals the product of the elements in its principal diagonal, $|\Delta| = \prod_{i=1}^{n} \delta_{ii}$.

13. The determinant of an orthogonal matrix **P** equals ± 1.

14. $|\mathbf{A}^{-1}| = \dfrac{1}{|A|}$.

15. $|\mathbf{I}| = 1$.

16. $|\mathbf{0}| = 0$.

Simultaneous Linear Equations

Consider the following system of simultaneous linear equations in which there are n equations and n unknowns:

$$a_{11}x_1 + a_{12}x_2 + \cdots + a_{1n}x_n = b_1,$$
$$a_{21}x_1 + a_{22}x_2 + \cdots + a_{2n}x_n = b_2,$$
$$\cdots\cdots\cdots\cdots\cdots\cdots\cdots\cdots\cdots\cdots\cdots$$
$$a_{n1}x_1 + a_{n2}x_2 + \cdots + a_{nn}x_n = b_n.$$

This system can be expressed in matrix form as

$$
\begin{bmatrix}
a_{11} & a_{12} & \cdots & a_{1n} \\
a_{21} & a_{22} & \cdots & a_{2n} \\
\vdots & \vdots & \ddots & \vdots \\
a_{n1} & a_{n2} & \cdots & a_{nn}
\end{bmatrix}
\begin{bmatrix}
x_1 \\ x_2 \\ \vdots \\ x_n
\end{bmatrix}
=
\begin{bmatrix}
b_1 \\ b_2 \\ \vdots \\ b_n
\end{bmatrix},
\tag{2.12}
$$

or in equation form as

$$\mathbf{Ax} = \mathbf{b}, \tag{2.13}$$

where **A** is the square matrix of known coefficients, **x** is the n-component column vector of unknown values $x_1, x_2, \ldots, x_n$, and **b** is the n-component column vector of known quantities. The task is to solve for the unknown quantities $x_1, x_2, \ldots, x_n$ in terms of known quantities in **A** and **b**.

Although there is a way of solving for the unknown quantities using determinants, known as Cramer's rule, we will instead consider how we might do this using matrix algebra. This is more practical. Suppose we multiply both sides of Equation 2.13 by $\mathbf{A}^{-1}$, we obtain

$$\mathbf{A}^{-1}\mathbf{A}\mathbf{x} = \mathbf{A}^{-1}\mathbf{b},$$

$$\mathbf{x} = \mathbf{A}^{-1}\mathbf{b}. \tag{2.14}$$

In other words, the column vector can be solved for by premultiplying both sides of the equation by the inverse matrix of $\mathbf{A}$.

If $\mathbf{A}$ is singular, that is, has a 0 determinant, then no solution is possible, because there will be no inverse matrix.

Treatment of Variables as Vectors

We indicated in Section 2.3 that the key idea of this book is that variables may be treated as vectors. In this section we intend to show how this may be done.

There are several ways to define a variable. A common definition is that a variable is a quantity that may take any one of a set of possible values. Another definition is that a variable is a property on which individuals of a population differ. For the purposes of this book, however, we will use the following definition, with qualification: a variable is a functional relation that associates members of a first set (population) with members of a second set of ordered sets of real numbers in such a way that no member of the first set (population) is associated with more than one ordered set of real numbers at a time. This definition is quite general for it includes both the familiar case of the unidimensional variable (involving a single quantifiable property or attribute) and the case of the multidimensional variable (involving simultaneously several quantifiable properties or attributes). Because most theoretical work, however, deals with relations among individual attributes, we shall hereafter usually mean by the term variable a unidimensional variable that as a functional relation associates each member of a population with only one real number at a time.

There are several ways to classify variables. For example, statisticians frequently classify variables according to whether they are discrete or continuous. A discrete variable can take on at most a countably infinite number of values, whereas a continuous variable can take on an uncountably infinite number of values. (A set is said to have a countably infinite number of members if it is infinite and each member of the set can be put in a one-to-one correspondence with members of the set of integers.) Mathematical operations on discrete variables involve only algebra, whereas on continuous variables they involve calculus. Discrete variables may be represented by N-tuples but continuous variables must be represented in other ways. A major concern is not whether variables are discrete or continuous but whether they involve finite or infinite populations.

Variables in Finite Populations

Variables in finite populations are treated as vectors in a unitary vector space of N-tuples. For example, suppose we have a finite population of five individuals whose scores on a variable are 3, 1, 1, 4, and 2, respectively. Let us represent these scores by a 5-tuple $(3, 1, 1, 4, 2)$. Each coordinate of the 5-tuple corresponds to an individual member of the population and takes on the values of the variable for that individual.

When more than one variable is defined for the members of a finite population, we represent the variables by N-tuples arranged in the form of $N \times 1$ column vectors (with N the number of individuals in the population. When the variables are dealt with collectively (as in the case of finding linear combinations of the variables), we transpose the column vectors to row vectors and arrange them as the rows of an $n \times N$ matrix (with n the number of variables). For example, let $\mathbf{x}_1, \mathbf{x}_2, \ldots, \mathbf{x}_n$ be n N-tuple column vectors standing for n variables $X_1, X_2, \ldots, X_n$ defined in a finite population of N individuals. By $\mathbf{X}$ we mean the $n \times N$ matrix, whose row vectors are the transposed variable vectors $\mathbf{x}_1, \mathbf{x}_2, \ldots, \mathbf{x}_n$.

$$\mathbf{X} = \begin{bmatrix} \mathbf{x}'_1 \\ \mathbf{x}'_2 \\ \vdots \\ \mathbf{x}'_n \end{bmatrix} = \begin{bmatrix} x_{11} & x_{12} & \cdots & x_{1N} \\ x_{21} & x_{22} & \cdots & x_{2N} \\ \cdots & \cdots & \cdots & \cdots \\ x_{n1} & x_{n2} & \cdots & x_{nN} \end{bmatrix}. \tag{2.15}$$

Consider now the linear combination

$$\mathbf{y}' = \mathbf{a}'\mathbf{X} = (a_1, a_2, \ldots, a_n) \begin{bmatrix} \mathbf{x}'_1 \\ \mathbf{x}'_2 \\ \vdots \\ \mathbf{x}'_n \end{bmatrix} = (y_1, y_2, \ldots, y_N). \tag{2.16}$$

An important matrix in factor analysis and structural equation modeling would be the matrix of variances and covariances among these variables:

$$\mathbf{S}_{XX} = \frac{1}{N} \left[\mathbf{X}\mathbf{X}' - \frac{1}{N}\mathbf{X}\mathbf{1}\mathbf{1}'\mathbf{X}' \right]. \tag{2.17}$$

Here $\mathbf{1}$ is the column sum vector consisting of all 1's. $\mathbf{X}$ is defined in Equation 2.15. $\mathbf{X}\mathbf{X}'$ is a square symmetric $n \times n$ matrix of sums of squares and cross-products among the variables, summed across individuals.

$$\mathbf{X}\mathbf{X}' = \begin{bmatrix} \sum_{i=1}^{N} X_{1i}^2 & \sum_{i=1}^{N} X_{1i}X_{2i} & \cdots & \sum_{i=1}^{N} X_{1i}X_{ni} \\ \sum_{i=1}^{N} X_{2i}X_{1i} & \sum_{i=1}^{N} X_{2i}^2 & \cdots & \sum_{i=1}^{N} X_{2i}X_{ni} \\ \vdots & \vdots & \ddots & \vdots \\ \sum_{i=1}^{N} X_{ni}X_{1i} & \sum_{i=1}^{N} X_{ni}X_{2i} & & \sum_{i=1}^{N} X_{ni}^2 \end{bmatrix}.$$

X1 is an $n \times 1$ column vector of sums of scores in each respective row of **X**. The column of variable means would be given by

$$\bar{\mathbf{X}} = \frac{1}{N}\mathbf{X1} = \frac{1}{N}\begin{bmatrix} x_{11} & x_{12} & \cdots & x_{1N} \\ x_{21} & x_{22} & \cdots & x_{2N} \\ \cdots & \cdots & \cdots & \cdots \\ x_{n1} & x_{n2} & \cdots & x_{nN} \end{bmatrix}\begin{bmatrix} 1 \\ 1 \\ \vdots \\ 1 \end{bmatrix} = \frac{1}{N}\begin{bmatrix} \sum_{i=1}^{N} X_{1i} \\ \sum_{i=1}^{N} X_{2i} \\ \vdots \\ \sum_{i=1}^{N} X_{ni} \end{bmatrix} = \begin{bmatrix} \bar{X}_1 \\ \bar{X}_2 \\ \vdots \\ \bar{X}_n \end{bmatrix}.$$

Another important matrix derivable from the matrix **X** is the symmetric $n \times n$ matrix $\mathbf{R}_{XX}$ containing the intercorrelation coefficients for pairs of variables in **X**. To obtain $\mathbf{R}_{XX}$ let $\mathbf{D}_X^2 = [\mathrm{diag}\mathbf{S}_{XX}]$. Here $[\mathrm{diag}\mathbf{S}_{XX}]$ means extract the diagonal elements of $\mathbf{S}_{XX}$ and place them in the principal diagonal of a diagonal matrix with zero off-diagonal elements. Then

$$\mathbf{R}_{XX} = \mathbf{D}_X^{-1}\mathbf{S}_{XX}\mathbf{D}_X^{-1}. \tag{2.18}$$

The matrix $\mathbf{D}_X^2$ has as its diagonal elements the variances of the respective variables in **X**.

Variables in Infinite Populations

Variables defined for a countably infinite population may be represented by N-tuples with infinitely many coordinates; however, not all operations defined on N-tuples with a finite number of coordinates apply to N-tuples with infinitely many coordinates. In particular, the definition given for the scalar product of N-tuples of finite order does not work on n-tuples of infinite order, because in the case of N-tuples of infinite order sums of infinitely many product terms are involved, and these normally do not converge to finite scalar values. Because of this difficulty and because of the additional difficulty that variables in uncountably infinite populations are not expressible as N-tuples, other notations have been sought to represent variables in infinite populations.

Mathematical statisticians have developed a useful notation for dealing with variables defined on infinite populations. They let a capital letter stand for a variable and use a corresponding lowercase letter to stand for a particular value of the variable. For example, X is a variable, and x is some particular value of the variable X. Mathematical statisticians have also adopted the convention of combining variables in this notation as if variables were elements of a scalar algebra. For example, U, V, W, X, Y, and Z are variables defined on a population; then the expression $U = V/W + XY - Z^2$ is a permissible expression and means that if u, v, w, x, y, and z are the respective real values of these variables for any arbitrary member of the population, then $u = v/w + xy - z^2$. Thus there is a 1 : 1 correspondence between expressions involving variables and expressions involving scalars, which reveals that variables form a field under the operations of addition and multiplication. An immediate consequence of this fact is that variables may serve as vectors in a vector space, with the operation of addition of vectors in the vector space corresponding to the operation of addition of variables in the field of variables.

However, a vector space of variables is not a unitary vector space—which we require for multivariate linear analysis—unless we can find a way to define for the vector space an operation having the properties of the scalar product of vectors. The simple product XY of two variables X and Y does not satisfy the requirements of a scalar product because the result is a variable and not a scalar. However, we might consider the possibility that the simple product XY is part of but not all the solution to the requirements for a scalar product. What we need to look for is an operation that takes a random variable and produces from it a unique number known as its scalar product. Since XY is a random variable representing a unique combination of the variables X and Y, applying this operation to XY will produce a unique scalar corresponding to this pair.

The operation we have in mind is the expected value operator, denoted by $E(\ )$, that obtains the mean of a random variable. Applied to a discrete variable (a variable that takes on only a finite number of distinct values), this operator resembles in many ways the operation for finding the mean of a variable for a finite population (or sample) by the method of grouping of individuals by scores. In the latter method, for each value on the discrete variable X, one obtains the proportion $p(x_i) = f_i/N$ of individuals in the finite population with value x_i on the discrete variable, where f_i is the number of individuals with value x_i and N is the number of individuals in the population (or sample). Then the mean is given as $E(X) = \sum_{i=1}^{m} x_i p(x_i)$, with m the number of possible values of X in the population.

Regardless, in the general discrete case with countably infinite populations, for each value x_i mathematical statisticians assume that there exists a non-negative number $p(x_i)$, which, roughly speaking, represents the proportion of the total population with value x_i of the variable X. Then the expected value of X, denoted $E(X)$, is given as

$$E(X) = \sum x_i p(x_i) \quad \text{for all } p(x_i) > 0. \tag{2.19}$$

In the case of an uncountably infinite population, it is not possible to obtain the proportions of the total population by a process of counting. In fact, if a value x_i on a variable X has only a finite or countably infinite individuals in an uncountably infinite population associated with it, then it still has, paradoxically, a zero proportion of the total population associated with it. We will not attempt to explain the basis for this paradox since it is based on one of the subtle features of measure theory, which is beyond the scope of this book (*cf.* Singh, 1959).

The expected value of a continuous random variable (a variable that can take on any of the values in a continuum of values on any interval or intervals of the real numbers) has to be defined differently. Given the continuous variable X, it is postulated that there exists a function $f(x)$, known as the density function of X, that allows one to determine the proportion $P(X \mid a < x \le b)$ of the total population with values x in any continuous interval $a < x \le b$ over which the variable is defined. Since this is a density function, the total area under the curve $f(x)$ equals unity. If one refers to the graph of the function $f(x)$ in Figure 2.7, the proportion $P(X \mid a < x \le b)$ corresponds to the area under the curve of the function $f(x)$ in the interval $a < x < b$. This area may be obtained by the integral calculus from the integral equation

$$P(X \mid a < x \le b) = \int_a^b f(x)\,dx.$$

The expected value of the continuous variable X is given as

$$E(x) = \int_a^b xf(x)\,dx. \tag{2.20}$$

Returning now to the problem of defining the scalar product **xy** of two vectors **x** and **y** representing two random variables X and Y defined on an infinite population, it would be possible to define

$$\mathbf{xy} = E(XY) \quad \text{and} \quad |\mathbf{x}| = \sqrt{E(X^2)}.$$

If X and Y are both discrete variables, then

$$E(XY) = E(XY) = \sum x_i y_i p(x_i, y_i) \quad \text{over all } p(x_i, y_i) > 0,$$

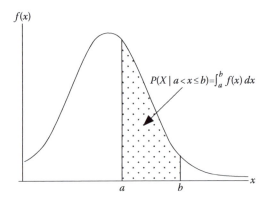

FIGURE 2.7 The probability that random variable X takes on values greater than a or less than or equal to b is given by the area under density function $f(x)$ between a and b on the x axis.

where $p(x_i, y_i) > 0$ is the probability in the population of having simultaneously the values x_i and y_i on the variables X and Y, respectively. If X and Y are continuous random variables, then

$$E(XY) = \int_{-\infty}^{+\infty} \int_{-\infty}^{+\infty} xy f(x, y) \, dx \, dy,$$

where $f(x,y)$ is the joint density function of X and Y.

But $E(XY)$ and $\sqrt{E(XX)} = \sqrt{E(X^2)}$ are not pivotal concepts in multivariate statistics. They do become pivotal if we consider

$$\mathbf{xy} = \text{cov}(X, Y) = E(XY) - E(X)E(Y) = E[(X - E(X))(Y - E(Y))],$$

where $\text{cov}(X, Y)$ denotes the covariance between variables X and Y. Here X and Y have their means subtracted from them before obtaining the products. The resulting transformed variables $X^* = X - E(X)$ and $Y^* = Y - E(Y)$ now have zero means, because $E(X^*) = E[X - E(X)] = E(X) - E(X) = 0$ and the same for $\hat{E}E(Y^*)$. Then $\text{cov}(X^*, Y^*) = E(X^*Y^*) - E(X^*)E(Y^*) = E(X^*Y^*)$. This suggests that without loss of generality in those cases involving linear algebra, we may assume that all variables under consideration have zero means. In much of this book, we will make this assumption (there will be exceptions and these will be noted). Consequently, we will assume that for variables having zero means $\mathbf{xy} = E(XY) = \text{cov}(X, Y)$ and

$$|\mathbf{x}| = \sqrt{\mathbf{xx}} = \sqrt{E(XX)} = \sqrt{E(X^2)} = \sqrt{\text{var}(X)}, \tag{2.21}$$

where $|\mathbf{x}|$ denotes the length of the vector $\mathbf{x}$ corresponding to the variable X, and $\text{var}(X)$ denotes the variance of the variable X. So, the length of a vector corresponding to a random variable is the standard deviation of the variable.

The point of all this discussion of how to define a scalar product between vectors representing random variables is to establish that most multivariate statistics implicitly assumes that the linear algebra of unitary vector spaces applies to its topics. Although the notation may not explicitly reveal the vectors involved, they are there in the random variables under operations of addition and scalar multiplication and the expected value of the product of two random variables.

Random Vectors of Random Variables

Mathematical statisticians do not readily acknowledge that the random variables they pack into what they call random vectors are themselves vectors. Nor do they explicitly indicate that much of multivariate statistics with random vectors of random variables is an application of linear algebra. But when they use their random-vector notation to obtain covariance matrices, they are using the scalar product of vectors, applied simultaneously to many vectors, each corresponding to a random variable. We will not dwell excessively on this fact beyond this point, since we will conform to convention and work with the random-vector notation. But it pays to recognize that the underlying structure of multivariate statistics is linear algebra.

As we may have already indicated, in mathematical statistics a random variable is a random real-valued quantity whose values depend on the outcomes of an experiment governed by chance. A single random variable is represented by an italicized, uppercase letter, with particular values of the variable represented by subscripted lowercase versions of the same letter. Several random variables are treated collectively as the coordinates of what is known as a random vector. For example, if $X_1, X_2, \ldots, X_n$ are n random variables with zero means, then the random vector $\mathbf{X}$ is defined as the column vector with coordinates being the random variables $X_1, X_2, \ldots, X_n$, that is,

$$\mathbf{X} = \begin{bmatrix} X_1 \\ X_2 \\ \vdots \\ X_n \end{bmatrix}.$$

Although this looks like an ordinary n-tuple, it is not because the elements of an ordinary n-tuple are constants, and here they are random variables. Assume now that the random variables have zero means. We may now wish to obtain a linear combination of these random variables $Y = a_1 X_1 + a_2 X_2 + \cdots + a_n X_n$ with the a's constant weights. This linear combination may be written in matrix notation as $Y = \mathbf{a}'\mathbf{X}$. The result Y, however, is not a constant but also a random variable. Now, consider the $n \times n$ matrix

$$\Sigma_{XX} = E(\mathbf{X}\mathbf{X}').$$

An arbitrary element σ_{jk} of the matrix Σ_{XX} has its value $\sigma_{jk} = E(X_j X_k)$ since the expected value of any matrix (in this case the expected value of the matrix $\mathbf{X}\mathbf{X}'$) is the matrix of expected values of the corresponding elements of the original matrix. The matrix $\mathbf{X}\mathbf{X}'$ is itself a random matrix of random variables for its elements:

$$\mathbf{X}\mathbf{X}' = \begin{bmatrix} X1 \\ X_2 \\ \vdots \\ X_n \end{bmatrix} (X_1 \ \ X_2 \ \cdots \ X_n) = \begin{bmatrix} X_1 X_1 & X_1 X_2 & \cdots & X_1 X_n \\ X_2 X_1 & X_2 X_2 & \cdots & X_2 X_n \\ \vdots & \vdots & \ddots & \vdots \\ X_n X_1 & X_n X_2 & \cdots & X_n X_n \end{bmatrix}.$$

Hence

$$E(\mathbf{XX'}) = \begin{bmatrix} E(X_1X_1) & E(X_1X_2) & \cdots & E(X_1X_n) \\ E(X_2X_1) & E(X_2X_2) & \cdots & E(X_2X_n) \\ \vdots & \vdots & \ddots & \vdots \\ E(X_nX_1) & E(X_nX_2) & \cdots & E(X_nX_n) \end{bmatrix}$$

$$= \begin{bmatrix} \sigma_1^2 & \sigma_{12} & \cdots & \sigma_{1n} \\ \sigma_{21} & \sigma_2^2 & \cdots & \sigma_{2n} \\ \vdots & \vdots & \ddots & \vdots \\ \sigma_{n1} & \sigma_{n2} & \cdots & \sigma_n^2 \end{bmatrix}.$$

Again, if $\mathbf{X}$ is an $n \times 1$ random vector of random variables $X_1, X_2, \ldots, X_n$ and $\mathbf{Y}$ is a $p \times 1$ vector of random variables $Y_1, Y_2, \ldots, Y_p$, then

$$\Sigma_{XY} = E(\mathbf{XY'})$$

is an $n \times p$ covariance matrix of cross-covariances between the two sets of variables with typical element $\sigma_{gh} = E(X_g Y_h)$. In effect, covariance matrices are matrices of scalar products of vectors.

Maxima and Minima of Functions

There are numerous occasions scattered throughout multivariate statistics, including multiple regression, factor analysis, and structural equation modeling, where we must find the value or (values) of an independent variable (or variables) that will maximize (or minimize) some function. For example, in multiple regression, we seek values of weights to assign to predictor variables in a regression equation that minimizes the average value of the squared differences between the predictive composite scores and the actual scores on the criterion. In principal components analysis, we seek the weights of a linear combination (under the restraint that the sum of squares of the weights add up to 1) of a set of variables that will have the maximum variance over all possible such linear combinations. In the varimax method of rotation, we seek values of the elements of the orthogonal transformation matrix that will rotate the factors so that the sum of the variances of the squared loadings on the respective factors will be a maximum. In maximum-likelihood factor analysis we seek values for the factor loadings, and unique variances that will maximize the joint-likelihood function. In structural equation modeling using maximum-likelihood estimation, we seek values of the free model parameters that will minimize the likelihood fit function.

In this section, we expect the student who is fresh to the topic of calculus to obtain nothing more than an intuitive understanding of what is involved in the solution of maximization and minimization problems by calculus. To obtain a more thorough background in this subject, the student should take a course in elementary calculus. Those sections in this book involving obtaining the solution to the maximization or minimization of some function will be marked with an asterisk in front of the title to the section heading and may be passed over by those who are mathematically challenged. Much of this book does not require calculus.

Slope as the Indicator of a Maximum or Minimum

Consider the graphical representation of the function $y = f(x)$ in Figure 2.8. The function portrayed in this figure is not designed to represent any particular function other than one having a minimum and maximum value over the range of the values of x depicted.

Now, consider the effects of drawing straight lines tangent to the surface of the curve at those points on the curve corresponding to the values of x_1, x_2, x_3, and x_4. We see, going from left to right along the x axis, that the slope of the line tangent at the point corresponding to x_1 is downward, indicating the function is descending at that point. At x_2 the tangent line is horizontal, and the function is at a local minimum. At x_3 the tangent line is sloping upward, indicating a rise in the function at that point. And at x_4 the tangent line is again horizontal, but this time at a maximum for the function. We see at this point that lines tangent to minima or maxima are horizontal. On the basis of this simple observation, calculus builds the methodology for finding points that correspond to maxima or minima or functions.

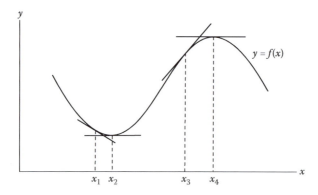

FIGURE 2.8 Graph of function $y = f(x)$ showing slopes of curve of function corresponding to different values of x.

An Index for Slope

Let us now consider quantifying what we mean by the slope of a straight line drawn tangent to a point on the curve of some function. Consider the graph of the function $y = f(x)$ in Figure 2.9. Let the point R be a point on the curve with coordinates $(x, f(x))$. Draw a line tangent to the curve at point R. Next, consider a line drawn from the point R to another point S on the curve a short distance away from the point R in the direction of increasing x. The coordinates of the point S are $(x + \Delta x, f(x + \Delta x))$, where Δx is some small increment added to x. Intuitively, we can see that if we make Δx smaller and smaller, the point S will come closer and closer to the point R. Also, the slope of the line passing through R and S will come closer and closer to the slope of the line tangent to the curve at R. Using this fact, let us first define the slope of the line passing through R and S as

$$\text{Slope(RS)} = \frac{f(x + \Delta x) - f(x)}{\Delta x}.$$

We see that this is simply the ratio of the side opposite to the side adjacent in a right triangle. Next, we will define the slope of the line tangent to the curve at point R, written as dy/dx, to be

$$\frac{dy}{dx} = \lim_{\Delta x \to 0} \frac{f(x + \Delta x) - f(x)}{\Delta x}. \tag{2.22}$$

Although to those not familiar with taking the limit of an expression, Equation 2.22 may appear to involve dividing by 0, such is not the case. Generally, we can simplify the right-hand expression so that the denominator Δx

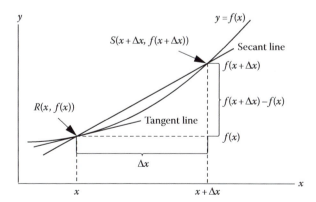

FIGURE 2.9 Graph of function $y = f(x)$ showing how the line tangent to the curve at point $R(x, f(x))$ may be approximated by the secant line drawn between points $R(x, f(x))$ and $S(x + \Delta x, f(x + \Delta x))$. The approximation improves as Δx approaches 0.

cancels out with quantities in the numerator before we take the limit when Δx approaches zero.

Derivative of a Function

An important point to realize about Equation 2.22 is that it also represents a function of the variable x. In this case, Equation 2.22 is a function relating the values of the slope of a line drawn tangent to the curve of the function $y = f(x)$ to the values of x corresponding to the points to which the line is drawn tangent. The slope function corresponding to the function $y = f(x)$ is known as the derivative of the function $y = f(x)$ with respect to x.

For example, consider the function $y = x^2$ shown in Figure 2.10. Let us look for the derivative of this function, using the definition in Equation 2.22:

$$\frac{dy}{dx} = \lim_{\Delta x \to 0} \frac{(x + \Delta x)^2 - x^2}{\Delta x}$$

$$= \lim_{\Delta x \to 0} \frac{x^2 + 2x\Delta x + \Delta x^2 - x^2}{\Delta x} = \frac{2x\Delta x + \Delta x^2}{\Delta x}$$

$$= \lim_{\Delta x \to 0} 2x + \Delta x = 2x.$$

Using the derivative, we can find the value of the slope of a line drawn tangent to any point on the curve $y = x^2$ corresponding to a given value of

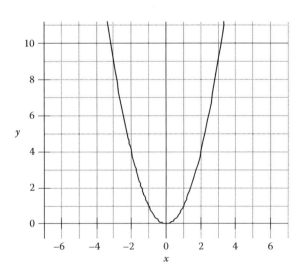

FIGURE 2.10 Graph of function $y = x^2$ which is minimized when $x = 0$.

x. Consider, for example, the slope of a line at the point on the curve where x equals 1. Substituting the value of 1 for x in the derivative $dy/dx = 2x$, we find that the slope is equal to 2 at this point. When x is 3, we find that the line drawn tangent to a corresponding point on the curve has a slope of 6.

But in the equation $y = x^2$, at what value of x is y a minimum? This can be found by setting the derivative equal to zero and solving for x.

$$\text{If } 2x = 0 \quad \text{then } x = 0.$$

When x is zero, the slope is zero, indicating that the tangent line drawn to the curve at the point where x equals 0 is horizontal. If we examine our graph for $y = x^2$, we will see that this corresponds to the minimum point on the curve.

Derivative of a Constant

Consider the function $y = k$, where k is some constant. By using Equation 2.22, we have $f(x + \Delta x) = k$ and $f(x) = k$, so that

$$\frac{dy}{dx} = \lim_{\Delta x \to 0} \frac{k - k}{\Delta x} = \lim_{\Delta x \to 0} 0 = 0.$$

Again, we do not have to divide by zero because $(k - k) = 0$ and $0/\Delta x = 0$. Geometrically the curve of $y = k$ is a horizontal straight line at altitude k. At any point on the line, the slope is 0.

Derivative of Other Functions

The derivatives for different functions are normally different themselves. As a consequence, one must determine for each function what its derivative is. This might seem to involve a bit of labor, especially if we had to apply the definition in Equation 2.22 to complicated functions. But fortunately this is not necessary, because most functions can be analyzed into parts and the derivatives of these parts taken separately and then added together. Moreover, it is possible to memorize tables of derivatives of common elementary functions that have already been derived using Equation 2.22. Some common derivatives are given in Table 2.1.

As an example of the use of Table 2.1, consider the problem of finding values of x that will either maximize or minimize the function

$$y = \frac{x^3}{3} - \frac{x^2}{2} - 2x + 1.$$

By rule 8 in Table 2.1, we see that the derivative of a function that is the sum of some other functions is equal to the sum of the derivatives of the other functions.

TABLE 2.1

Derivatives of Common Functions

	Function	Derivative
1.	$y = k$	$\dfrac{dy}{dx} = 0$
2.	$Y = x$	$\dfrac{dy}{dx} = 1$
3.	$Y = kx$	$\dfrac{dy}{dx} = k$
4.	$y = ax^n$	$\dfrac{dy}{dx} = anx^{n-1}$
5.	$Y = e^x$	$\dfrac{dy}{dx} = e^x$
6.	$y = a^x$	$\dfrac{dy}{dx} = a^x \ln a$
7.	$y = \ln x$	$\dfrac{dy}{dx} = \dfrac{1}{x}$

If $u = u(x)$ and $v = v(x)$ are functions of x, then for

8.	$y = u + v$	$\dfrac{dy}{dx} = \dfrac{du}{dx} + \dfrac{dv}{dx}$
9.	$Y = uv$	$\dfrac{dy}{dx} = u\dfrac{dv}{dx} + v\dfrac{du}{dx}$
10.	$y = \dfrac{u}{v}$	$\dfrac{dy}{dx} = \dfrac{v(du/dx) - u(dv/dx)}{v^2}$
11.	$y = \sin x$	$\dfrac{dy}{dx} = \cos x$
12.	$y = \cos x$	$\dfrac{dy}{dx} = -\sin x$
13.	$y = \tan x$	$\dfrac{dy}{dx} = \sec^2 x$

Hence

$$\frac{dy}{dx} = x^2 - x - 2.$$

To find the points of x that correspond to maximum or minimum values of y, we need to set the derivative of y equal to zero, and then solve for x, that is, solve

$$x^2 - x - 2 = 0.$$

The solution to this is obtained by factoring, so that we have

$$(x + 1)(x - 2) = 0.$$

We see that solutions that will make the left-hand side of the equation equal to zero are $x = -1$ and $x = 2$. In other words, at $x = -1$ and $x = +2$, y is either a maximum or a minimum.

When a function is known to have only a maximum or a minimum, there is no question of whether one has found either a maximum or a minimum when he or she finds an x that makes the derivative of the function equal to zero. However, in those cases where both maximum and minimum solutions exist, ambiguity exists.

Several methods may be used to resolve this ambiguity: (1) By direct computation, one can compute values of the function $y = f(x)$ for values in the vicinity of x_m, where x_m is a value of x at which the derivative of the function equals zero. If x_m leads to a value for y that is less than the values of y in the vicinity of x_m, then x_m is where the minimum is located. If the value of y at x_m is greater than the values of y in the vicinity of x_m, then x_m is likely a maximum. (2) A more certain method to use is based on finding the second derivative of $y = f(x)$. The second derivative is simply the derivative of the function that is the derivative of $y = f(x)$. One then finds the value of the second derivative when $x = x_m$. If the second derivative is positive, then x_m corresponds to a minimum y. If the second derivative is negative, then x_m corresponds to a maximum y.

To illustrate the use of the second derivative, consider the problem thus cited in which

$$\frac{dy}{dx} = x^2 - x - 2.$$

The second derivative is

$$\frac{d^2y}{dx^2} = 2x - 1,$$

where d^2y/dx^2 is the symbol for the second derivative. Substituting -1 for x into the equation for the second derivative, we have

$$2(-1) - 1 = -3.$$

Since the value of the second derivative when $x = -1$ is negative, y must be a maximum when $x = -1$. Substitute $x = 2$ into the equation for the second derivative:

$$2(2) - 1 = 3.$$

Since the value of the second derivative is positive when $x = 2$, y must be a minimum when $x = 2$.

A word of caution is necessary regarding using zero slope as a sufficient indication of a maximum or minimum for a function. Some functions will have zero slope where there is neither a maximum nor a minimum; for example, the function $y = x^3$ has a zero derivative when x equals 0, but the function is at neither a maximum nor a minimum at that point. So, zero derivative of the function at some point is only a necessary but not a sufficient condition that there exists a maximum or a minimum at that point. A sufficient condition for a maximum or a minimum is that the second derivative does not equal zero at the point where the function has zero slope.

Partial Differentiation

In multivariate techniques such as structural equation modeling, we are concerned with functions not of a single independent variable but of many independent variables. Thus we may be interested in knowing how to treat change in the function in connection with changes in the independent variables. This may be done by allowing only one variable at a time to change while holding the other variables constant and by observing the degree of change in the value of the function. In this connection it is possible to obtain the derivative of the function in the direction of one of the independent variables, treating all other independent variables as constants. Such a directional derivative is known as a partial derivative. For example, suppose we wish to know the rate of change of the function $y = 3x^2 + 12xz + 4z^2$ when we change x while holding z constant. We can obtain the partial derivative of y with respect to x as

$$\frac{\partial y}{\partial x} = 6x + 12z.$$

Here we obtained the derivative in the usual way, but we treated the other independent variable, z, as a constant for the differentiation. On the other hand, the partial derivative of y with respect to z is

$$\frac{\partial y}{\partial z} = 12x + 8z.$$

Again, we now treated x as if it were a constant while taking the derivatives of the terms with respect to z. So, there is very little new to learn to obtain partial derivatives.

Maxima and Minima of Functions of Several Variables

In Figure 2.11, we illustrate a function $z = f(x, y)$ of two independent variables x and y with a relative maximum. Consider that at a maximum or a minimum for the function $z = f(x, y)$, there must be some point (x, y) on the plane surface defined by the x and y axes, where the slopes of the function in the directions of the x and y axes are zero, respectively. In other words, a necessary condition

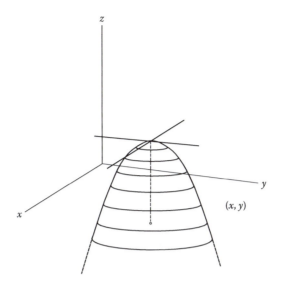

FIGURE 2.11   Graph of a function of two variables illustrating point (x, y) at which function $z = f(x, y)$ is a maximum. Note that tangents to surface of function in directions of x and y axes at point (x, y) are horizontal, indicating zero slopes in those directions.

for a maximum or a minimum is that at some point (x, y)

$$\frac{\partial z}{\partial x} = 0,$$

$$\frac{\partial z}{\partial y} = 0.$$

To establish the sufficient conditions for a maximum or a minimum, define the Jacobian matrix of second derivatives of z evaluated at (x, y) as

$$\mathbf{J} = \begin{bmatrix} \dfrac{\partial^2 z}{\partial x^2} & \dfrac{\partial^2 z}{\partial x \partial y} \\ \dfrac{\partial^2 z}{\partial y \partial x} & \dfrac{\partial^2 z}{\partial y^2} \end{bmatrix}.$$

A sufficient condition that there exists a maximum or a minimum at this point (x, y), where $\partial z/\partial x = 0$ and $\partial z/\partial y = 0$, is that the determinant $|\mathbf{J}| \neq 0$. If $|\mathbf{J}| < 0$, then z is a relative maximum at (x, y); on the other hand, if $|\mathbf{J}| > 0$, then z is a relative minimum at (x, y).

For example, investigate the relative maxima or minima of the function $f(x, y) = 2x^2 + y^2 + 4x + 6y + 2xy$. Taking the partial derivatives of $f(x, y)$

with respect to x and y and setting them equal to zero, we obtain

$$\frac{\partial f}{\partial x} = 4x + 4 + 2y = 0,$$

$$\frac{\partial f}{\partial y} = 2y + 6 + 2x = 0$$

or

$$4x + 2y = -4,$$
$$2x + 2y = -6.$$

Solving this system of simultaneous linear equations for x and y, we obtain $x = 1$ and $y = -4$ as the point where $f(x, y)$ is at a possible extremum. The second derivatives of $f(x, y)$ are

$$\frac{\partial^2 f}{\partial x^2} = 4, \quad \frac{\partial^2 f}{\partial x \partial y} = 2,$$

$$\frac{\partial^2 f}{\partial y \partial x} = 2, \quad \frac{\partial^2 f}{\partial y^2} = 2.$$

Hence the Jacobian matrix is

$$\mathbf{J} = \begin{bmatrix} 4 & 2 \\ 2 & 2 \end{bmatrix},$$

whose determinant is $|\mathbf{J}| = 4$. Hence, with the determinant of the Jacobian matrix positive, an extremum must exist and it is a minimum.

Constrained Maxima and Minima

Up to now we have considered maximizing (or minimizing) a function of the type $z = f(x, y)$, where x and y are independent of one another. We shall now consider maximizing (minimizing) a function of the type $z = f(x, y)$, where x and y must maintain certain dependent relationships between them as defined by one or more equations of the type $g(x, y) = 0$. The most general and powerful method for finding solutions for x and y that maximize $f(x, y)$ under constraints is the method of Lagrangian multipliers. Assuming there is one equation of constraint $g(x, y) = 0$, we first form the function

$$F(x, y) = f(x, y) + \theta g(x, y),$$

where θ is an unknown multiplier (to be determined) multiplied by the equation of constraint. Then we form the equations

$$\frac{\partial F}{\partial x} = \frac{\partial f}{\partial x} + \theta \frac{\partial g}{\partial x} = 0,$$

$$\frac{\partial F}{\partial y} = \frac{\partial f}{\partial y} + \theta \frac{\partial g}{\partial y} = 0,$$

$$\frac{\partial F}{\partial \theta} = g(x,y) = 0,$$

which we then solve for x, y, and θ.

For example, maximize the function $f(x,y) = 10 - 2x^2 - y^2$ subject to the constraint $x + y - 1 = 0$. First, form the equation $F = 10 - 2x^2 - y^2 + \theta(x + y - 1)$.

Then

$$\frac{\partial F}{\partial x} = -4x + \theta = 0,$$

$$\frac{\partial F}{\partial y} = -2y + \theta = 0,$$

$$\frac{\partial F}{\partial g} = x + y - 1 = 0.$$

Solving this system of equations for x, y, and θ, we obtain $x = 1/3$, $y = 2/3$, and $\theta = 4/3$. In other words, $f(x,y)$ is a maximum at $x = 1/3$ and $y = 2/3$ when x and y are constrained so that $x + y - 1 = 0$.

The method of Lagrangian multipliers just described may be generalized to functions of more than two independent variables and problems with more than one equation of constraint. In general, given a function $f(x_1, x_2, \ldots, x_n)$ to maximize subject to the constraints defined by the equations $g_1(x_1, x_2, \ldots, x_n) = 0, \ldots, g_m(x_1, x_2, \ldots, x_n) = 0, m \leq n$, form the function

$$F(x_1, x_2, \ldots, x_n) + \theta_1 g_1(x_1, x_2, \ldots, x_n) + \cdots + \theta_m g_m(x_1, x_2, \ldots, x_n).$$

Then obtain the equations

$$\frac{\partial F}{\partial x_1} = \frac{\partial F}{\partial x_2} = \cdots = \frac{\partial F}{\partial x_n} = 0,$$

which may be combined with the m equations of constraint to form a system of $n + m$ equations in $n + m$ unknowns which may then be solved. In factor analysis and structural equation modeling, we find that the solutions for such systems of equations require iterative, numerical methods instead of direct algebraic manipulation.

3

Causation

Historical Background

After completing work on confirmatory factor analysis and analysis of covariance structures around the end of 1968, Karl Jöreskog had laid the groundwork for a methodological development that was to have considerable impact on the behavioral and social sciences in the years to follow. His interests turned to structural equation modeling, a topic that had its first origins in the field of genetics in the work of Sewell Wright (1921, 1931, 1934) on path analysis, but which had been extended by biometricians (Turner and Stephens, 1959) and sociologists (e.g., Blalock, 1964, 1969; Duncan, 1966; Land, 1969; Heise, 1969; Costner, 1969) and approached independently from the theory of regression by econometricians (e.g., Klein, 1953, 1969; Wold and Jureen, 1953; Koopmans and Hood, 1953; Goldberger, 1964; Fisher, 1966). Jöreskog's contribution to this literature was to merge the latent variable idea of common factor analysis with the traditional theory of systems of linear structural equations of measured variables, to produce a new and more general model along with an efficient algorithm for the estimation of its parameters. By the fall of 1970, he had completed a paper on his new system and presented it at a conference cosponsored by the Social Science Research Council and the Social Systems Research Institute of the University of Wisconsin at Madison, Wisconsin, November 12–16, 1970. This paper, published as Jöreskog (1973), was the first of many papers from his and others' laboratories that was to be concerned with the new field of structural equation modeling with latent variables. The conference also brought key figures concerned with path analysis, structural equations modeling, and causal modeling with correlational data from many fields and gave new impetus to the development of this methodological area. Because of the generality of Jöreskog's model and the practicality of the computing packages that he made available for its implementation, much of the direction of this field in the years to follow were to be in terms of this model. A major consequence of Jöreskog's model and the computer programs available for its implementation was an awakened interest of researchers who traditionally conducted correlational research in the issues of causality and the study of causal models, because the new methodology made this seemingly possible.

A question naturally arises, why did such interest not develop among this kind of researcher earlier? Why was the emphasis for so long primarily upon exploratory analysis and correlation and not on hypothesis testing and causality? It is not easy to determine precisely why this was the case, although it is clear that, at its inception, the earliest form of structural equation modeling, path analysis (Wright, 1921), met considerable hostile resistance. Niles (1922) attempted to discredit Wright's contrast of causality with correlation with copious citations from Pearson's *The Grammar of Science* (Pearson, 1911). Pearson, who provided the maximum-likelihood estimate of the correlation coefficient and the seminal work on multiple correlation, was also a noted philosopher of science, perhaps the last of the great nineteenth-century British empiricists and a forerunner of the logical positivist movement that sprang up later in Vienna. Pearson regarded causality as association and believed that the physics of his day was moving away from determinism to probabilism. As with most nineteenth-century empiricists, he stressed induction (generalizing from particulars) as opposed to the use of hypotheses and hypothesis testing (see Mulaik, 1985, 1987). He regarded correlation as the new replacement of causation, with correlations less than unity indicative of the imperfect or probabilistic nature of causation. Niles (1922) thus argued that the distinction between causality and correlation was nonsense, because in the new way of thinking causality *is* correlation. He also rejected path analysis because he believed it required one to formulate causal hypotheses totally *a priori*, and he believed there was no point at which one could stop in tracing back the causes of things.

That did not dissuade Wright from persisting in developing his new technique. Wright (1923) argued that causality implied directionality, whereas correlation did not. He argued that researchers had the knowledge needed within an area to formulate causal hypotheses, so hypotheses are not generated purely *a priori*. And he believed it was possible to isolate a portion of the universe to study the causal influences on and within such a limited system. Other geneticists were also not dissuaded from exploring Wright's path analytic methods. But it is not clear why psychologists and sociologists at this time did not discover this methodology and seek to incorporate it into their own research. Perhaps they did not read the genetics journals.

There were other influences that inhibited the development of a hypothesis-testing methodology designed to study causal relations using correlation and covariance. Although by the beginning of the twentieth century physicists were developing hypothesis-testing approaches to science, there were still strong beliefs in some fields of science, held over from the nineteenth century, that science is basically inductive and should avoid hypotheses and stick to description and careful generalizations from "the facts" (Laudan, 1980; Mulaik, 1985). And then, just before and after World War II, when so many members of the logical positivist movement, then centered in Vienna, migrated to English-speaking universities to escape Naziism, there arose among philosophers of science the claim that science had abandoned causality. Some claimed causality is a metaphysical concept, whereas others claimed causality is simply determinism in the form of functional relationship and no longer compatible with the probabilism of the new quantum physics (Schlick, 1932/1959). Causality was also displaced by logical implication in the efforts of logical empiricists to express all scientific relations in logical form (Hempel, 1965). Not the least influence was a feeling that the behavioral and social sciences were still in an exploratory mode, with insufficient knowledge on which to base causal hypotheses. There was also the maxim *Correlation does not imply causation*, which because of Pearson's influence on the field of correlational statistics became the received wisdom, no doubt with the encouragement of experimentalists.

Causation among the Ancients

The idea of causation, of course, is ancient, being closely associated with the abandonment of mythic forms of explanation and the rise of science among the Greeks. Aristotle regarded causation as explanation and synthesized the major forms of explanation developed by his predecessors into four forms of causation: (1) *material*, (2) *formal*, (3) *efficient*, and (4) *teleological*. Causal explanations explain something by showing how it is dependent on something else. Explaining things in terms of *material causes* involves showing how certain properties of things are due to substances of which they are made or the effects of certain substances. For example, a statue is heavy and hard because it is made out of stone. Ice forms at the top of milk containers at temperatures below 0°C because milk contains water. A man dies because his wife puts arsenic in his food. *Formal causes* explain things in terms of their forms. An arrow penetrates the flesh because of the shape of its wedge-like point, which concentrates the force at a point and moves aside the flesh as it penetrates by the principle of the inclined plane. A fist-sized rock moved by a similar force as imparted to an arrow does not penetrate flesh, because, due to its shape, the rock's force is spread out over a greater area that results in insufficient force at any point to penetrate the flesh. The shape of a wing causes lift when air moves across it. Many psychological explanations today invoke formal causation: Some people get higher grades in school because they are more intelligent. Intelligence is a form of behavior as measured by tests in which a person demonstrates the ability to correctly infer the nature of a rule from seeing simply some instances of the rule (Guttman, 1965). Some people are prone to anxiety attacks because they have more labile parasympathetic nervous systems. Efficient *causes* explain things in terms of the agents or events that make the things take on the forms they have: A slab of marble looks like a man because a sculptor hewed at the stone until it took the shape of a man. The tree fell because it was struck by lightning. The boat turned to the left because the helmsman pushed the tiller to the right. Some dogs salivate whenever a bell rings because they have been conditioned to do so by giving them meat powder immediately after ringing the bell. *Teleological* explanations explain things in terms of the final forms or states that the things are changing into: An acorn develops as it does because its final state is to be a tree. A statue is made to achieve the final state of a pleasing representation of a man for others to see. A man goes into a restaurant because he wants to eat.

Aristotle's writings were lost to the West during the Dark Ages after the fall of Rome and were found again only during the later Middle Ages. Aristotle had rejected the purposeless view of nature held by his atomist predecessors to argue that the universe is like a living, growing, developing thing, which was to be explained not just materially, formally, and in terms of efficient causes, but teleologically, in terms of its final ends and purposes. During the Renaissance other Greek writers were studied, and platonist/atomist and, subsequently, mechanistic forms of thought reemerged to challenge Aristotelian thought, and these rejected especially teleological explanations and favored efficient causation as the basic form of explanation.

Causation in the Seventeenth Century

Descartes

The rationalist French philosopher René Descartes (1596–1650) held that the mind gains knowledge by means of the processes of analysis and synthesis, which he called intuition and deduction. Analysis or intuition breaks things down into clear and distinct component elements or parts, and synthesis or deduction forms composites of them by putting parts together. He sought certain knowledge by analyzing philosophical problems into fundamental and certain ideas, and then synthesizing them by a "deduction" involving a continuous and uninterrupted action of the mind that joins them successively together into complex truths. Descartes further held that while many ideas arise solely from experience or imagination, some ideas do not, but rather are innate, being stimulated or caused to appear before the mind by experience. The ideas of extension, solidity, quantity, and mobility of substances he held are innate to the mind. In fact, reason, the laws by which thinking proceeds, is innate, and the principle of causation, of noncontradiction, that nothingness cannot be the efficient cause of anything; the ideas of geometry are all innate.

However, Descartes' theory of causation is not well developed compared with that of philosophers who followed him. Descartes is also a major founder of the school of Continental rationalist philosophers such as Spinoza, Leibniz, and Wolf, who followed him. The rationalist school sought certain knowledge that could be deduced logically from fundamental, self-evident truths.

Locke

In England, a younger contemporary of Descartes, John Locke (1632–1704), while admiring and even adopting some features of Descartes' method of analysis and synthesis, rejected Descartes' ideal of certain and incorrigible knowledge. Locke, as a physician, regarded human knowledge as verifiable and corrigible by experience, but rarely, if never, absolutely certain, while still probable to a degree by experience. Locke rejected outright Descartes' idea of innate ideas. He sought then to develop an alternative theory of how ideas and knowledge are obtained from experience and the role of reasoning from experience in acquiring knowledge. As a consequence, Locke is considered to be the founder of the school of British empiricism, which includes Locke, George Berkeley, David Hume, James Mill, John Stuart Mill, and even Karl Pearson.

For Locke (1694/1905/1962), all of our ideas have their origins in experience. Reason then operates upon our experience-derived ideas to achieve knowledge. His method then shows by analysis the origins of our ideas in simple ideas of sensory and reflective experience, and then by synthesis how more complex ideas are derived from the simple ideas. Some of his attention is directed to showing how what Descartes regarded as innate ideas are derived from experience. He also focuses on an analysis of the "powers of the mind," which is a kind of faculty theory of mind. Our interest here, however, is principally upon Locke's view of causation. This is closely joined with Locke's view of "power," which is a fundamental concept in his system of thought.

The idea of power, Locke says (1694/1905/1962, p. 128), contains in it a relation between substances and things. There are two forms of power: active and passive. Active power is the potential that an agent's acting in certain ways or a substance's having certain properties will make certain kinds of changes in certain other things. Passive power is the potential to change in a certain way when operated on in a certain way by an agent or a certain substance. Thus power inheres in things or substances. However, our idea of power is nothing more than an inductive generalization from what we have observed in the past, "that like changes will for the future be made in the same things by like agents, and by the like ways" (p. 127). But because Locke believes that some simple ideas such as extension, size, form, and motion reflect corresponding properties in external things, which thus are the causes of these simple ideas, he believes that power also may be a simple idea caused by the acts and reactions of external things. Nevertheless, for power involving action, we have the clearest idea of the power that produces action: For the beginning of motion, by reflecting on what occurs in ourselves, we find that by willing it we move parts of our bodies at rest. This Locke believes is a clearer and more distinct source for the idea of power than gained by observing objects (Locke, 1694, p. 130). Locke thus wants to reduce the idea of power to our own experience of the ability to will our bodies to act in certain ways.

An associated idea of causation is *substance*. The word means "that which stands under." This is a central concept of the rationalists because substance is supposed to be that to which qualities and properties such as color and weight inhere. Since powers are qualities and properties, they inhere in substances. But people in Locke's time had little knowledge derived from experience of the substrate of things and how the properties and qualities are supported by the "substance" underlying them. So substance effectively, for Locke, was "the supposed, but unknown, support for those qualities we find existing, which we imagine cannot exist *sine re substante*, 'without something to support them' " (Locke, 1694/1905/1962, p. 195). But once we have the general notion of a "substance," we have the idea of particular substances, such as gold, a man, horse, blood, etc. Locke then says "It is the ordinary qualities observable in iron or a diamond, put together, that make the true complex idea of those substances ..." (p. 195). And one "... has no other idea of those substances than what is framed by a collection of those simple ideas which are to be found in them ..." (p. 196). At best, for Locke, the notion of a substance was something in thought on which to hang the qualities and to provide for a convenient, shortened way to think about collections of simple ideas. But there was no substance to "substance."

Power for Locke was the major part of our ideas of complex substances. Passive powers of substances were capacities to be changed by other substances. Thus the capacity for iron to be drawn toward a magnet was a passive power of iron. A magnet had the active power to draw iron to it. In short, an active power functions as a cause and a passive power as an effect. Furthermore, these powers are simply the potential to make sensible changes in sensible qualities in other objects or to be changed by changes in the sensible qualities of other objects. But the simple idea of a power of a specific substance resides in the changes observed in the sensible qualities of a thing (Locke, 1694/1905/1962, p. 199). (For Locke, perception of *change* in sensible qualities itself is a simple idea given by the senses, on which point, we shall see, he differs from successor empiricist philosophers.) And so causation is grounded in sensible changes in qualities. Substances are "nothing else but a collection of a certain number of simple ideas, considered as united in one thing" (p. 206).

Also important for Locke was that things do exist independently of our minds and cause the sensible ideas we have. He held that there were three kinds of qualities of physical things. Primary qualities, such as extension, magnitude, quantity, shape, and motion, or rest, are given to us as simple ideas. By means of the senses, these are directly and immediately given properties of things as they truly are. Then there are secondary qualities, such as color, taste, warmth, smells, and sounds, which are produced by primary qualities of things. These are sensed qualities, but they do not have immediate and exact counterparts in things, and are merely effects on the senses of the primary qualities of things. Finally, there are powers to make changes in bulk, figure, texture, and motion of another body.

Eighteenth-Century Empiricists

Berkeley

George Berkeley (1685–1753) was an English cleric who was appalled at the degree to which the various atheists of his time depended on the so-called existence of material substance while holding that it is difficult to understand how it could be created out of nothing by God's divine command. He sought to show that nothing exists independently of mind. Furthermore, he rejected both Descartes' and Locke's dualism of mind and matter. All we have given to us is in the mind and to think of something beyond as an independent world is unnecessary. There is but one reality, and to be is to be perceived. As for things that we do not perceive but which we believe nevertheless exist, we are justified in so believing by our belief that God perceives them when we do not, which is a hypothesis that accounts for all the facts of experience. However, beyond his preoccupation with the Divine, in general, Berkeley's system was simply Locke's stripped of an external reality. All we have are ideas.

Hume

It fell to David Hume (1711–1776) to work out in unflinching detail the logical implications of Locke's empiricism with Berkeley's addendum that there is no need to consider an external reality, and Hume's exclusion of God's mind (Hume, 1739, 1777). The fundamental reality is mental, and the atomic elements of that reality are the vivid *impressions* of sense or reflection. *Ideas* in turn are but less vivid copies of impressions. (Hume avoids saying where ideas are kept or come from. They just appear at some points in the mind.) Complex ideas are composites of simple ideas. We observe furthermore that complex ideas occur in certain categories that suggest the existence of uniform principles of how the mind functions. Hume distinguished three kinds of principles involved in forming three general categories of complex concepts: (1) resemblance, (2) contiguity, and (3) cause and effect. Together they represented different forms of association of ideas: (1) association or grouping of simple ideas into single (complex) ideas because of similarity; (2) association of ideas because they occur together in space; and (3) association of distinct ideas because they seem to regularly follow one another in time.

For the empiricists, causality was a learned association between certain kinds of events. Causality is not a necessary connection. John Locke argued that the idea of causality arises initially in our experience where we learn to associate our actions with certain consequences and then we observe "constantly" that certain actions produce certain results and generalize that the same actions will produce the same results in the future. In fact, causality, David Hume argued, is a kind of illusory connection because there is nothing in experience that corresponds to the causal connection. There is just the mysterious association of one kind of event with another that follows it on a regular basis that leads to our expecting the second kind of event whenever we experience the first. But logically, there is no necessity that the preceding event will always be followed by the second kind of event. Some other kind of event might follow. Thus Hume's concept of causality is an event–event concept, wherein the cause, so-called, is an event that precedes the effect event in time. An event is some kind of complex of sense impressions that occurs at a given point in time.

Similarly, the idea of substance, Hume held, was also an illusion, for there is no necessity to posit an external reality beyond what is given to us immediately in our minds. Substances, things, are simply certain kinds of collages of simple ideas that we experience regularly and come to expect will thus co-occur in a similar way in the future. In fact, the idea of the self seems to be nothing more than the introspected collection of ideas and impressions that we regularly experience. Nowhere do we experience in our minds an entity that is the self. In this way Hume pushed empiricism to a skeptical conclusion, that there is no external world, only the mind's construction of the world through associative processes, and there is no necessity in our ideas about the world. And there is no "self." Consequently, later empiricist philosophers tended to debunk the ideas of substance, causality, and of the self, and to attempt to drive the idea of causation out of science.

Immanuel Kant

Toward the end of the eighteenth century, Immanuel Kant (1724–1804) in Prussia, although trained in the rationalist tradition, and functioning as a professional philosophy professor, came to deeply admire the works of Isaac Newton on mechanics. Newton had claimed not to feign hypotheses for his explanations of physical phenomena, but rather used experiment and the testimony of the senses. Rationalist "natural philosophers" or scientists, on the other hand, tended to proceed by formulating systems grounded in fundamental self-evident truths from which they would seek to deduce the phenomena observed. The physicist Isaac Newton obviously succeeded way beyond anything rationalist philosophers had produced, in using simple basic assumptions and experimental findings in understanding and predicting a broad range of physical phenomena that held up with experiment. Newton's system was one that combined reason with experience, on the assumption that the world was understandable with reason. But Hume had driven a wedge between reason and experience, and the challenge was to find how reason could operate with experience (Jones, 1952).

At age 57, in 1781, Kant published *Critique of Pure Reason*, arguing that the way of pure reason was a failure in understanding the world. He sought to reconcile rationalist and empirical thought in arguing that we do not know things as they are in themselves independent of the mind. Rather we know things only in terms of *a priori* categories or schemas by which we synthesize (join together, combine) what is given to us by the senses. Causality is an *a priori* schema by which the mind organizes and synthesizes experience by showing how certain attributes of things are dependent on other attributes of things. However, causal connections are not logically necessary connections. Logic demonstrates necessity by establishing identity under the principle of contradiction. The conclusion of a syllogism can contain nothing more than what is in its premises. But an effect is more than or different from its cause. The necessary connection in causation is an aspect of the synthesizing schema of causation, but the necessity is not in the material provided by the senses, but in the *a priori* synthetic form provided by the mind. This form presumes a schema whose components are the attributes of the causal substance or object and the attributes of the affected substance or object, and their relation is one in which the effect is dependent on the cause for its existence.

Thus we can reason about the implications of causal connections, using the necessity of the *a priori* synthetic form in the reasoning. But the conclusions would be corrigible by experience because the component of the reasoning about the world introduced by experience via the senses is necessarily incomplete, limited, and conditioned, simply because the knower is incapable of perceiving the whole of existence, past, present, or future from a finite point of view.

Watkins (2005) argues from a comprehensive review and understanding of Kant's works and the works of his contemporaries and predecessors that, contrary to some commentators on Kant in the English-speaking world, Kant did not write a refutation of Hume in the *Critique of Pure Reason*. The German-speaking world of Kant's time did not take Hume's skeptical stance seriously, and no one felt the need to write a refutation of Hume. This was the case for Kant, who wrote for a German-speaking, professional, rationalist philosophical audience. Kant simply offered his view of how pure reason alone was ineffective in understanding the world and yet reason and experience together can work.

In fact, Watkins (2005) argues that Kant's views on causality are so different from Hume's that there is no common ground from which to write a refutation relevant to the other. Whereas Hume takes an event–event succession of sensations in time as a given, and regards causation as a familiar, regular succession of events, Kant treats causation in a manner rather like Locke, as involving powers of substances to produce certain effects in certain other substances. The powers are grounded in the essential yet empirical nature of the substances. The essential nature of a substance concerns those attributes without which the substance would not be what it is. This contrasts with accidental attributes that can change without changing the substance's essential nature.

Kant differs from Locke regarding the concepts of substances having attributes as a synthetic *a priori* category on which the human understanding depends for its reasoning about experience. The way in which humans think about the world depends on the use of this category. This category can only be used as a framework to reason about experience. It cannot be used as do the rationalists, apart from experience. The idea of substance is not useless or unnecessary, because without it we cannot reason about things and objects in the world. In fact, the whole set of synthetic *a priori* concepts was necessary to have a concept of the world as a unity for the knower.

Kant was also intrigued with concepts that came in threes. In fact, his famous table of categories of objective judgment has four major classes: (1) of quantity, (2) of quality, (3) of relation, and (4) of modality. In each of these classes, there are three categories. Kant's explanation of his categories in the *Critique of Pure Reason* is sparse, and many scholars have puzzled over their manner of formation in threes. My understanding of them is based on a footnote in Kant's (1790) *Critique of Judgment*, which is a clearer statement of the principle of the categories' occurrence in threes.

> That my divisions in pure philosophy almost always turn out tripartite has aroused suspicion. Yet that is the nature of the case. If a division is to be made *a priori*, then it will be either *analytic* or *synthetic*. If it is analytic, then it is governed by the principle of contradiction and hence is always bipartite (*quodlibet ens est aut A aut non A* [Any entity is either A or not A]). If it is synthetic, but is to be made on the basis of *a priori concepts* (rather than, as in mathematics, on the basis of the *intuition* corresponding *a priori* to the concept), then we must have what is required for a synthetic unity in general, namely, (1) a condition, (2) something conditioned, (3) the concept that arises from the union of the conditions with its condition; hence the division must of necessity be a trichotomy. (Kant, p. 198; 1790/1987, p. 38)

For Kant, following Descartes, the mind functions in terms of analysis and synthesis. Analysis breaks things down into more fundamental distinct components or concepts, whereas synthesis puts components and concepts together into new compositions and concepts. Kant's threes or triples just represent the different aspects of synthesis. First, there are categories considered by themselves, without reference to other categories. This is a first-level concept. Second, given one category, one considers another category as it is distinct from or contrasted with the first. This is a second-level concept. It represents a form of synthesis, in that the two categories must be considered together to note their distinctness. Third, one considers a pair of seconds joined or synthesized together into a new whole to be treated as a unit. The act of joining these together is a distinct act of the mind from that of forming simple pairs of distinct categories or even from considering individual categories. It involves a third-level concept, a synthesis. Consider an example in linguistics: Suppose we have a *noun phrase*, then a *verbal phrase*, and next we note they are distinct. Then consider their further synthesis in a *sentence*.

Kant (1787/1996) considered the triple in the class of *relation* that contains causality: First, one begins with an attribute of some object without consideration in awareness for any other attribute or object. Kant draws upon Aristotelian philosophy that divides an object's attributes into *substance* (the essential attributes without which the object would not be what it is) and accidents (attributes that may change or be changed without destroying the object's essential nature). Then one considers another attribute of the same or some other object that depends on the original attribute for its state or existence. This requires simultaneously being aware of both attributes and their distinction, as might occur in vision when one sees simultaneously two distinct objects and their respective attributes. This is what originally causality concerned, a relation of dependence between attributes of objects (note that attribute and object stand in a relation also, known as *inherence*). Kant then went on to form a third-level concept from these pairs in the idea of "community," which today we would call a system. This again requires being able to represent all these pairwise relations simultaneously in awareness and how they are interrelated. Consider a community of objects acting on one another's attributes by mutual and reciprocal causation.

Whereas Hume took time and succession in time for granted, Kant held that the categories under the mode of *relation* are essential to determinations of time. Kant held that the sense of change in time was grounded in the substance–attribute relation. The essential or enduring substrate of attributes that do not change determines a *substance*, whereas accidents are changes in some of the attributes of the substance that are not essential to the substance, without which it would not be what it is. Change is sensed in the contrast of the change with the enduring permanence of the substance. Thus, time determination depends on something permanent against which changes can be compared.

Kant held that all changes in time occur according to the law of cause and effect. Hence the determination of objective succession in time depends on relations of cause and effect between the attributes of substances. For example, a clock consists of parts of a mechanism made of enduring and relatively unchanging materials, which nevertheless change in their motions or states according to causes governing the mechanism of the clock. Against the clock we are able to compare objectively the order in which events in appearance occur.

Finally, Kant asserted that "all substances as they can be perceived in space as simultaneous are in thorough-going interaction" (Kant, p. A211). Again mutual interactive causality in a community of substances determines simultaneity.

Whereas Hume's concept of causation involves certain events being succeeded regularly in time by other events, Kant held that causes are not events but powers of substances (here thought of as objects) to produce effects in other substances. Powers are derived from the essential natures of the substances. Furthermore, unlike Hume's concept of successions of events in time, effects occur simultaneously with their causes and not successively in time. Kant (1787) says "The majority of efficient causes in nature are simultaneous with their effects, and the temporal sequence of the latter is occasioned only by the fact that the cause cannot achieve its entire effect in one instant. But in the instant in which the effect first arises, it is always simultaneous with the causality of its cause, since if the cause had ceased to be an instant before, then the effect would never have arisen" (pp. A203, B248). The effect of a change in motion on one billiard ball when struck by another in motion (the cause) is simultaneous with the cause. A billiard ball placed on a pillow causes, by the force of gravity due to its weight, the effect of an indentation in the pillow that is simultaneous with the cause. The die press forces the die into sheet metal, which simultaneously conforms its shape to the die.

Kant also gives the mind more freedom in imagination than Hume to formulate causal hypotheses rather than being passively driven by processes of association based on the order in which sensory impressions are given to the mind.

I think Kant's introduction of synthetic *a priori* concepts was a fundamental insight gained from studying Newton's *Principia* that provided laws of motion. The first law, of inertia, states that an object at rest will remain at rest, and an object in motion will remain in motion in a straight line with constant velocity, unless acted upon by some external force. We have no experience of objects at absolute rest, nor of those in motion without forces acting on them. So this concept does not come from experience. Yet Newton's law of universal gravitation argues that the force of attraction between every pair of bodies in the universe is directly proportional to the product of their masses and inversely proportional to the square of the distance between them. So every body in the universe has a force of attraction acting between it and every other body. For Kant the law of inertia represented an *a priori* principle by which in combination with other principles based on experience and quantities also derived from experience, we may understand the motions of objects. So what Kant teaches us is that reasoning about the world involves both experience and synthesizing concepts introduced *a priori* that provide a framework or scaffold on which experience is built.

Something analogous occurs in our modeling. To represent phenomena by a model, we must introduce *a priori* a minimum set of constraints on the entities of our models so that they may be put in correspondence with observed entities and from which empirical values may be estimated. When this occurs, the number of observed parameters in the phenomenon equals the number of estimated parameters in the model. But any number of such models may be constructed that all reproduce the same observed entities. So, such models cannot be tested for lack of fit, because they necessarily always fit. Hence additional constraints must be imposed *a priori* on the model's entities so that a possibility may exist that the model does not conform to the observed entities. Some of these additional constraints may be taken from prior experience. But there will remain some constraints that by themselves cannot be proven true or false, but simply provide a framework within which additional constraints may be inserted that together with the original constraints may constitute a model that can be tested against observed phenomena.

Causation among Nineteenth-Century Empiricists

British scientific thought in the nineteenth century was basically empiricist and inductive (generalizing from particulars) and mostly unaware of Kant. The British empiricist John Stuart Mill (1874) revived Francis Bacon's method of eliminative induction in developing four methods of discovering causation. Karl Pearson, the founder of modern multivariate statistics at the end of the nineteenth century, was perhaps the last leading nineteenth-century British empiricist. Trained as a physicist, and influenced by the empiricist Austrian physicist Ernst Mach, Pearson (1892/1911) wrote a popular book, *The Grammar of Science*, on the philosophy of science from an empiricist standpoint. Pearson emphasized the idea of causation as association, and linked the new idea of correlation with causality, because correlation was an index of association. At the turn of the twentieth century, Bertrand Russell in England further emphasized causality as association.

Causation in the Twentieth Century

In Germany Kant's idealism evolved into forms he would have hardly recognized as his own. Toward the end of the nineteenth century, however, Austrian philosophers and scientists were more sympathetic with Humean empiricism than German idealism. They initially stressed induction and association. Ernst Mach regarded causation as having been replaced in science by the use of mathematical forms that stressed functional interrelationships among variables such as between the temperature, pressure, and volume of a gas. In Vienna, after World War I, a group of ex-NeoKantian Austrian philosophers of science were impressed with Russell's *Principia Mathematica*, which attempted to derive all mathematics from logic. What impressed them was the fine detail with which one could express relations in experience with the forms of logical propositions. They sought to express all scientific concepts in logical form. This led, as we have already noted, to their reliance on the logical form of material implication, for example, IF A, THEN B, as the relation for expressing contingent dependency relations in experience, but qualifying that by saying that no necessity is inherent in this use of "logical implication"; the logical connections are (as Kant might have argued) provided by the logical language by which concepts are expressed. This group of philosophers became the logical positivists and then the logical empiricists. Because many of them were Jewish, they almost all emigrated from Austria to England and the United States after the rise of Fascism in Germany and Austria in the 1930s, eclipsing pragmatic and realist forms of thought in philosophy of science circles extant in those countries. We have already noted that the logical empiricists held critical views of the causality concept. However, in the late 1960s, logical empiricism underwent severe criticism and philosophers of science since then have diverged into several schools.

Perception of Causation

Hume's Doctrine

David Hume (1777) developed an introspectionist psychology which was very influential in the nineteenth and twentieth centuries. For him the sole reality was what appeared in the mind as vivid impressions of sense and the fainter ideas based upon them. Try as we might to discern what joined ideas together either in space or through time, we never discover an impression of the link or connection between them. Objects are merely familiar collages of impressions or their corresponding ideas arranged in a certain way in space. Causality is a familiar succession of kinds of events through time. Both joinings, either in space or in time, are regarded as associations, which are forms of synthesis. Hume likened association to gravity, which draws objects together but acts at a distance without any perceived link or connection. We can apply the concept of gravity without the need for the idea of a substance linking the objects together, since there is none, nor any impression on which it would be based. Similarly we can apply the concept of association without the need for a link between the ideas that is the basis of the association. Associations simply are driven by the order, arrangement, or successions by which impressions and their corresponding ideas in thought are given in the mind. Causal associations are formed and strengthened through repetition in experience, such as event-kind A's always being followed by event-kind B. We may be aware of an impression of familiarity that accompanies a collage of impressions encountered in the past or a sequence of certain kinds of events experienced previously. The familiarity will lead us, in the case of successions through time, to judge that causation has occurred and that, given that A has occurred, B will follow. But, Hume argued, there is no logical necessity for B to follow A, only that it always has in the past. We can always conceive, logically, that something else may follow A when it occurs. So, causation for Hume was a familiar, regular succession of events in experience that had no more to it than a familiar association and no logical necessity of occurring.

Contemporary Cognitive Psychology

Contemporary cognitive psychology regards empiricism's phenomenalist psychology as limited in a number of ways. To begin with, phenomenalism assumes that everything that occurs in the mind is given in conscious awareness. For the empiricist, knowledge begins with sensory impressions, which become ideas when later the impressions are retrieved in weaker images from memory. The sensory impressions visually are like patches and small blobs of color positioned in space and moving through time. One thinks of the later nineteenth-century impressionist paintings by Monet and Pissarro with their short strokes of light and color, or of the pointilist technique of Georges Seurat using tiny dots of paint. The mind forms these impressions into objects by associating them with similar collages of light and color with which the observer previously became familiar through repetitious experience of them.

J. J. Gibson

Twentieth-century psychologists such as J. J. Gibson (1950, 1966) rejected the impressionist account of empiricism, arguing that we are immediately and directly conscious of solid, invariant objects with textured, three-dimensional surfaces, and edges. There is no such thing as sensation as a conscious awareness of each receptor cell's detection of a stimulus. Rather perception is the gathering and processing of information provided in the physical stimulus, beginning at the receptor, organizing it, and synthesizing it so that we perceive objects and their interactions. So, we are aware of objects and not sensations. Perceptual processing, in fact, most mental functions, occurs outside of conscious awareness. The concept of the *cognitive unconscious*, that mental operations occur beyond the knower's direct awareness, is now a central concept of cognitive psychology. Consciousness begins with attention to the contents of short-term working memory (Baars, 1997).

Next, empiricism's associationist psychology does not endow the animal nervous system with a sufficiently rich set of analyzers and synthesizers that are tuned, practically at birth, to features of the environment that the organism uses to survive. Empiricism was proposed before the theory of evolution and took no account of how organisms' perceptual processes have evolved so that the organisms dealt more efficiently with their physical environments from the time of their births.

As for causality, we shall argue that we directly and immediately perceive causal connections between things. We do that by perceiving quantities in magnitudes and motions of causes that are conserved in the effects of causal exchanges.

Michotte

Michotte (1946/1963) is regarded as the psychologist who first systematically studied the perception of causes experimentally, using animated displays of colliding objects. Michotte's displays were crude mechanical representations of motion and collision of objects. Nevertheless, he was able to create the appearance of one object moving and colliding with another and the second object moving off as a result. By adjusting the timing between when the first object contacted the second and the second object moved off, he was able to elicit or fail to elicit a judgment of causation in his observers (Twardy and Bingham, 2002). Contrary to Hume's assertion that causality involves an association formed by a repeated regular succession, observers often declared causation occurred on the first trial, and in other cases, as with a long delay between contact of the first object with the second object and then the moving off of the second object, no judgment of causation was made, regardless of repeated presentations of the sequence (Twardy and Bingham, 2002).

Twardy and Bingham

Drawing upon Dowe's (1999, 2000) conserved-quantity (CQ) theory of causation, Twardy and Bingham (2002) argue that what perception works with to perceive causal connections is the information of conserved quantities in the exchanges of causal and effect events: "According to the CQ theory, *causal interactions* are marked by the exchange of CQs between or among *causal processes*. A causal process is a (space–time) trajectory of an object, be it a photon or a baseball. CQs are whatever quantities are actually conserved in nature, and these are taken to be those indicated by current physics, such as charge, energy, momentum, and angular momentum" (Twardy and Bingham, 2002, p. 956). They then cite as an example, ". . . a moving billiard ball has energy and momentum in proportion to its speed. It will also have other CQs such as angular momentum . . . . During a collision, the trajectories of two billiard balls briefly intersect, and the objects (processes) exchange energy and momentum. That constitutes a causal interaction" (p. 957). The senses take in information about these quantities and perception processes them. Visually what is seen in a causal event involving two objects is that something in the object regarded as cause is *transferred* to the object that receives the effect. It can be a component of the causal object's trajectory, or velocity, or shape, or mass (in inelastic collisions) to name a few cases. In structural equation modeling, the structural coefficient is multiplied with the value of the causal variable to obtain the component that is "transferred" to the object bearing the effect. However, Dowe (2000) remains agnostic about the transfer of any stuff in one object to the other. Sums of quantities are just conserved. Hence his physical theory is not a transference theory but has other aspects similar to one.

   This is not to argue, Twardy and Bingham (2002) indicate, that our perceptual processes have a built-in precise knowledge of Newton's mechanics. Rather our perceptual systems are sensitive to aspects of the physical determinants of causal interactions, such as energy and momentum, at least to a degree of approximation. Observers can discriminate forms of motion in the trajectories of objects. "The form of the trajectories by which objects exchange energy and momentum over time constitutes the structure of an event" (Twardy and Bingham, 2002, p. 957). Momentum is a vector quantity.

To see how a component of an object's attribute is transferred to another in a causal exchange, consider that a first billiard ball A with a mass of 1 unit moving at a constant velocity of 1.3 units at an angle of 270° collides off-center with a second ball B with a mass also of 1 mass unit, moving somewhat in the opposite direction at 104° with a velocity of 1.7 units. The second ball abruptly moves off to the upper right at 36° with a velocity of 1.4 units. At the same time the first ball moves off toward the lower left at 201° and with an increased velocity of 1.6 units. Something has changed. The first ball has changed direction and picked up 3 units of velocity from the second ball while the second ball also changed direction and has lost 3 units of velocity. So, we will say, "The first ball caused the second ball to move toward the upper right at a somewhat slower speed than it moved before the collision." But we are also authorized to say, "The second ball caused the first ball to be deflected toward the lower left at a somewhat faster speed." Which object is cause and which is effect depends on your focus of interest.

What is exchanged in this collision is some momentum. Momentum is mass × velocity. (Velocity is the rate at which an object's position changes with time, roughly "speed," like miles per hour. Acceleration is the rate with which velocity changes with time. Starting at 0 mph, pressing on the gas pedal will make a car accelerate from 0 to 60 mph in a short interval of time. The acceleration at any point in time is how much the velocity is changing at that point. Mass is a measure of an object's resistance to acceleration, and when measured against the force of the earth's gravity, it is measured in units of weight. Force is equal to mass × acceleration.)

To see how a component of momentum is transferred in the above example, consider the diagram in Figure 3.1.

Figure 3.1 is a diagram of the collision between two balls, A and B, represented by the circles. The vectors shown are momentum vectors. With equal and unit masses for the balls, the lengths of the momentum vectors are directly proportional to the velocities with which the balls are moving in the directions indicated by the vectors. The exchange of momentum in the collision between the two balls takes place on the axis of collision on the line between the centers of the two balls. Each ball contributes a vector of momentum to the other ball equal to the projection of the initial momentum vector onto the axis of collision. (We show the projections although they are also shown again added to the reactive momentum of the other ball.) So P_{BA} is the momentum vector given by A to B, by projecting P_{A1} onto the axis of collision. P_{AB} is the momentum vector "given" by B to A, and is the projection of P_{B1} onto the axis of collision. (The recipient is listed first, then the donor.) But by Newton's third law—to every action there is an equal and opposite reaction—each ball generates a negative momentum vector corresponding to the component of its momentum gained by the other ball. This effectively subtracts from one ball the momentum gained by or transferred to the other. So, we combine the

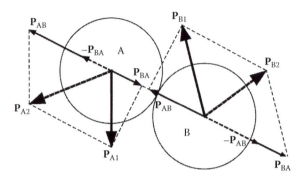

FIGURE 3.1 Physics of collision between two balls of equal mass, moving somewhat in opposite directions. The first ball A moves initially down in the direction of the vector $\mathbf{P}_{A1}$ with a velocity equal to its length. The second ball B moves up and to the left in the direction given by $\mathbf{P}_{B1}$ with a velocity equal to its length. The exchange of momentum between the two balls is on the axis of collision connecting their centers. Ball A then moves in the direction of $\mathbf{P}_{A2}$, whereas B moves in the direction of $\mathbf{P}_{B2}$.

momentum vector gained by the other ball with the negative reaction vector of the ball in question to get a vector equal to their sum along the axis of collision. So, for ball A the vector $\mathbf{P}_{AB}$ is added to $-\mathbf{P}_{BA}$, and the resulting vector is then added to the initial vector of momentum $\mathbf{P}_{A1}$ to obtain the resulting vector $\mathbf{P}_{A2}$, the new direction and momentum of A. (By the parallelogram law of vector addition, we can construct the resultant vector which points in the direction in which the ball will then move.) Similarly for ball B we add $-\mathbf{P}_{AB}$ to $\mathbf{P}_{BA}$ to obtain a vector that we in turn add to $\mathbf{P}_{B1}$ to get the new momentum and direction $\mathbf{P}_{B2}$ for B. Causally, the effect of A's collision on B is $\mathbf{P}_{BA}$, whereas the effect of B's collision with A on A is $\mathbf{P}_{AB}$. Because in each case the effect constitutes a transfer of a component vector of momentum to the other ball, it must also be subtracted from the original ball.

Wolff

Kinetic models of perceived causation seek to map visible physical properties of an event onto perceptual judgments, for example, shapes, sizes, positions, trajectories, points of contact, velocities, and accelerations of entities in a situation. These are given in the case of the collision of two moving balls by the velocity and directions of their motion. These quantities are easily perceived in most instances, although they may not by themselves yield exact causal knowledge. Dynamic models concern the "invisible properties," for example, underlying energies and forces (mass × acceleration) and momentum (mass × velocity). People also perceive forces, momentums, and energies, at least approximately (Wolff, 2007). So, there is much information in the physical world involving causality that is available to perception.

But Wolff (2007) argues that any model of the perception of causality (he prefers just a dynamic model, which I think is too limited, since kinetic information is often sufficient) must map the physics to perceptual categories. He considers four causation-related perceptual categories: CAUSE, ENABLE, PREVENT, and DESPITE. I accept these as genuine concepts. However, I have reservations about the manner in which he analyzes these into more fundamental components, although I think he is on the right track when he seeks to understand them in terms of objects interacting. The major problem is that Wolff (2007) analyzes these concepts in terms of just two objects, an affector and a patient, whether the patient initially is tending toward the endstate to be produced by the cause, whether affector and patient are in concordance in their forces, and whether the result approaches the endstate or not. ENABLE, PREVENT, and DESPITE, in my view, based on Lakoff and Johnson's (1999) event-structure schemas, concern movements of an object along a path toward a prespecified goal of a sentient being, where movement is a metaphor for causal action and the path is whatever requires specific causes to be applied to reach the goal. But these three concepts also require the presence or absence of third objects that provide forces which Wolff's account lacks. ENABLE is instanced by sentences such as "A crank enabled him to close the window" (Wolff, 2007, p. 84) or "John's help enabled Bill to build his boat." PREVENT is instanced by "The fence *prevented* the aliens from entering the country" or "The use of doxycycline *prevents* an infection with malaria." DESPITE is instanced by "The river flooded the town despite the dikes" (Wolff, 2007, p. 88), "The wind caused the boat to capsize despite their reefing the sails," or "The aliens entered the country despite the fence."

Lakoff and Johnson

Lakoff and Johnson (1999) would analyze PREVENT as someone's doing something ahead of time to provide counterforce against the movement of the object to its goal, say, by erecting a barrier in its path or deflecting the movement into a different path. DESPITE would be the inadequacy of the counterforce to stop the movement of the object to the goal. ENABLE has many schemas to which it could apply: A bridge or boat could enable one to cross a stream, a ladder enables one to climb a fence, a handle enables one to crank a window up, and a key unlocks a gate through which one must pass. Here the enabler is a third object that does something (cause) that makes it possible for the causal efforts of attaining the goal to function. Here the enabler is a nonlinear moderator of the affector's causation. Or enabling via helping could be another object providing extra force (cause) to overcome any counterforce (difficulties, barriers, and deflectors). Here helping is additive in its effect. Or enabling may be accomplished by allowing or letting, that is, some third object removing any barriers, so that pre-existent causes may function.

Wolff (2007) tries to represent causation by a resultant of forces approaching an endstate, equating the endstate of CAUSE with the endstates for the other concepts, which are prespecified. For Wolff, *a cause is an object (affector) whose motion or force alters the initial motion or location of another object (patient) to approach a specific endstate.* But causes can occur whether or not there is a *prespecified* endstate to attain. There just has to be an altering of motion or location of the patient in a regular way. Regularity is not the same as a prespecified goal to be attained. Furthermore, the causal object may be altered itself in its motion away from the endstate for the patient. We see this in the example of the collision of two billiard balls. This reflects Newton's third law that to every action there is an equal and opposed reaction. Consideration of this law is not found in Wolff's account, but it is part of the effect on the patient.

Wolff (2007) next considers how perceptual concepts of physical causation can be mapped to social and psychological situations analogically, which leads us to Lakoff's metaphor theory of causal concepts.

Contemporary metaphor theory (Lakoff, 1987, 1993; Lakoff and Johnson, 1980, 1999; Lakoff and Nuñez, 2000) would then argue that the perceptual schema of a physical causal exchange can become the basis of a metaphor for conceptually joining in working memory percepts from memory and from graphic materials to form concepts of "causal connections." The schema involves a cause and an effect and something exchanged between them. Lakoff and Johnson (1999) argue that causality is a "radial concept," meaning that it has a central core schema that is extended in different ways in different concepts of causation. We will discuss this in more detail later.

Causality

Is Causality Material Implication?

Empiricist philosophers since Hume (1739/1969, 1777/1975) had analyzed causes and effects in terms of binary events. If A occurs, then B occurs. If A does not occur, B does not occur. This fostered analyses of causation along the lines of logical implication, which works with the binary true/false system. The binary treatment of causation was also accompanied by the belief that all events are logically independent of one another. David Hume, the empiricist, had argued that logically there is no logical necessity that any kind of event in experience must accompany or follow another. Events are logically independent. It was on the basis of such an idea that Wittgenstein (1922/1978) together with Russell (1918, 1919) formulated the theory of logical atomism, the idea that fundamental elements of experience can be represented by elementary propositions that are either true or false and logically independent of one another. Connections between atoms of experience were to be supplied by association, which, following Hume, does not represent a logically necessary connection. This was the key idea underlying the later logical empiricism of the Viennese positivists who believed relations of empirical dependence between scientific propositions must be expressed in the *form* of *material implication*, but without the notion of logical necessity inherent in the form.

Nevertheless, there are paradoxes if one uses "logical implication" as the form of the causal relation. "If A then B" implies logically "If not B, then not A." For example, "If it is raining, then the ground is wet" implies "If the ground is not wet, then it is not raining." If one treats material implication as causality, then "The ground's not being wet *causes* it not to rain" is a consequence, and this seems to contradict our sense of causal order and agency here. Furthermore, what instances correspond to "not wet" and "it is not raining?" If we assume all events and attributes are logically independent, "not wet" can be any attribution other than "wet," say, "rocky," and "it is not raining" can be "it is windy." Hence, "If the ground is rocky, then it is windy" is an instance of the same relation. But our sense that there are constraints on the way in which we categorize events not represented in the formal logic makes us suspect its use here for causality.

Wittgenstein (1975) finally came to the realization of a fundamental mistake in logical atomism: We do not describe experience in terms of logically independent atomic propositions. If we did, then we would be able to say, "The bar is 45 cm long" and "The bar is 42 cm long," at the same time. But when we state that "The bar is 45 cm long," we imply at the same time that the propositions, "The bar is 41 cm long," "The bar is 42 cm long," and so on are all necessarily false. Their truth or falsity is not independent of the truth of "The bar is 42 cm long" and so on. If we say "The ball is red" is true, we at the same time imply 'The ball is green," "The ball is blue," "The ball is yellow," and so on are false. In other words, attributes in our languages come joined logically in sets of mutually exclusive categories, frequently having more than two categories or values, and to any object we assign only one of the possible values in the set of values, thus constituting a variable. Within such classes of attributes that constitute a variable, the varying attributes are not mutually independent.

Is Causality a Functional Relation between Variables?

With the decline of logical empiricism in the late 1960s, the attitudes of philosophers of science became more tolerant of the concept of causality. Bunge (1959) reawakened interest in causality among philosophers of science by his thorough survey of the topic. Simon (1952, 1953, 1977) and Simon and Rescher (1966) argued that causality does not take the form of logical implication but rather of an asymmetric functional relation between variables. A functional relation is a relation between two sets, a first set and a second set, wherein to each element in the first set there is assigned one and only one member from the second set (Shapiro and Whitney, 1967, p. 53). Typically in science, the two sets correspond to the sets of values taken on by the variables.

Thus one could designate Y as a dependent or effect variable and X as an independent or causal variable and write $y = f(x)$, which states that a value y of the variable Y is a function of the value x of the variable X.

The implication I drew from this (Mulaik, 1986, 1987) was that, in support of Simon's view, causality, which concerns relations of dependence between attributes of things in the world, must be expressed between variables and not just specific values of variables. The quantification of attributes leads us to treat the attributes of objects as quantities, varying across objects, and hence as variables. Within variables there are logical dependencies among values of the variable. Between variables there are only contingent dependencies. The attention philosophers have given to the relations between specific pairs of events is a limited perspective. The whole of causality involves a functional relation between variables—the form for expressing relations of dependence between variables. It is true that in a functional relation there are relations or links between the elements of the domain and the range, but there are numerous such links, and none of them is the functional relation, but a component of the functional relation, which integrates them in a higher-level synthesis that concerns the dependence of the variable of the range on the variable of the domain. It is also important to see the big picture, that "causality" is a relation expressed in a language of objects, where according to the "grammar," so to speak, of this language, objects are bearers of properties, and causality concerns functional relations of dependence between the attributes of objects, with attributes grouped into logically inter-dependent families known as variables (Mulaik, 1987, 2004). Contemporary science thus expresses causal relations as functional relations between variables. In this book, however, we focus principally on those causal relations that are linear relations.

Still, Why Variables?

Some may still demur at the idea that we need to discuss causality in terms of variables. Aristotle, and also Hume, Mill, Kant, and the logical empiricists, and many modern philosophers of science, did not treat causality in terms of variables.

We discussed earlier how Kant formed concepts in threes. His tripartite formation of inherence, causality, and community was one way in which attributes of objects could be organized in three categories. But this is not the only three-level concept that we can form by synthesizing distinct attributes. The idea of a variable is itself a third-level synthesis. First there is an attribute. Then there is another attribute distinct from it such that if the first attribute describes an object, the second attribute cannot also do so at the same time. If something is 100 kg in weight, it cannot also be 101, 95, or 99 kg, and so on. That attributes fall into sets of mutually exclusive application is just the way our brains work in categorizing things. Wittgenstein (1953) held that this is just a matter of *a priori* "grammar" that lays down the rules of how the attributes are to be applied. There is no need, he held, for any further explanation. In other words, this is how we do it, how we play the game. A variable is a set of attributes such that no two members of the set can apply or describe the same object at the same time. We say the members of this set are mutually exclusive with respect to applying to another.

The nineteenth-century philosopher of science C. S. Peirce, whom many philosophers believe was way ahead of his time in logic and scientific method, drew heavily on Immanuel Kant's (Kant, 1787/1996) insight that many concepts come in threes. Considering the set of mutually exclusive attributes or properties as a *whole* is a third, in Peirce's terminology. It is a synthesis of all those pairs of attributes that stand in the relation of mutual exclusivity with respect to applying to some object. A variable is a higher-level concept than the concept of a simple attribute, or a simple pair of attributes that stand in a relation of mutual exclusivity. A variable concerns all such pairs simultaneously. A comprehensive logic of causality must recognize the way in which attributes are joined into sets of mutual exclusive attributes with respect to their application to objects.

Sciences begin, Peirce held, by making qualitative distinctions: Object A does or does not have a certain quality. This functions at the level of second-level concepts. But for a science to advance, it has to attain the level of thirds as well. This occurs "... when, no longer content with such rough distinctions, we require to insert a possible half-way between every two possible conditions of the subject [or object] in regard to its possession of the quality indicated by the predicate. Ancient mechanics recognized forces as causes which produce motions as their immediate effects, looking no further than the essentially dual relation of cause and effect. That was why it could make no progress with dynamics" (Peirce, 1931, 1.359). Modern physics developed only when the concept of a quantitative variable came into physics to deal with dynamics, which involves third-level, continuous quantities.

So, if there are variables, distinct variables are different sets of mutually exclusive attributes or quantities that may themselves be applied simultaneously to the same object. But once we have come to regard a variable as a unity in thought, we may regard a variable then as a first-level concept in a new hierarchy of syntheses. A functional relation between variables is then a second-level concept, a synthesis of pairs of variables, such that given two variables, if a value of a first variable is assigned to an object, it maps by a causal relation to one, and only one value of the second variable also assigned to the object. A third level involving a second level of synthesis would be a linear causal model involving functional and correlational relations among sets of variables, including reciprocal causal relations between them in some cases.

Causality and Counterfactuals

According to Wikipedia, the philosopher David Lewis (1973, 1979, 2000) held that causal statements are to be understood as counterfactual statements.

A counterfactual statement is a statement about something that is imagined or nonfactual. Because causal relations are often stated as occurring between events, for example, "A is a cause of B," we interpret this as implying "If A occurs then B occurs" (an imagined antecedent A and consequence B). We furthermore take this as allowing us to argue that "If A had occurred, then B would have occurred," even though A in fact had not occurred. "If A had occurred, then B would have occurred" is a counterfactual statement.

Regarding a causal relation as a functional relation between variables provides a perspicuous view not only of all the possible values of the causal variable in the first set of the functional relation and all the possible values of the effect variable in the second set, but also of the "causal links" between these values in these two sets as given by the mapping between the elements of these sets by the functional relation. Now, no object can take on more than one attribute value of a variable at a time. If A and B are two attribute variables, then at any instance we should observe only one value of both A and B to occur. If the variables A and B are attribute variables related by a functional relation in which the value A_i maps to B_j, then suppose we observe A_i occur, we should then expect to observe B_j occur. Suppose another pair in the functional relation is (A_h, B_k); then even though neither of them has occurred, are not a fact of existence, we would still be licensed by the functional relation between A and B to say "If A_h had occurred, then B_k would have occurred. In fact, all pairs (A_h, B_k) in the mapping of the functional relation are potential components of a counterfactual statement "If A_h had occurred, then B_k would have occurred," given some other pair (A_i, B_j) in fact has occurred.

Now, it is my contention that philosophers in their analysis of causality and counterfactuality have been all too wedded to the schema of a relation between binary variables (A, not A) and (B, not B), perhaps because functional relations are mathematically challenging for them. But then they wish to assert the negative counterfactual statement "If A had not occurred, then B would not have occurred." This is true as far as it goes, as long as A and B are binary variables. But can this binary paradigm apply to the cases when A and B are not binary variables but multivalued variables, connected by a functional relation in which it just happens that more than one value of A maps to the same value of B? Consider a sine function. Could we then say something like "If A_i had not occurred, then B_j would not have occurred," if another value of A, A_m maps to B_j as well? Being able to state this negation counterfactually depends on the nature of the functional relation between the variables in question and whether or not B_j is linked to more than one value of the causal variable. So, negative counterfactual statements are not the same as positive ones and are not as generally true counterfactually.

What we have considered so far concerned counterfactuality in connection with relations between values of variables of a functional relation between multivalued variables. Problems arise in understanding causality if one tries to impose binary relations on such cases. One must have the perspicuous view of all the possible values of a variable that rules out other values not considered.

But another, analogous problem arises if one fails to consider causality counterfactually within a system of variables, where more than one variable may be the cause of a given variable. Structural equation modeling considers systems of variables wherein it is possible for more than one variable to be a cause of a given variable. Again one must have the perspicuous view of a closed, self-contained system of variables, because one cannot establish a functional relationship between two variables if one cannot control for all other immediate causes of the effect variable than the causal variable in question. If more than one value of an effect variable B occurs in the presence of a given value of the causal variable A, either A is not really a causal variable for B, or there is some other variable U, unmeasured or unobserved, that is also a cause of B. We cannot then state counterfactually that "if A had been A_i then B would have been B_j." B might then be $B_j + B_n$, and this varies with the value of B_n, and this in turn with the value U_g of U, because of the causal relation between U and B that maps U_g to B_n.

Consequently, an assumption, often implicitly made, but it should be explicitly evaluated for empirical validity when feasible, is that our models involve closed, self-contained systems of variables. Attempts to make this happen involve closing off the effects of extraneous causal variables by shielding, by holding extraneous variables constant when not including them in the model, by manipulation of causal variables so that causes are independent of any other unmeasured effects, by randomization of the occurrence of value of causal variables (again to achieve independence), and by isolation. By including otherwise extraneous causal variables explicitly within a model, one can also control for their effects mathematically by conditioning on the extraneous variables when studying the relation between a focal cause and focal effect.

Probabilistic Causality

A common approach to probabilistic causality among philosophers of science today is that causes increase the probability of effects (Dowe and Noordhof, 2004). But this reflects a lack of a full understanding of probability theory and a failure to consider causality in terms of variables. If some event increases in probability, then necessarily the probability of its complement decreases. So, causes both increase and decrease the probability of events, depending on one's focus. It is more important to know the probability of an event given the occurrence of some cause. But how might we reconcile that with causality as a functional relation between variables? There are those who still argue that because causality as a functional relation implies determinism, causality has now been abandoned because modern science now views the world probabilistically. That argument too can be defeated by showing how functional relations can be combined with probability. Simon (1977) gave a hint of how to do this when he wrote, "... we can replace the causal ordering of the variables in the deterministic model by the assumption that the realized values of certain variables at one point or period in time determines the probability distribution of certain variables at later points in time" (p. 54).

Exploiting Simon's hint, I argued (Mulaik, 1986) that probabilistic causality can be expressed using functional relations in the following way: Causal variables determine the probability distributions by which the values of dependent variables occur. Max Born (1951) had a similar idea for retaining causality in quantum physics. Thus with causality a functional relation between two sets, the first set, the domain, contains the values of a causal variable, the second set, the range, contains as its "elements" probability *distributions* defined on the dependent variable, so that for each value of the causal variable there is assigned to it one and only one probability distribution from the set of probability distributions in the second set. This is illustrated in Figure 3.2. Thus, even though pure determinism is ruled out in a probabilistic world, so that causes no longer determine individual events, there is still enough determinism to determine the probabilities with which events occur. A more formal "epistemological" definition of probabilistic causality follows, adapted from Mulaik (1986):

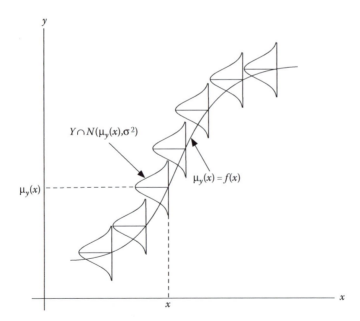

FIGURE 3.2 An illustration of an example of a probabilistic causal relation wherein a causal variable x determines by a functional relation $\mu_y(x) = f(x)$ the mean of a normally distributed probability distribution, and thereby the distribution itself (assuming invariant variance), by which a dependent variable y *occurs*. (Adapted from Mulaik, S. A. (1993). Objectivity and multivariate statistics. *Multivariate Behavioral Research, 28,* 171–203.)

Definition: A variable X, representing states of nature, is said to be a prima facie probabilistic cause of a variable Y, also representing states of nature, if—given certain background conditions C establishing causal direction and order, mediating mechanisms and connections, relevance criteria, closure, stability, and (where appropriate) the form of the joint probability distribution of the system of variables—there exists a function $P : X \to \mathbf{P}(Y)$ that assigns to each value x of X a unique probability distribution function $P_x(y)$ in the set $\mathbf{P}(Y)$ of probability distribution functions defined on Y and the following conditions hold: In the domain of the relation there exist at least two distinct elements x_i and x_j corresponding to two distinct probability distributions $P_{x_i}(y)$ and $P_{x_j}(y)$ in the range of the function and the states corresponding to the values of X occur before or simultaneously with the states represented by the values of Y.

This definition of probabilistic causality is broadly applicable. It can apply to both formal and efficient causation, but efficient causation usually requires additional assumptions involving the specification of agents and how *changes* in the states of agents, reflected in changes in certain variables describing these agents, produce changes in the effect variables or their probability distributions. It will also include not only cases of linear causal modeling, where the disturbance or error variables introduce probabilistic elements into the dependent variables, but also other cases in psychological measurement. In fact, my inspiration for the idea that causal variables determine probabilities was the Rasch model of item response theory. In that case the subject ability and item difficulty parameters are variables that determine the probability distribution of a binary response variable, "the answer is correct vs. the answer is incorrect." (*Note*: the Rasch model and other IRT models are models of *formal* causation and not efficient causation, although under some circumstances variables such as item difficulty and subject ability might be altered as effects of other causal variables).

The function $P_x(y)$ is realized in the matrix $\mathbf{P} = [p_{ij}]$ of Markov transition probabilities, where $p_{ij} = p(y_j \mid x_i)$ is the conditional probability of y_j given x_i. Mathematicians have studied Markov transition probabilities extensively in connection with sequences of random variables $X(1), X(2), \ldots, X(n-1)$ known as *Markov processes*. In such sequences the probability of a value of a variable $X(j)$ in the sequence is stochastically dependent on only the value of the immediately preceding variable in the sequence, and none preceding that. This property is known as the *Markov property* and is given by the equation

$$p(x(j) \mid x(j-1)) = p[x(j) \mid x(j-1), x(j-2), \ldots, x(1)], \qquad (3.1)$$

which shows the independence of $X(j)$ from all those variables preceding $X(j-1)$.

As an example, consider two discrete random variables X and Y, where X is measured at time t and Y is measured at time t', $t < t'$. We will assume that X takes on four values and Y takes on five values. A matrix of transition probabilities for the variables X and Y is given by

$$\mathbf{P} = \begin{array}{c|ccccc} & Y_1 & Y_2 & Y_3 & Y_4 & Y_5 \\ \hline X_1 & p_{11} & p_{12} & p_{13} & p_{14} & p_{15} \\ X_2 & p_{21} & p_{22} & p_{23} & p_{24} & p_{25} \\ X_3 & p_{31} & p_{32} & p_{33} & p_{34} & p_{35} \\ X_4 & p_{41} & p_{42} & p_{43} & p_{44} & p_{45} \end{array} \cdot$$

An element in this matrix p_{ij} is a *transition probability* which is the conditional probability of Y_j given X_i has occurred. The elements in each row of $\mathbf{P}$ must add up to unity. Thus there is a functional relation between the values of X and the conditional probability distributions of Y given X.

The following is a useful property of Markov processes that is relevant to the concept of probabilistic causality. If X is a probabilistic cause of Y, and Y is a probabilistic cause of Z, then X is a probabilistic cause of Z, that is, probabilistic causality is *transitive*. For example, if X is a probabilistic cause of Y, this implies the function $P : X \rightarrow \mathbf{P}(Y)$. Similarly, if in turn Y is a probabilistic cause of Z, this implies the function $Q : Y \rightarrow \mathbf{Q}(Z)$. PQ is then the function $PQ : X \rightarrow \mathbf{PQ}(Z)$, where PQ is realized in the matrix product $\mathbf{PQ}$. This result implies that $\overline{pq}(z_k \mid x_i) = \sum_j p(y_j \mid x_i)q(z_k \mid y_j) = \sum_j p_{ij}q_{jk}$. This is a well-known result for Markov processes. It generalizes to causal series such as

$$X \xrightarrow{\mathbf{P}} Y \xrightarrow{\mathbf{Q}} Z \xrightarrow{\mathbf{R}} V \xrightarrow{\mathbf{S}} W,$$

with W a function of X given by $PQRS : X \rightarrow \mathbf{PQRS}(W)$, where $\mathbf{P}, \mathbf{Q}, \mathbf{R}$, and $\mathbf{S}$ are Markov transition matrices containing the probabilities by which each succeeding variable's values occur given a certain value of the preceding causal variable occurs.

When is a variable causally independent of another variable? We will say that, conditional on any other direct causes of Y, Y is causally independent of X, if for each value x of the variable X, the conditional probability distribution $p(y \mid x)$ is the same distribution as $p(y)$. In other words, variation in x makes no difference to the probabilities with which y occurs. This implies, in the case of Markov transition probability matrices, that such matrices will have rank 1, since the elements in each row will be the same.

So far, we have indicated that a causal variable affects the probability distribution with which an effect variable occurs. There are many ways by which this might be modeled. For example, if the form of the probability distribution of the effect is given by background conditions (e.g., the distribution of the effect is normal), then a causal variable may determine some or all the values of the parameters of the distribution, so that the parameters of the distribution are functions of the causal variable. The causal effect is shown in the variation in the values of the parameters of the mean and/or variance of the effect variable. In linear models with normal distributions for the effects, a unit change of the causal variable will produce a specified change in the mean of the effect distribution.

As I indicated in Mulaik (1986), the concept of Markov processes is useful for elucidation of probabilistic causation between two variables, but does not provide the whole story about probabilistic causality. I noted: "Variables in nature can be connected together in complex causal networks in which a variable can be the combined effect of numerous independently acting causes and in turn the cause of numerous other variables. The simple model of a Markov process, which involves a simple sequence or chain of 'singly connected' variables ... is inadequate as a representation of the complexities involved in such networks. The value of the analysis in terms of a Markov process is to show how causality between a cause variable and an effect variable operates, if one isolates the two variables in question from varying causal influences of other variables" (p. 323).

But in cases where a single variable Y is the effect of each of k other causal variables, $X_1, X_2, \ldots, X_k$, which we will call, after Pearl (2000), the *Markovian parents* of Y, we may treat the causal variables jointly as a single variable. In this case the causal variables span a space of k dimensions, and the points in this space, whose coordinates $(x_1, x_2, \ldots, x_k)$ are given jointly by the respective values of each of the k variables, are then the elements of the domain of a functional relation that maps these points onto a set of *conditional* probability distributions $f(y \mid x_1, x_2, \ldots, x_k)$ defined on the variable Y, such that at least two of these probability distributions are distinct.

Local Independence

In the early development of probabilistic models, authors frequently introduced an assumption known as the *conditional independence assumption* (Anderson, 1955, 1959) or the *axiom of local independence* (Lazerfeld and Henry, 1968), or the *local independence assumption* (Lord and Novick, 1968). Rarely in these cases did these authors explicitly describe these models as causal models. Perhaps this was because they wrote in an era when causality was a suspect concept. The focus of these discussions was furthermore on error distributions rather than probabilistic causality. Nevertheless, by the definition of probabilistic causality just formulated, their models were causal models, and the local independence assumption (which I will call it here) is both a necessary and sufficient condition for probabilistic causality (within a closed system of variables).

LOCAL INDEPENDENCE THEOREM Let X and $Y_1, \ldots, Y_q$ constitute a closed system of variables, implying that they are independent of variables outside the system. A necessary and sufficient condition within this system that X is the sole common cause of the variables $Y_1, \ldots, Y_q$, respectively, is that

$$f(y_1, \ldots, y_q \mid x) = f_1(y_1 \mid x) \cdots f_q(y_q \mid x)$$

for all values x of X, where $f_j(y_j \mid x), j = 1, \ldots, q$, is the conditional probability distribution for the variable Y_j given $X = x$, and there exists for each variable Y_j at least two values of X, x and x', such that $f_j(y_j \mid x) \neq f_j(y_j \mid x')$.

PROOF We will first prove the necessity of the condition. We begin with the case where $q = 2$ and prove that it is true. Then we extend the result by induction and recursion to the case where $q > 2$.

Given X and $Y_1, \ldots, Y_q$ constitute a closed system of variables and $q = 2$. If by hypothesis X is the sole cause of Y_1 and Y_2, respectively, then

$$f(y_1 \mid y_2, x) = f_1(y_1 \mid x) \tag{3.2a}$$

and

$$f(y_2 \mid y_1, x) = f_2(y_2 \mid x) \tag{3.2b}$$

because there exists for each variable Y_j a single functional relation $F : X \to F(Y_j), j = 1, 2$, that maps each value of X, x, into one and only one distribution $f_x(y_j) = f_j(y_j \mid x)$, and there exists for each variable Y_j, $j = 1, 2$, at least two values of X, x and x', such that $f_j(y_j \mid x) \neq f_j(y_j \mid x')$. It would be a contradiction of $X's$ being the sole probabilistic cause of Y_j if more than one probability distribution for Y_j were associated with a given value of x.

Note now that by the definition of conditional probability, we may write

$$f(y_1 \mid y_2, x) = \frac{f(y_1, y_2, x)}{g_2(y_2, x)}, \tag{3.3a}$$

$$f(y_2 \mid y_1, x) = \frac{f(y_1, y_2, x)}{g_1(y_1, x)}, \tag{3.3b}$$

where $g_1(y_1, x)$ is the joint density of Y_1 and X at (y_1, x), and $g_2(y_2, x)$ is the joint density of Y_2 and X at (y_2, x). Again, because of the definition for conditional probability,

$$f_1(y_1 \mid x) = \frac{g_1(y_1, x)}{h(x)}, \tag{3.4a}$$

$$f_2(y_2 \mid x) = \frac{g_2(y_2, x)}{h(x)}, \tag{3.4b}$$

where $g_1(y_1, x)$ and $g_2(y_2, x)$ are joint densities of Y_1 and X at (y_1, x) and Y_2 and X at (y_2, x), respectively, and $h(x)$ is the marginal density of x.

Substituting the right-hand side of Equations 3.3a and b for the left-hand side of Equations 16.2a and b, and the right-hand side of Equations 3.4a and

3.4b for the right-hand side of Equations 16.2a and b, respectively, we obtain

$$\frac{f(y_1, y_2, x)}{g_2(y_2, x)} = \frac{g_1(y_1, x)}{h(x)}, \tag{3.5a}$$

$$\frac{f(y_1, y_2, x)}{g_1(y_1, x)} = \frac{g_2(y_2, x)}{h(x)}. \tag{3.5b}$$

Dividing both sides of Equation 3.5a by $h(x)$ and multiplying both sides by $g_2(y_2, x)$, we obtain

$$\frac{f(y_1, y_2, x)}{h(x)} = \frac{g_1(y_1, x)g_2(y_2, x)}{h^2(x)},$$

or after distributing the $h(x)$ on the right to each term on the right,

$$f(y_1, y_2 \mid x) = f_1(y_1 \mid x)f_2(y_2 \mid x).$$

Similarly dividing both sides of Equation 3.5b by $h(x)$ and multiplying both sides by $g_1(y_1, x)$, we obtain

$$\frac{f(y_1, y_2, x)}{h(x)} = \frac{g_1(y_1, x)g_2(y_2, x)}{h^2(x)}$$

or

$$f(y_1, y_2 \mid x) = f_1(y_1 \mid x)f_2(y_2 \mid x),$$

which proves the case for $q = 2$, which we may extend by induction and recursion to the case where $q > 2$, thus proving the theorem. Another term for this result is "conditional independence."

Next we prove sufficiency: Assuming again a closed system of variables and given for all x,

$$f(y_1, \ldots, y_q \mid x) = f_1(y_1 \mid x) \cdots f_q(y_q \mid x).$$

This implies that the variables $y_1, \ldots, y_q$ are mutually conditionally independent given values of X, which in turn implies for the subset of variables $y_2, \ldots, y_q$, obtained by eliminating y_1 from $y_1, \ldots, y_q$, that

$$f(y_2, \ldots, y_q \mid x) = f_2(y_2 \mid x) \cdots f_q(y_q \mid x).$$

Thus we may divide both sides of

$$f(y_1, \ldots, y_q \mid x) = f_1(y_1 \mid x) \cdots f_q(y_q \mid x)$$

by $f(y_2, \ldots, y_q \mid x)$ to obtain

$$f(y_1 \mid y_2, \ldots y_q, x) = f_1(y_1 \mid x).$$

In fact, we may cycle each of the Y variables into the first position so that we may write for the jth variable

$$f(y_j \mid y_1, \ldots, y_{j-1}, y_{j+1}, \ldots, y_q, x) = f_j(y_j \mid x),$$

which shows that each Y variable is conditioned only by X and is conditionally independent of the other Y variables. So, there is for each Y_j one and only one distribution $f_j(y_j \mid x)$ conditional in any way on x. If further there exists for each variable Y_j at least two values of X, x and x', such that $f_j(y_j \mid x) \neq f_j(y_j \mid x')$, then there exists a functional relation $F : X \to \mathbf{F}(Y_j)$ that maps at least two values, x and x', of the variable X, each to a unique distribution for Y_j, implying that X is a probabilistic cause of Y_j. Furthermore, because each variable Y_j is conditionally independent of the other Y variables, it cannot be a probabilistic effect of them. Because the system of variables is closed, there can be no other variables that are causes of Y_j than variable X. So, X is the sole common cause of the variables $Y_1, \ldots, Y_q$.

The result just proved extends easily enough to the special case where $Y_1, \ldots, Y_q$ represent a stochastic sequence of observations of the same variable. A multivariate extension of this theorem also follows easily:

Theorem on local independence as a necessary and sufficient condition that within a closed system of variables, one subset of variables contains the sole causes of another distinct subset of variables. Given a closed system of variables $X_1, \ldots, X_p$, $Y_1, \ldots, Y_q$, a necessary and sufficient condition that $X_1, \ldots, X_p$ are the sole probabilistic causes of $Y_1, \ldots, Y_q$ is that for all x

$$f(y_1, \ldots, y_q \mid x) = f_1(y_1 \mid x) \cdots f_q(y_q \mid x), \qquad (3.6)$$

where $x = (x_1, \ldots, x_p)$ is a specific joint realization of the variables in $X_1, \ldots, X_p$, and there exists for each variable Y_j at least two realizations of the X variables, x and x', such that $f_j(y_j \mid x) \neq f_j(y_j \mid x')$.

PROOF The proof follows immediately from the previous theorem by replacing x by x.

Implications for factor analysis. The implications of local independence for factor analysis is that in certain special circumstances, the assumptions of the lack of correlation between the unique factors may represent a local independence assumption, primarily in the case where all variables, manifest and latent, of the model have joint multivariate normal distributions. In the common factor model, $\mathbf{Y} = \mathbf{\Lambda X} + \mathbf{\Psi E}$, if $\mathbf{X}$ and $\mathbf{E}$, the common and unique factor variables, respectively, have a joint multivariate normal distribution, say, with means of zero (for convenience) and variance–covariance matrix

$$\mathrm{var} \begin{bmatrix} \mathbf{X} \\ \mathbf{E} \end{bmatrix} = \begin{bmatrix} \mathbf{\Phi_{XX}} & \mathbf{0} \\ \mathbf{0} & \mathbf{I} \end{bmatrix},$$

then by elementary theorems of multivariate statistics, $\mathbf{Y}$ has a multivariate normal distribution with a null mean vector and variance–covariance matrix

$$\Sigma_{YY} = \Lambda \Phi \Lambda' + \Psi^2.$$

The conditional distribution of $\mathbf{Y}$, given $\mathbf{X} = \mathbf{x}$, is then a multivariate normal distribution with variance–covariance matrix

$$\mathrm{var}(\mathbf{Y} \mid \mathbf{x}) = \Sigma_{YY} - \Lambda \Phi \Lambda' = \Psi^2,$$

which is a diagonal matrix. Variables having multivariate normal distributions whose variance–covariance matrices are diagonal are jointly independent. Requiring in factor analysis under multivariate normality that the unique factor variance–covariance matrix is diagonal amounts to assuming that the observed variables are independently distributed if conditioned on their common factors, which is local independence. Because the common factor model is often applied to variables that do not have multivariate normal distributions, requiring in these cases the variance–covariance matrix of the observed variables to be diagonal when one partials out the common factors amounts to a weaker form of independence known as *linear experimental independence* (Lord and Novick, 1968, p. 45). This does not establish causality. But in such situations the common factor model may be used simply as an analogue of a causal model. On the other hand, if the residual variables are uncorrelated after the common factors are partialed out and have a joint multivariate normal distribution, then this implies independence, and causality may be invoked. So, even if the manifest variables do not have multivariate normal distributions, if partialing out the common factors produces uncorrelated residual variables that have multivariate normal distributions, this may allow one to establish causality for the common factors. These conclusions will be extended to the properties of the disturbances of structural equation models (SEMs), which are analogues of the unique factors of common factor analysis.

Is Causation Manipulation?

We have already considered how Locke (1694/1905/1962) regarded causation as reducible to the experience of willful movement of parts of one's body as the basis for the concept of causation. Since in willing actions of our bodies we come to manipulate things in the world, this concept of causation has been extended to the idea that only those things in the world that can be in principle manipulated should be regarded as causes. Experimenters especially favor this view of causation because they tend to believe that only by experimentation involving manipulation can causality be demonstrated. It is true that we frequently regard inferences of causality from experiments to be well founded, but I argue that the validity of such inferences is not a result of

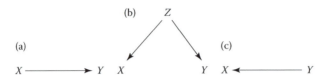

FIGURE 3.3   Three forms of causation that produce a correlation between X and Y.

manipulation *per se* but because manipulation of a causal variable frequently, but not always, interrupts the effects of an unmeasured extraneous cause on both the causal variable and the effect variable.

The illusion of causation can occur between two variables X and Y, when X is correlated with Y. But X may not be a cause of Y but rather X and Y may have a common cause Z, which accounts for their correlation. Or X may be an effect of Y. This is illustrated in Figure 3.3, which shows three possibilities for the causal basis for a correlation between X and Y.

The three possibilities as shown in Figure 3.3 are what we would have to consider if we simply observe variables X and Y in nature and we are not ourselves the causes of variation in X. The ambiguity cannot be resolved if this is all we have to go on.

Suppose, however, as diagrammed in Figure 3.4, we introduce another variable M representing the actions of an experimenter manipulating X. It is implicit that M is the cause of X; in fact, it may be a perfect cause in that the correlation between the experimenter's actions and the variable X is unity. (We have also introduced an error random variable that operates on Y in the variable E_Y). In Figure 3.4a, we illustrate how most experimenters who think manipulation is causation believe manipulation is the nature of causation. The manipulation is an intervention in the world that brings the variation in the causal variable under the total control of the experimenter by the act of manipulation. No longer, it is thought, is X a function of some other causal variable Z that may just also be a cause of Y. The act of manipulation breaks any causal connection that may otherwise in nature exist between some other variable Z and X. Hence, it is concluded, if X and Y are correlated, then X is a probabilistic cause of Y. But consider the situation illustrated in Figure 3.4b. The experimenter's manipulations, represented by the variable M, is a perfect cause of an intervening variable Z, which is also a probabilistic cause of Y as

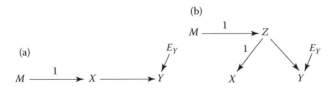

FIGURE 3.4   Two situations in which an experimenter manipulates a causal variable perfectly.

well as a perfect cause of X. X thus is perfectly correlated with Z and M. X will then be correlated with Y, but X is not a cause of Y. How could this be?

Consider the following possible scenario. Suppose a pediatrician believes that administering premature neonate infants drops of a weak saline solution to their eyes at periodic intervals reduces the number of problems in the development of their eyes usually experienced in the nursery. The neonates happen to be kept in incubators and breathe pure oxygen that is fed into their incubators. The doctor thus decides to perform an experiment. He divides a large group of premature neonate infants into two groups, completely at random. To one group he administers the weak saline solution to their eyes several times a day; the other group gets nothing. He finds that administering the weak saline solution produces a significant effect in reducing the number of developmental problems with the infants' eyes. He concludes that the saline solution is the cause of reduction in eye problems.

But another doctor notes that there is an unmeasured common cause of both the administering of the saline and the reduction in eye problems: When the first physician administered the saline solution, he opened a door to the incubator to get at the infant. That let outside air into the incubator that mixed with the oxygen already in the incubator. This did not occur with the control group that received no eye drops. So, the second physician performs another experiment with a placebo. The placebo variable is now Z', opening the door to the incubator periodically without administering eye drops. In this experiment there is a correlation between X and Y, but the interpretation is that Z' is the true cause of Y: Opening the door periodically introduces normal air with nitrogen and less oxygen into the incubator. Figure 3.5a illustrates this case. But even this experiment does not prove that administering the saline eye drops has no effect, only that opening the incubator to outside air has an effect.

In Figure 3.5b we show how both Z and X may be causes of reducing eye problems. Another experiment with incubators having rubber gloves in the sides to allow personnel to manipulate the infants without opening the incubators is then performed. One group of infants chosen at random receives the eye drops, the other does not. If there is an effect of the saline eye drops, this condition will support the eye drops as a cause of reducing eye problems.

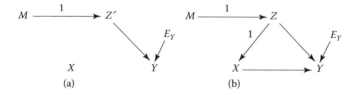

FIGURE 3.5 In condition (a) Z' opens the incubator door with no eye drops X given. Condition (b) illustrates how opening the door and administering eye drops may both have effects on reducing eye problems.

But this experiment should be paired with another condition in which some other placebo solution is administered to more convincingly show/test that the saline solution per se is a cause of reducing eye problems.

The point to make is that manipulation per se does not guarantee causality. What manipulation can often do is bring the variation in the causal variable under experimenter control and break the influence of the effects of other causes on the experimental causal variable. But as we have seen in the above example, there may be perfect correlation between manipulation and the putative cause, and yet the manipulation is itself introducing an extraneous cause that is not controlled. Medical researchers are trained to look out for such effects of manipulations and to control for them with placebo groups or other forms of experimental control. But causation is in the nature of things in the world and not in just human manipulations, which also happen to be causes in the world. Correct inference of causality requires control or reasonable grounds for assuming that extraneous common causes are not present, which, for example, astronomers and geophysicists often assume but do not manipulate.

Woodward (2001) lists conditions for a manipulation M on a causal variable X with respect to Y to be what it is for X to cause Y. (I have changed his "intervention I" to "M"):

(M1) M must be the only cause of X, that is, as with Pearl (2000) the intervention must completely disrupt the causal relationship between X and its previous causes so that the value of X is set entirely by M.

(M2) M must not directly cause Y via a route that does not go through X as in the placebo example (Figure 3.5a).

(M3) M should not itself be caused by any cause that affects Y via a route that does not go through X.

(M4) M leaves the values taken by any cause of Y, except those that are on the directed path from M to X to Y (should this exist), unchanged (Woodward, 2001, p. 8).

Several authors (Pearl, 2000; Spirtes et al., 2000; Woodward, 2007) consider an ambiguous situation that can arise. Suppose there are two directed causal paths from X to Y, and the effect of one causal path cancels the effect of the other path. In such a case one would not detect that X has an effect on Y. We illustrate this in Figure 3.6.

The canceling of parallel effects is a possibility. For example, in an experiment one may administer a poison in some food which unknowingly also contains the antidote. Spirtes et al. (2000) point out that in general, where one has not systematically caused the administration of one treatment to be accompanied by another treatment that specifically cancels the first treatment's effect, the likelihood of encountering such cases has a Lebegue measure

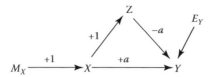

FIGURE 3.6 A case with two directed paths from X to Y in which the effect of one path is the negative of the other, causing them to cancel the total effect of X on Y.

of zero, that is, effectively zero likelihood. The implication is, "Don't worry about it." Treat the variable X as having no effect on Y. Still they concede that this may not seem satisfactory. So, for the general case (Figure 3.7a) where all variables are probabilistic, to retain a consistency between probabilistic and graphical representations, which they call the "faithfulness condition," they interpret such situations in general as in Figure 3.7b.

Of course, that will not work in the manipulative case where Z is perfectly correlated with M and X, since Z is perfectly correlated with X and cannot be correlated nonzero with Y if X is correlated zero with Y. Experimenters will see readily that the solution is to disentangle X and Z by finding a manipulation of Z that is unconnected with X, and vice versa for X with respect to Z. These manipulations are shown in Figure 3.8.

An assumption made by manipulationists is that for any cause X of some effect Y, one can always find a direct manipulation of X that is independent of any other causes of Y. This is almost a metaphysical assumption about the nature of reality. But it more likely rests on an induction from considerable successes in doing so in the past in the progress of science. This, I think, is why manipulationist concepts of causation are so popular among scientists. But there is no logically necessary reason that humans will always be able to find manipulations that are direct causes of causes studied without also causing other causes of the same dependent variable. But nothing says one should not try to find such causes.

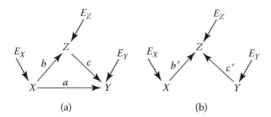

FIGURE 3.7 In (a) where $a = -bc$ the direct and indirect effects of X on Y cancel, leaving no effect of X on Y. To maintain consistency between statistical/probabilistic formulations of causal relations with graphical representations, (a) may then be interpreted as (b), where Z has nonzero correlation with X and conditioning on Z makes X and Y dependent.

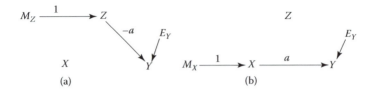

FIGURE 3.8 Two experiments that disentangle X and Z. In (a) M_Z causes only Z and not X. In (b) M_X causes only X and not Z.

But, in my view, "manipulation" in science is only a metonym for causation—identifying something salient that accompanies causation in their experimental studies with causality itself. As Woodward (2001, 2007) points out, one of the problems philosophers have identified with the reductive doctrine that causality *is* manipulation is that it is circular. Manipulations are causes. So one is trying to explain causation in terms of something that is already a cause. Another criticism of reductive manipulationism is that it is, as Woodward (2001) puts it, "unacceptably anthropocentric or at least linked much too closely to the practical possibility of human manipulation" (p. 2). Causes certainly occurred between objects in the universe before man appeared and began manipulating things on earth. And causes are now occurring where no human is now, such as effects of meteors striking the surface of Jupiter. And humans did not manipulate an asteroid to strike the earth and create conditions leading to the demise of the dinosaurs. Nor did their manipulations cause Vesuvius to erupt, or the San Francisco earthquake of 1906 to occur. There is also a sophomoric quality to the reductive concept that "causality" *is* "manipulation," wherein one seeks to make a challenging and outrageous claim against common sense. This seems true of Locke (1694) and Hume (1739, 1777).

The argument here is not that manipulation is an ineffective method of determining causation. Actually, it is generally the best way. But its effectiveness is in the manipulation's breaking causal connections with other variables that are also causes of the effect variable. But it is not infallible. And in circumstances where one cannot manipulate causes, inferences to causality are only provisional, and stronger only to the extent that one can reasonably show that the causal variables in the model are free of the effects of unmeasured/unmodeled variables.

Conditions for Causal Inference

Background Conditions

The events of everyday experience are rarely isolated from a universe of causal influences that produce them. By seeking to comprehend the dependencies between different kinds of events by means of the concept of functional relationship, researchers have learned over the centuries a number of general conditions that must obtain for one to have prima facie evidence of causal relations. The most important of these are the following (James, Mulaik, and Brett, 1982; Mulaik and James, 1995):

(1) Through experience and reasoning, we must isolate variables into *closed self-contained systems* that exclude extraneous and irrelevant influences of other variables that may mask causal connections or spuriously suggest them. Similar conditions of closure are made for systems describing laws of conservation in physics.

(2) We must focus our attention on the *coupling* between what we will regard as the independent and dependent variables to demonstrate that unmeasured mediating causal variables between them are free from interference, so that the independent variables are not shielded from the dependent variables by interruptions in the mediating pathways. This requires explicit attention to the *mediating mechanisms* by which causal and effect variables are linked.

(3) We must correctly specify the *causal direction* between variables. Sometimes this is established by the temporal order of measured events. Other times it is established by manipulation of the causes. In other situations, directionality is recognized by the dependent variables being a relational property between objects that is altered by changes in the attributes of the objects. For example, the lift of a wing is a relational property between moving air and the shape of the wing. Holding the velocity and direction of the air constant, we can change the lift of a wing by altering its shape. We cannot change the shape of the wing by altering lift.

(4) Since causal relations are functional relations between attributes of objects, this means that the objects studied must be *causally homogeneous*, that is, for each object the causal relation between a given pair of variables must be the same functional relation. One must also specify the relevant *environmental context* in which the attributes are defined and observed.

(5) Because the effects of causes frequently change gradually rather than instantly, one must come to identify when to measure the effects. One must identify the points of *stability* or *equilibrium* in which the causal system settles down after a change in a causal variable is introduced into the system. If one ignores this, but measures the effects haphazardly, one may not get repeatable results, or biased results.

(6) With probabilistic causal systems, when one does not work with the immediate probabilities themselves, but with parametric probability distribution functions, one must *specify the forms of the distributions*

and the parameters involved. Frequently researchers with little experience with experimentation or causal modeling are unaware of these conditions. James et al. (1982) were specifically concerned with calling structural equation modelers' attention to these or similar conditions, but Mulaik (1986, 1987), Mulaik and James (1995), and others (e.g., Bollen, 1989; Heise, 1975) have discussed them as well.

Nonlinear Causation

Although this book concerns *linear models* of causation, something must be said about nonlinear causation. We have asserted that causal relations are expressed by functional relations between causal and effect variables. Causal models are linear if they represent causal relations between variables by linear functions. A nonlinear causal model contains functional relations between variables that are not linear. It is easy to define what a linear function is. A linear function $f(x)$ is one that has both of the following properties:

1. Additivity: $f(x + y) = f(x) + f(y)$.
2. Homogeneity: $f(bx) = bf(x)$.

Examples: $y = ax + b$,

$$y = b_1 x_1 + b_2 x_2 + \cdots + b_n x_n,$$

$$y = a_1 + a_2 x + a_3 x^2 + \cdots + a_k x^{k-1}.$$

The latter is nonlinear in the functions of x but linear in its parameters $a_1, a_2, a_3, \ldots, a_k$. We may see this by writing this function as

$$y(x) = a_1 X_1(x) + a_2 X_2(x) + a_3 X_3(x) + \cdots + a_k X_k(x).$$

The functions $X_i(x)$ can be extremely nonlinear functions of x. Linear, in this case, only refers to the parameters.

The possibilities for nonlinear functions are endless.

$$y = a \, \exp(-bx) + c,$$

$$y = x^5 + x - 1,$$

$$y = \sin x + b \, \exp(cx),$$

$$y = a + bxz + x^2 + z^3,$$

$$y = \frac{2}{x} + 3xz + bwz + \frac{1 - x}{\cos v}.$$

Mathematicians and physicists are interested in nonlinear functions because many phenomena in physics display nonlinear behavior. But modeling nonlinear phenomena is difficult. To begin with it is not always obvious what independent variables to consider and how they are to interact. Physicists have found that a method known as differential equations is useful in modeling nonlinear phenomena such as a pendulum. Sometimes the effect variables are monotonic rather than linear functions of causal variables. A monotonic function is always increasing (or decreasing) with respect to the increase of the independent variable. In such a case, a linear function represented by a straight line may be a close approximation to the monotonic function.

Moderator variables have been used in regression to treat cases where the relation of the independent variable to the dependent variable varies with variation in a third moderator variable. A simple example would be a regression equation such as

$$y = b_0 + b_1 x + b_2 z x + e = b_0 + (b_1 + b_2 z)x + e,$$

where y is a dependent variable, x is an independent variable, z is a moderator variable, and the b's are regression coefficients. We see that the middle term reduces to the term on the right, and $(b_1 + b_2 z)$ becomes a regression coefficient on x that varies with different values of z. So, for all cases with the same value for z, a straight line with slope $(b_1 + b_2 z)$ will describe the relationship of x to y. This is not a serious problem if x, y, and z are observable and there are numerous cases for each distinct value of z. We can study the group homogeneous in value in z separately and estimate the regression coefficient for that group. The problem becomes more difficult if z is not quantitative but qualitative. In these cases we can still segregate the cases into groups homogeneous for the quality of the moderator variable and estimate the regression coefficient. Sometimes this is done using dummy variables to represent each of the qualitatively different groups. In structural equation modeling where the moderator is a qualitative variable that we can segregate cases on, we can model relations between the latent independent and latent dependent variables separately in each group, meaning we estimate the structural coefficients separately in multiple groups. This is the safest method to use.

There have been efforts to formulate ways of treating interactions in SEMs. An excellent work with chapters by leading researchers in this field is the edited work of Schumacker and Marcoulides (1998). These methods have not caught on, because they are complex to implement and have difficulties in implementing estimating algorithms and often yield seriously biased standard errors for model parameters. To limit the length of this book, I have not included a chapter on interaction and nonlinear effects. But the Schumacker and Marcoulides (1998) book is the best reference now available on this topic.

Science as Knowledge of Objects Demands Testing of Causal Hypotheses

What may be new about linear causal modeling with structural equations for students is that the emphasis is on forming causal models and testing them. Whereas most statistics taught to beginners involve exploratory statistics—discovering a mean, discovering whether a difference is significantly different from 0, discovering that a correlation differs significantly from zero—the statistics of structural equation modeling involve testing for specific values or patterns of correlations among observed variables dictated by a linear causal model. (We will only concern ourselves in this book with linear models of causation, although for the physical world we may often need to formulate nonlinear models.) To provide the student with a rationale for hypothesis testing, I offer the following explanations.

The common practice of asserting a substantive hypothesis that there will be a nonzero correlation between two variables but testing whether the estimated correlation differs from zero is not a very strong way of using statistics. For example, if one believes there may be a relationship between two variables, showing that it is significantly different from zero does not lend much support for any particular substantive theory, since any number of substantive theories may assert that a nonzero correlation exists between the variables. Showing that the correlation is not zero only supports all such theories. But science seeks to develop specific theories for substantive phenomena and to rule out rival theories. And so, at some point we need to begin testing hypotheses for specific values or specific patterns of values that are at best limited to only one theory, or at worst to only a few theories.

But a problem in testing specific values or specific patterns of values for parameters may occur to the reader. Where do these specific values or specific patterns come from? In many respects they can come from our general knowledge about things and how they are related or, more frequently, *not* related. In other instances, they may come from previous studies and are then to be applied to a new context as invariants.

Abduction, Deduction, and Induction

The researcher should realize that scientists conduct research by repeatedly cycling through three phases as the research progresses. The philosopher, logician, and physical scientist Charles Saunders Peirce (1839–1914) was perhaps the first to articulate how scientists actually function through three phases, reflecting his first-hand experience as a scientist and his deep knowledge of the history of science. He named these three phases *abduction*, *deduction*, and *induction*. In *abduction* the scientist is confronted with new phenomena which beg for an explanation or a theory to account for them. The scientist then proceeds to develop a hypothesis to account for all of the known pertinent phenomena in question at that point. This may involve developing a number of hypotheses and eliminating all but one by how well they are able to reproduce the known phenomena in question. Sometimes abduction is described as "inference to the best explanation." As a logician, Peirce held that the inference in abduction had the following form:

> The surprising fact, C, is observed;
> But if A were true, C would be a matter of course,
> Hence, there is reason to suspect A is true. (CP, 5.189; Forster, 2001)

Peirce does not claim that A is now indeed the unique or true explanation of C. That would be to commit the fallacy of affirming the consequent. There may be many reasons for C instead of A. But the scientist has reasons to believe that A is the reason for C. Thus A becomes his/her hypothesis to explain C. There is still a need to test the hypothesis.

The second phase in the scientific cycle according to Peirce is *deduction*. Given prior knowledge and hypothesis A, the scientist deduces logically some new consequence that might be observed. The inference may be of two kinds (Forster, 2001):

Necessary deduction:

> All M's are P.
> All S's are M.
> Hence, all S's are P.

Probabilistic deduction:

> $p\%$ of M's are P.
> $S_1, \ldots, S_n$ is a large sample of M.
> Hence, probably and approximately $p\%$ of $S_1, \ldots, S_n$ are P.

It is essential that the consequence deduced from the hypothesis be something not in the knowledge or data previously known without the hypothesis to serve as a premiss to deduce it from that knowledge.

The third phase in Peirce's scientific cycle is *induction*. This is reasoning from a sample to a population. Today, behavioral and social scientists know this as statistical inference based on the results of a statistical test. The researcher collects or uses data not used in formulating the hypothesis or the deduction to test the hypothesis. The data collected should not be merely another random sample from the same population from which came the data used in hypothesis formation, but data collected in a new context, sometimes with other variables not observed before. But it is not sufficient that the data conform to the hypothesis, but rather that one be able to infer that it would conform in the population from which the data are sampled.

Science is the Study of Objects

There is another rationale for why scientists form and test hypotheses. This rationale demands of scientists throughout history less expertise in formal logic and deduction than required by Peirce's logical analysis. It also does not require a formal deduction after formulating a hypothesis. The hypothesis can simply assert invariants.

Scientists generally accept that scientific knowledge is (1) acquired by the senses, (2) is unbiased, (3) is intersubjective, (4) is repeatable, (5) is falsifiable (or disconfirmable), as well as possessing other characteristics that are often blends of these five (Mulaik, 2004). Now, what I contend is that these characteristics of scientific knowledge all follow from an unconscious application of a metaphoric schema taken from object perception: Science is knowledge of objects (Mulaik, 1994, 2004).

That this may be a metaphor may not be immediately obvious. George Lakoff and Mark Johnson (1980, 1999), Lakoff (1987, 1993), Lakoff and Nuñez (2000), and Lakoff and Turner (1989) have argued that abstract thought is metaphoric, with metaphors taken from embodied perception and action. Metaphors are mappings from components of a source domain to components of a target domain, so that relationships between the components in the source domain may be analogously applied to make inferences in the target domain. By "embodied" it is meant that we as humans move and observe as bodies within the world with coordinated perceptual and motor schemas adapted over millions of years of evolution that integrate what is given by the senses with muscular action that permits us most of the time to successfully move among and manipulate objects in the world.

These perceptual and motor schemas, I hold, are applied rapidly—often within milliseconds, in perception—to integrate the data coming in from sensory neurons (Blumenthal, 1977). Or they may be applied to *percepts* (Jaynes, 1990/1976) received in working memory from perceptual processes within at most a few seconds of one another. In this respect, humans share much in common with most mammals and other vertebrates.

But humans, I contend, also have the capability to form concepts. Concepts are formed and integrated in working memory from percepts retrieved from long-term memory that have been excerpted (Jaynes, 1990/1976) from perceptual experience, sometimes hours, days, and years apart in time and/or from widely spaced locations, sometimes hundreds and thousands of miles apart in space. Or the percepts may be recorded as signs or writing and taken in through perception and then into working memory. The concept is not based on a percept that one simultaneously perceives in a single glance of all its components organized in a whole. The metaphor is what, in working memory, provides the organization or structure that combines all the percepts into a synthesized whole of distinct parts, constituting the concept.

When Columbus argued that the earth is round, he had not seen the earth as a whole to know this. He had a number of observations, such as seeing first the tops of ships coming toward him over the horizon, then later the hulls. Then there was the round shape of the shadow of the earth on the moon during a lunar eclipse. The sun being overhead at noon at the equator and increasingly lower in the sky as one goes north is another consequence of the earth's being round. The existence of round bodies in the heavens, such as the moon and sun, also suggested the roundness of the earth. But the concept of a ball, a sphere, integrates all these observations into a concept of the round earth.

Leeuwenhoek looks through his water-drop-lens microscope and sees tiny things moving around in some water he is focused on and calls them *little animals*, assimilating them thus metaphorically to the domain of ordinary living animals seen with the naked eye. They are living things. He does not see the "little animals" with his naked, unaided eyes.

Again, Tycho Brahe's observations of Mars, used by Kepler to determine that the orbit of Mars is an ellipse, were gathered over many years from locations of the earth in space that were often hundreds of thousands of miles apart. The metaphor of an ellipse, taken from drawings in which the ellipse was perspicuously perceived as a whole, integrated all these observations into an orbit.

Brown (2003) cites a metaphor in chemistry in the following sentence: "[Cell] Membranes contain *channels* that are permeable to hydrogen ions and other positive ions." This is an example of the use of a conduit metaphor (Lakoff, 1987). A conduit guides the movement of something through space. Usually a conduit has boundaries that confine what is transported through it within the conduit. A *channel* is a particular kind of conduit, usually between bodies of water. Here hydrogen and other positive ions are confined to the *channel* conduits as they move from the outside to the interior of a cell.

Most metaphors, according to Lakoff and Johnson, are schemas of embodied action and perception—such as paths followed in space, conduits, locations, obstacles in paths, movement of objects—used to synthesize or integrate experience into complex concepts. So, human knowledge is conditioned on and mediated by prior schemas of embodied perception and action. In other words, a schema of perception or embodied action becomes a metaphor when used to blend or join in thought diverse elements of experience not perceived jointly in a single percept, but conceived in thought. By mapping the components of the schema to the diverse elements, we can then conceive of them as the whole suggested by the schema. The mapping of course has to be constrained by what Lakoff (1993) calls the "invariance principle": "Metaphoric mappings preserve the cognitive topology (that is, the image-schema structure of the source domain [e.g. in perception], in a way consistent with the inherent structure of the target domain" (Lakoff, 1993, p. 215). For the object schema, invariants of the image schema of an object must be mapped to invariants of the target domain, different points of view must be mapped to different sources of observation at different times and locations, independence of the observer from the object in the perceptual domain must be mapped to the independence of the observer scientist in the target domain, and so on. We will now consider this more explicitly.

Objects as Invariants in the Perceptual Field

Having said that, why is science's concern with objects built around a metaphor? Most abstract scientific concepts are not based on direct percepts of objects corresponding to them. Atoms, molecules, cells, channels in cell membranes, intelligence, and cooperation are regarded as objects or attributes of objects by the scientists concerned with them, but no one sees these directly and immediately by unaided perception. The use of instruments to observe these involves complicated theories of how the instruments work to reveal the objects through lenses, meters, and imaging screens. All these theories work at the conceptual level and incorporate numerous metaphors. So, these are conceptual objects, not perceived objects.

What then is a perceived object, if it is the basis of science's conceptual objects? J. J. Gibson (1950) has provided what I take to be an appropriate answer here. According to Topper (1983), Gibson (1950) drew upon the mathematical concept of an invariant. In topology "… a series of transformations can be endlessly and gradually applied to a pattern without affecting its invariant properties. The retinal image of a moving observer would be an example of this principal" (Gibson, 1950, pp. 153–154). Topper (1983) paraphrases Gibson as asserting that the retinal image is a projection of a solid object onto the curved surface of the retina of the eye. As the observer or the object or both move around in space with respect to each other, the image will change. But within the image there will remain invariant properties that are analogous to those in the geometry of transformations (Topper, 1983).

But there is another aspect of object perception. While objects may be invariants in the perceptual field, they are still regarded as *other to* and *independent* of the observer. This is possible, Gibson (1950) held, because the observer also simultaneously gains information in perception about the actions and position of the observer with respect to the object and its surroundings. He called this *proprioception*, in contrast to the perception of objects, which he named *exteroception*. As the observer moves with respect to the object, the visual image changes in regular ways. As the observer approaches the object, both the object and the immediate background loom out from a point toward which the observer is moving. As the observer moves away from the object, the image of the object and objects in its immediate vicinity shrink toward a point. On the other hand, if the object approaches the subject, only the object looms, but not the objects in the surrounding vicinity, while retaining invariant relations among its components. Moving the head to the right causes the image of the object to move to the left. Thus one has information in perception that the invariant object is distinct from the observer because it remains invariant despite the regular changes to the image produced by the actions and motions of the observer.

Objects are also intersubjective, because each observer can ask another whether he/she observes the same thing and acts consistent with that view. The other's report is further evidence bearing on invariance. The observer can also move around, and observe from different places whether the same thing is perceived. The observer may close his/her eyes for a few seconds, and then look again, to see whether the object remains invariant and endures through time.

Thus, I believe scientists unconsciously came to demand that one establish the objectivity of one's concepts because they unconsciously applied the metaphor taken from the schemas of object perception to their concepts. They did this, maybe naively, because objects they held are what are real, and the object conceived, although not directly observed, was still regarded as another object in the world. (It is a trap to fall into to believe that what we know about objects is incorrigible). Just as objects are invariants when seen from different points of view, scientific objects must be invariants when observed from different points of view in different laboratories and with different observers and means of observation. But observations are limited in time and from a few points of view, so they may not take in the whole object. Future observations from new points of view may be inconsistent with the concept formed of it.

Scientific concepts must also be free from systematic artifacts due to the methods, conceptions, and prejudices of a specific observer or group of observers. In other words, the concept of the scientific object must not be contaminated with effects due exclusively to the observer.

For example, when cold fusion was announced by researchers at the University of Utah in 1989, physicists all over the world scrambled to replicate the phenomenon in their laboratories. At Georgia Tech a week later, scientists reported having conducted an experiment that yielded evidence in support of cold fusion. But a few days later, the scientists retracted their claim, saying that a neutron detector they had adapted for the experiment unexpectedly yielded inflated readings of neutrons (a sign hypothesized of cold fusion) because of the confounding influence of temperature in the experiment on the instrument. Thus an artifact of observation rendered "subjective" conclusions.

Implications for Structural Equation Modeling

Structural equation modeling involves all three phases of scientific practice. Abduction occurs during review of prior knowledge and formulation of hypotheses. Deduction and induction are used when the researcher formulates and tests a hypothesis about how a set of theoretical constructs or latent variables contains the causes of another set of observed variables or indicators. The researcher may develop the theory in the abstract involving anticipated latent variables derived from general knowledge and then derive and/or construct observed indicators (variables) of the hypothesized common factors as effects of them. The resulting model with certain parameters fixed to specified values represents a hypothesis asserting invariant quantities, whereas other parameters are left freely estimated conditional on the fixed parameters, to fill the unknown blanks in the model. The values of the fixed parameters may be derived in some cases deductively from theory, but frequently they will be values estimated for these parameters in some other contexts.

Summary and Conclusion

Science is the knowledge of objects. Objects are invariants in experience independent of observers, bearing attributes and/or properties. Attributes are categories that come in classes of two or more categories such that no more than one member of the class may be assigned to an object at any one time. A variable is a symbol in thought that at any given time takes on the value of only one of the members of the attribute class that it represents. Causal relations have the form of functional relations between variables. A functional relation involves two sets, a first set and a second set, such that each member of the first set is assigned to one and only one member of the second set (but more than one member of the first set may be assigned to a given member of the second set). Probabilistic causal relations involve functional relations between a set of causal attributes and a set of distinct probability distributions defined on an effect-attribute variable such that at any given time a value of the causal variable is mapped to a value of the effect variable with a probability density given by the corresponding probability distribution. Physical causal interactions involve exchanges of quantities (attributes) between objects or parts of objects that conserve the total quantity in the system (when no outside forces act). Animal and human perceptual systems are sensitive to physical quantities in the causal exchanges between things and thus perceive causality in those exchanges. The schema of a causal relation involving cause and effect is so minimal that it easily becomes a metaphor at the basis of numerous causal concepts. Causes metaphorically can be forces, essences, substances, intentions, agents, settings, reasons, ideas, actions, instruments, goals, attractors, ends, barriers, and thwartings (Lakoff and Johnson, 1999). Causal relations being properties of objects must be invariants for certain kinds of objects and different times. Hypotheses about objects assert invariants, and testing such a hypothesis tests the invariance in representative samples of objects. Hypotheses about causes assert invariant functional relations between variables. "Science is knowledge of objects" is frequently a metaphor involving a metaphoric object.

4

Graph Theory for Causal Modeling

Directed Acyclic Graphs

Recent developments, summarized in Spirtes et al. (1993, 2000) and Pearl (2000), have attempted to sharpen statements of background principles for studying causation in mathematical terms, using graph theory. Their aim was to demonstrate correspondences between graphs representing causal relations among variables and the joint probability distributions of these variables. They further sought to determine in graph theoretic terms conditions under which causal relations could be demonstrated.

A graph consists of vertices (or nodes) and connecting links between them known as edges. Graph theorists have developed correspondences between graphs and systems of variables, wherein each vertex in a graph corresponds to a random variable, with the edges or links between them representing their dependencies. In our present discussion, we will consider graphs representing causal dependencies between variables. Since causes are presumed to be asymmetric relations, we will represent them in graphs by arrows drawn from causal variables to effect variables. Such graphs that have asymmetric links between nodes are directed graphs. A graph can be as simple as a Markov chain (Figure 4.1).

Graphs can also have branches and take on a tree-like structure (Figure 4.2).

Directed graphs are also distinguished according to whether they are cyclic or acyclic. *Cyclic graphs* have paths or series of variables connected by arrows that point back to earlier variables in the series. Acyclic graphs do not. *Cyclic*

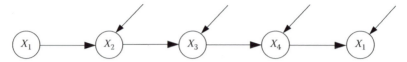

FIGURE 4.1 A simple directed graph of a Markov chain of variables. Nodes are random variables and edges or links indicate causal dependencies between variables. In this Markov chain, only adjacent variables have direct causal dependencies between them. Diagonal arrows represent disturbance or error variables.

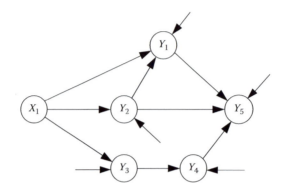

FIGURE 4.2 A directed graph with branches and disturbances.

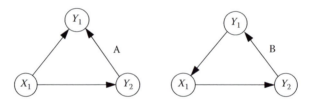

FIGURE 4.3 Directed acyclic and cyclic graphs. A is acyclic and B is cyclic.

graphs are also known as *nonrecursive models. Acyclic graphs* correspond to what are known as *recursive models* (Figure 4.3).

Graph theory can be used to examine independence between variables. Two variables X and Y are said to be independent if $f(x,y) = f(x)f(y)$ for all x and y. An equivalent expression of independence is that X and Y are independent if $f(x \mid y) = f(x)$. We will use the notation $X \perp Y$ to mean that variables X and Y are independent.

Conditional Independence

Given random variables X, Y, and Z,

$$X \perp Y \mid Z \Leftrightarrow f(x,y \mid z) = f(x \mid z)f(y \mid z) \Leftrightarrow f(x \mid y,z) = f(x \mid z).$$

The symbol $\Leftrightarrow$ may be read as "if and only if." This means that once Z is known to have a certain value, then no additional information about X may be obtained from Y. This is a variant of the local independence condition.

In the Markov chain of Figure 4.1, $X_3 \perp X_1 \mid X_2$ (X_3 is independent of X_1 given X_2) because $f(x_3 \mid x_2, x_1) = f(x_3 \mid x_2)$ for all values of x_1, x_2, and x_3. In a Markov chain all influence on a variable is mediated through that variable immediately preceding it in the chain. Once one conditions on the immediately preceding variable, all prior variables to it are independent of the successor variable. We can further note that for the variables in the Markov chain of Figure 4.1, $X_4 \perp X_1, X_2 \mid X_3$ and $X_5 \perp X_1, X_2, X_3 \mid X_4$. This suggests that we may test for a Markov chain in variables having a joint multivariate normal distribution, using partial correlations, because the partial correlations $\rho(x_3, x_1 \mid x_2)$, $\rho(x_4, x_1 \mid x_3)$, $\rho(x_4, x_2 \mid x_3)$, $\rho(x_5, x_1 \mid x_4)$, $\rho(x_5, x_2 \mid x_4)$, and $\rho(x_5, x_3 \mid x_4)$ should all equal zero, indicating the conditional independence of the variables in question.

The joint distribution of the variables in a Markov chain can be expressed as the product of conditional distributions of adjacent variables in the chain. For the Markov chain in Figure 4.1,

$$f(x_1, x_2, x_3, x_4, x_5) = f(x_1)f(x_2 \mid x_1)f(x_3 \mid x_2)f(x_4 \mid x_3)f(x_5 \mid x_4).$$

This demonstrates that the joint distribution of a set of variables in a directed acyclic graph (DAG) can be derived from certain sets of conditional distributions suggested by the graph. More about this is discussed later.

Markov Condition

We turn now to directed causal graphs in which variables may have more than one cause.

We will first develop some terminology for describing a DAG. This terminology draws heavily upon kinship relationships. Variables that are immediate causes of an effect variable, that is, variables whose arrows leave them and point directly to the effect variable, are known as the *parents* of the variable. For example, in Figure 4.4 variables X_1 and X_2 are parents of Y_1, variables Y_1 and X_3 are parents of Y_2, and X_1 and Y_2 are parents of Y_4, while Y_2 is also a parent of Y_3. A variable that is an immediate effect of a causal variable is called a *child* or *daughter* of the causal variable. In Figure 4.4, Y_1 is a child of X_1 and X_2; Y_4 is a child of X_1 and Y_2, but Y_4 is not a child of Y_1 or X_2; and Y_3 is a child of Y_2. An *ancestor* of a variable is any variable that is in a directed path to the variable in question. X_1, X_2, Y_1, and X_3 are ancestors of Y_2. Next, X_1, X_2, X_3, Y_1, and Y_2 are ancestors of Y_4. A *descendent* of a variable is any variable in a directed path leading from the variable in question. Y_2, Y_3, and Y_4 are descendents of Y_1 and Y_1, Y_2, Y_3, and Y_4 are descendents of X_1. All parents are ancestors and all children are descendents.

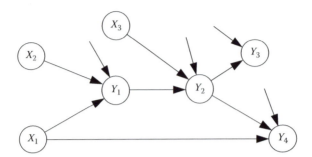

FIGURE 4.4 DAG with variables having more than one cause used to illustrate the Causal Markov condition.

Pearl and Verma (1991), Pearl (2000), and Spirtes et al. (2000) described the *Markov condition* as a fundamentally important condition for establishing independence relations in a DAG of causal relations between random variables. Data satisfying this condition are necessary for both exploratory and confirmatory analyses. In the exploratory case, which is developed by Spirtes et al. (2000), who have developed an algorithm for the discovery of causal structure from covariances among variables, variables must have covariance structures satisfying this condition to be able to establish causal relations. In the confirmatory context, this condition must be satisfied by the hypothesized model graph so that one can identify conditional independence conditions in the model that follow from a causal structure among the variables and that may be put to the test against the data. Furthermore, the *Markov condition* may not be satisfied by any arbitrary set of variables, for it may not be possible in any case to subdivide the variables into three proper subsets such that a first subset is independent of a third subset given values of the variables in a second subset. Thus, all common causes of variables within a set of variables to be considered must be included within the set for the Markov condition to apply. Moreover, the population of subjects must be *causally homogeneous* as mentioned previously. Pearl (2000) provides the following statement for the Markov condition.

The Markov Condition

"A necessary and sufficient condition for a probability distribution P to be Markov relative [to] a DAG G is that every variable be independent of all its nondescendents (in G) conditional on its parents" (p. 19). To this I add: *Every variable with parents is unconditionally dependent on its parents.*

The Markov condition implies that variables will be unconditionally dependent on their parents but conditionally independent of all other nondescendent variables, conditional on the parents.

In Figure 4.4, $Y_2 \perp X_1, X_2 \mid Y_1, X_3$ (Y_2 is independent of X_1 and X_2 conditional on Y_1 and X_3). Also, $Y_4 \perp Y_1, X_2, X_3, Y_3 \mid X_1, Y_2$ (Y_4 is independent of $Y_1, X_2, X_3,$ and Y_4 conditional on Y_2 and X_1). When the variables have a joint multivariate normal distribution, these independence conditions imply that the following partial correlations are zero: $\rho(Y_2, X_1 \mid Y_1, X_3)$, $\rho(Y_2, X_2 \mid Y_1, X_3)$, $\rho(Y_4, Y_1 \mid X_1, Y_2)$, $\rho(Y_4, X_2 \mid X_1, Y_2)$, $\rho(Y_4, X_3 \mid X_1, Y_2)$, and $\rho(Y_4, Y_3 \mid X_1, Y_2)$. An important consequence of this is that if $\mathbf{V}$ represents the variables corresponding to nodes in a DAG, then for all values $\mathbf{v}$ of $\mathbf{V}$, for which $f(\mathbf{v}) \neq 0$, the joint density function $f(\mathbf{V})$ of variables satisfying the Markov condition is given by (Spirtes et al., 2000, p. 12)

$$f(\mathbf{V}) = \prod_{V \in \mathbf{V}} f(V \mid \text{parents}(V)). \tag{4.1}$$

Here parents(V) is the set of variables that are parents of V. $f(V \mid \text{parents}(V))$ means the density of V conditional on the (possibly empty set of) parents of V. If the set of parents is empty, we simply take $f(V)$.

d-Separation Criterion

The Markov condition for a DAG gives rise to a criterion by which we can determine from the graph what mutually dependent random variables (represented as the nodes of the graph) can be made to be independent by conditioning on a certain set of other variables in the graph. The criterion also allows us to identify what initially mutually independent variables will become mutually dependent if we condition on a certain other set of variables. This criterion is known as the *d-separation criterion*. Pearl (2000) states the criterion as follows:

> A path is said to be d-separated (or blocked) by a set of nodes Z if and only if
>
> - p contains a chain $i \rightarrow m \rightarrow j$ or a fork $i \leftarrow m \rightarrow j$ such that the middle node m is in [set] Z, or
> - p contains an inverted fork (or collider) $i \rightarrow m \leftarrow j$ such that the middle node m is not in Z and such that no descendant of m is in Z.
>
> A set Z [of variables corresponding to nodes in the DAG] is said to d-separate [a set of variables] X from Y if and only if Z blocks every path from a node in X to a node in Y. (pp. 4–17)

In Figure 4.4, variables X_1 and X_2 are d-separated from Y_2 by Y_1, variables Y_3 and Y_4 are d-separated by Y_2, and variables Y_1 and Y_4 are d-separated by the set $\{X_1, Y_2\}$. A path from X_1 to X_2 is blocked by the collider Y_1. Conditioning on Y_1 will make the initially independent X_1 and X_2 become mutually dependent. Paths between X_1 and X_3 are blocked by Y_2, so that if we note that initially X_1 and X_3 are mutually independent, conditioning on Y_2 will make them become mutually dependent. The path between X_2 and X_3 is also blocked by the collider Y_2, so that conditioning on Y_2 will make the initially mutually independent X_2 and X_3 become mutually dependent.

The importance of the d-separation criterion is that it allows us to infer hypotheses as to when variables that appear initially to be mutually dependent will become independent when we condition on certain other variables. At the same time it also allows us to identify from the graph which initially independent variables will become dependent if we condition on certain colliders dependent on them. These hypotheses can be tested empirically. When the variables have a joint multivariate normal distribution, we can test these hypotheses using partial correlation, testing to see if correlations between variables become zero when we condition on common causes of them or intermediate causes between them.

Pearl (2000) notes that one of the paradoxes illustrated by the d-separation criterion is that variables such as music ability and intelligence, which may be independent in the general population, may become mutually dependent (negatively) in a student population of graduate students who are selected for having high grades in undergraduate music courses, which may be due to both high intelligence and high music ability.

Minimality Condition

A subgraph G' of a graph G is a graph that contains some or all of the vertices of G and some or all of the connected paths between pairs of vertices in G, such that any path missing in G must be missing in G'. Paths that exist in G may be missing in G'. However, no directed path between corresponding vertices in G' may be missing in the original G (effectively G' is nested within G).

If the graph is a causal graph with vertices and paths representing variables and causal influences, respectively, then each direct causal connection represented by a path in the graph will prevent either an independence or a conditional independence relation between variables that otherwise would result. So, the issue is what is the minimal set of directed paths that must be inserted within a graph that will allow it to represent the independence and conditional independence relations between the variables that empirically obtain? Spirtes et al. (2000) give the following.

Causal Minimality Condition

Let G be a causal graph with vertex set $\mathbf{V}$ and P a probability distribution on $\mathbf{V}$ generated by G. $\langle G, P \rangle$ satisfies the minimality condition if and only if for every proper subgraph H of G with vertex set $\mathbf{V}$, the pair $\langle H, P \rangle$ does not satisfy the Causal Markov condition.

Faithfulness Condition

Faithfulness concerns the fidelity with which a probability distribution P and a graph G conform to one another in representing dependence and independence relations. Consider the graph in Figure 4.5.

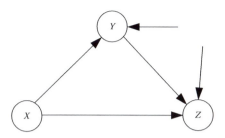

FIGURE 4.5   The graph may not be faithful if the direct effect of $X \to Z$ is canceled by the indirect effect $X \to Y \to Z$.

Ordinarily one would expect X and Z to be unconditionally dependent. But one may encounter empirical distributions in which the direct effect of X on Z is canceled by the indirect effect of X on Z through Y. Y in this case reverses the effect of X, making it a negative effect on Z. In this case, X and Z are unconditionally independent but conditionally dependent condition-alon Y. X and Y are unconditionally dependent, as are Y and Z. The Causal Markov condition when applied to this graph does not imply that X and Z are unconditionally independent. The Minimality condition is also not satisfied. No subgraph obtained by deleting a path in the graph will produce a graph that satisfies the Causal Markov condition with respect to the empirical distribution. For example, if the path $Y \to Z$ is deleted, conditioning on X will not leave Y and Z independent with respect to the empirical distribution. If we remove the path $X \to Z$, which would imply that X is independent of Z conditional on Y, we find instead that X and Z are dependent conditional on Y. G is not faithful to P.

The Faithfulness Condition

If G is a causal graph and P a probability distribution implied by G, then together G and P satisfy the Faithfulness condition if and only if every conditional and unconditional independence relation true in P is implied by the Causal Markov condition applied to G (Spirtes et al., 2000).

The Faithfulness condition requires that if one variable has causal influences on another variable over several parallel routes, these influences in different routes must not cancel one another in their combined effect. Otherwise, the Causal Markov condition will not be satisfied and cannot be used to infer causal relations. Spirtes et al. (2000) assert further that Faithfulness and the Markov condition imply Minimality, but Minimality and the Markov condition do not imply Faithfulness. Fortunately, Faithfulness is relatively common and allows us to use graph theory to infer causal relations from probability distributions.

5

Structural Equation Models

Basics of Structural Equation Models

The student who, at this point, may be familiar only with exploratory uses of statistics, such as the descriptive statistics of means, regression, and exploratory factor analysis, will now be introduced to a new way of using statistical models. The emphasis will be on testing hypothesized models in which certain "overidentifying" constraints on the model's parameters have been imposed. The aim is to test whether models with these constraints fit data to which they are applied. The closest example to this in the statistics of means would be a test of the null hypothesis that a mean is equal to a specific value prespecified by the researcher. Exploring is when you simply estimate the mean of a variable without any constraints on it other than those minimally necessary to make the estimate possible. In regression we estimate regression coefficients without placing constraints on them. In exploratory factor analysis we estimate factor loadings and factor correlations, with only minimal constraints necessary to achieve unique values for them. Hypothesis testing involves prespecifying certain parameters and performing tests of these values. We will not dwell on hypothesis testing now, but it is important that the student keep this generally in mind as we begin considering structural equation models.

In most cases, in addition to hypothesized constrained values for parameters, there may also be estimated parameters in structural equation models, because the researcher has no idea what values to constrain them to. But the models need to be completed (by having values for all parameters) to be able to test them. A test of a model consists of comparing a reproduced variance–covariance matrix for the observed variables derived from the model's parameters (both constrained and estimated) to the observed variance–covariance matrix, and noting any lack of fit. So, we complete the incompletely specified model by estimating the unknown parameters conditional on the constrained parameters. But the estimated parameters are not the principal aim of a structural equation model. This is because their values are conditioned on the constraints imposed on other parameters in the model, and whatever values are obtained for them by estimation techniques will be biased if the constraints are not properly specified. Thus the reproduced variance–covariance matrix

$$\Sigma = \mathbf{G}\mathbf{B}^{*-1}\mathbf{\Gamma}^{*}\mathbf{\Phi}\mathbf{\Gamma}^{*'}\mathbf{B}^{*'-1}\mathbf{G}'$$

is compared to the sample covariance matrix **S** (which is estimated without overidentifying constraints) and determining whether the difference between these matrices is more than could be accounted for by chance. But we need to work to a point where the above matrices and concepts have meaning for us.

Path Diagrams

To begin with, the complexity of structural equation models demands a perspicuous way of representing them. Sewell Wright (1921), the founder of structural equation modeling, invented the *path diagram* as a way of showing a structural equation model. A structural equation model expressed in path diagram form is shown in Figure 5.1. A path diagram is also known as a directed graph. *Manifest* variables in the diagram are shown by squares (sometimes rectangles) with variable labels written within them. *Latent* variables are shown by circles (sometimes ellipses). *Direct causal paths* are represented by one-headed arrows pointing from causal variable to effect variable. Covariances between pairs of variables are shown by a two-headed curve. Almost always covariance is shown only between exogenous variables. *Exogenous* variables in the diagram have one-headed arrows pointing away from them, and none pointing to them, and represent causal inputs into the system of variables. *Endogenous* variables have arrows pointing to them and are variables within the system that are the effects of exogenous variables or causes of other endogenous variables within the system. Manifest exogenous variables are represented by x's. (There are no manifest exogenous variables in the model of Figure 5.1.) Latent exogenous variables are designated by x's. Manifest endogenous variables are denoted by y's.

Associated with each direct causal path is a *structural coefficient*, which represents the direct causal *effect* of the cause on the effect variable. The effect represents how much a unit change in the causal variable has on the effect

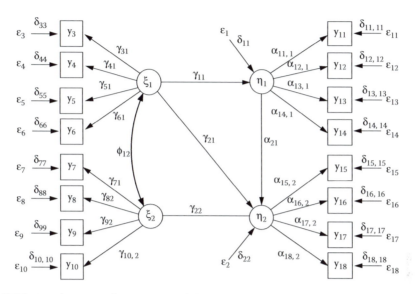

FIGURE 5.1 A structural equation model represented by a path diagram. Squares are manifest variables. Circles are latent variables. Disturbances, although latent variables, are represented without circles. One-headed arrows represent causal paths. Two-headed curves represent covariance. Associated with each causal path is a structural parameter.

variable or proportionally how much of the quantity of the causal variable is transferred to the effect variable. The structural coefficients of paths from exogenous to endogenous variables are g's. The structural coefficients for paths from endogenous to other endogenous variables are a's. When no arrow exists between a pair of variables where such an arrow could possibly exist, it means that there is no causal connection between the variables, and the corresponding structural coefficient is zero. Thus what is left out of a path diagram is very important. Indicating that a variable is *not* a cause of another variable is often the way we impose both identifying and testable constraints upon our models.

Disturbances are designated by e's. Structural coefficients of paths from disturbances to their respective endogenous variables are denoted by d's. Subscripts on structural coefficients of causal-path arrows begin on the left with the number label of the variable pointed to, followed by the number label of the variable from which the causal-path arrow originates.

Figure 5.1 has two latent exogenous variables and two latent endogenous variables. The two latent exogenous variables are correlated, as indicated by the curve with arrows at each end, with a covariance of f_{12}. Latent exogenous causal variable ξ_1 is a cause of both η_1 and η_2. Latent exogenous variable ξ_2 is a cause of η_2 also. Each of the latent variables, both exogenous and endogenous, has four manifest *indicator* variables. The distinction between indicators and the latent variables they are indicators of is analogous to the distinction between the use of readings on several different kinds of thermometers, for example, mercury-tube, alcohol-tube, thermocouple, thermistor thermometers, and the physical concept of temperature, of which the thermometers are indicators. It is a good practice to have at least four indicators for each latent variable to overidentify the latent variables of the model and permit tests of closure and/or self-containment. Note also that each endogenous variable, whether manifest or latent, has a *disturbance variable* pointing to it. Disturbance variables represent extraneous influences such as errors of measurement and random shocks that are combined with the effects of exogenous and/or endogenous variables on a given endogenous variable. Disturbances are analogous to unique factors in common factor analysis or errors of measurement in classical test theory. However, disturbances may contain both systematic and unsystematic error. They are usually assumed to be mutually uncorrelated and uncorrelated also with the exogenous variables. Technically disturbances are also exogenous variables, but whatever is contained in them is not of focal interest in contrast to the explicitly named exogenous variables.

Requiring the disturbances to be mutually uncorrelated imposes an analogue of the local independence condition wherein conditioning on the explicit exogenous variables leaves the conditioned endogenous variables mutually independent. The only variation left in the endogenous variables after conditioning on the explicit exogenous variables is due just to the disturbances. Requiring the disturbances to be uncorrelated with the exogenous variables implies that there are no other hidden *relevant* causes, not explicitly represented in the model, and permits unbiased estimation of the structural coefficients. In other words, the model represents a conception of external reality, and disturbances and their properties are supposed to hold in reality. When these assumptions are violated, the model may be compromised and yield misleading inferences when seemingly confirmed against data. (*Note*: Disturbances are not residual variables. Residual variables are formed when one partials from a set of variables what can be predicted in them from other variables. They are the result of a mathematical operation.) In linear models, residuals are necessarily uncorrelated with the predictor variables on which the partialled components are based. Disturbances, on the other hand, represent other causes of the variables not explicitly represented in the model otherwise, and subjunctively it is possible in some cases to imagine their being correlated with the explicit causal variables within the system and with each other. The constraints imposed on disturbances, that they are mutually uncorrelated and are uncorrelated with exogenous variables of the system, must be satisfied in the real-world situation represented by the model to achieve a closed system of variables in which causal relations can be inferred and structural coefficients estimated without bias. Residual variables become equivalent to disturbances when these constraints are satisfied. But if the constraints are not satisfied in the situation represented by the model, for example, there are hidden relevant causes in the disturbances that are correlated with the exogenous variables, then the residuals are not true disturbances and parameter estimates are likely biased (see James, Mulaik, and Brett, 1982, pp. 71–80).

Note in Figure 5.1 how the variables are numbered. The exogenous and the endogenous variables are numbered separately. The numbering begins in each case with latent variables and is followed by manifest variables. There are only two exogenous variables, ξ_1 and ξ_2. But there are 18 endogenous variables, $\eta_1, \eta_2, y_3, \ldots, y_{18}$. Disturbance variables are labeled with the same number as the endogenous variables they point to.

From Path Diagrams to Structural Equations

Most researchers begin formulating a structural equation model by drawing a path diagram. That allows them to see the various relationships between the variables of the model in a perspicuous way. But each path diagram can be converted into a system of linear equations, and because most computer programs today require that one specify the model to be tested by entering the system of equations, one line at a time, it is important that you learn how to convert a path diagram into a system of linear equations. The process is rather simple. What you have to do is focus on the endogenous variables of the system, for there will be a linear equation for each endogenous variable. We do this with the model in the path diagram of Figure 5.1.

Begin with the first endogenous variable in the system of equations, η_1. Identify the arrows coming into η_1. In this case there is an arrow from ξ_1 and from ε_1. There are no other arrows coming to η_1. Now, associated with each arrow is a structural coefficient. The arrow from ξ_1 has the structural coefficient γ_{11}. The arrow from ε_1 has the structural coefficient δ_{11}. We now write down a structural equation for η_1 as:

$$\eta_1 = \gamma_{11}\xi_1 + \delta_{11}\tilde{\varepsilon}_1.$$

Note that there is a term on the right for each arrow coming into η_1. The first term is for the arrow from ξ_1 to η_1, and represents the product of the structural parameter γ_{11} with the variable ξ_1, which is the origin of the arrow. The second term is for the arrow from ε_1 to η_1, and represents the product of the structural parameter δ_{11} with ε_1. Again we multiply the structural parameter with the variable that is the origin of the arrow and add that product to the equation.

Next, let us consider the equation for η_2. Write η_2 on the left of the equation. This time look for an endogenous variable's input into η_2. There is an arrow from η_1 to η_2. Multiply the structural coefficient on this arrow, α_{21}, with the variable that is the origin of this arrow, η_1, and write the product down as the first term of the equation on the right. Next look for any arrows from exogenous variables to η_2. There are two, from ξ_1 and ξ_2. Begin with the lowest numbered exogenous variable. Note the structural coefficient γ_{21} for the arrow from ξ_1 to η_2. Multiply γ_{21} with ξ_1 and add this product as the third term of the equation. Do the same with the arrow from ξ_1 to η_2. Finally note the arrow from the disturbance ε_2 to η_2. Multiply the structural coefficient δ_{22} with the variable ε_2 that is the origin of this arrow. Add the product to the equation. You should have

$$\eta_2 = \alpha_{21}\eta_1 + \gamma_{21}\xi_1 + \gamma_{22}\xi_2 + \delta_{22}\varepsilon_2.$$

The third endogenous variable is y_3. It has only two arrows coming into it, one from x_1 and the other from e_3. Multiplying structural coefficients with the variables that are the origins of these arrows and adding the products yields

$$y_3 = \gamma_{31}\xi_1 + \delta_{33}\varepsilon_3.$$

We would next write down each of the equations for the endogenous variables on the left, y_3 through y_{10}. (We will postpone doing so here.) Variables y_{11} through y_{18} yield slightly different results since they are indicators of latent endogenous variables η_1 and η_2. For example,

$$y_{11} = \alpha_{11,1}\eta_1 + \delta_{11,11}\varepsilon_{11}.$$

We are now ready to see the full system of equations for the 18 endogenous variables:

$$\eta_1 = \gamma_{11}\xi_1 + \delta_{11}\varepsilon_1$$

$$\eta_2 = \gamma_{21}\xi_1 + \gamma_{22}\xi_2 + \alpha_{21}\eta_1 + \delta_{22}\varepsilon_2$$

$$y_3 = \gamma_{31}\xi_1 + \delta_{33}\varepsilon_3$$

$$y_4 = \gamma_{41}\xi_1 + \delta_{44}\varepsilon_4$$

$$y_5 = \gamma_{51}\xi_1 + \delta_{55}\varepsilon_5$$

$$y_6 = \gamma_{61}\xi_1 + \delta_{66}\varepsilon_6$$

$$y_7 = \gamma_{72}\xi_2 + \delta_{77}\varepsilon_7$$

$$y_8 = \gamma_{82}\xi_2 + \delta_{88}\varepsilon_8$$

$$y_9 = \gamma_{92}\xi_2 + \delta_{99}\varepsilon_9$$

$$y_{10} = \gamma_{10,2}\xi_2 + \delta_{10,10}\varepsilon_{10}$$

$$y_{11} = \alpha_{11,1}\eta_1 + \delta_{11,11}\varepsilon_{11}$$

$$y_{12} = \alpha_{12,1}\eta_1 + \delta_{12,12}\varepsilon_{12}$$

$$y_{13} = \alpha_{13,1}\eta_1 + \delta_{13,13}\varepsilon_{13}$$

$$y_{14} = \alpha_{14,1}\eta_1 + \delta_{14,14}\varepsilon_{14}$$

$$y_{15} = \alpha_{15,2}\eta_2 + \delta_{15,15}\varepsilon_{15}$$

$$y_{16} = \alpha_{16,2}\eta_2 + \delta_{16,16}\varepsilon_{16}$$

$$y_{17} = \alpha_{17,2}\eta_2 + \delta_{17,17}\varepsilon_{17}$$

$$y_{18} = \alpha_{18,2}\eta_2 + \delta_{18,18}\varepsilon_{18} + \delta_{18,18}\varepsilon_{18} \tag{5.1}$$

Formulas for Variances and Covariances in Structural Equation Models

Algebraic Methods

As in factor analysis, we soon see that in structural equation modeling, the model is not fit to individual scores but to the variance–covariance matrix for the observed variables. In order to get the reproduced variance–covariance matrix from the model, we need expressions that derive the variances and covariances among the variables of the model from the model equations for the observed and latent endogenous variables. Although, for some purposes, this is easily done in matrix algebra, there are still numerous situations where it is useful to know how to do this in ordinary algebra and, even more easily, by path tracing rules. We first consider purely algebraic methods. Consider a simple causal model with four manifest variables. There are two endogenous variables in this model, Y_1 and Y_2, and two exogenous variables, X_1 and X_2. There are two model equations:

$$Y_1 = \gamma_{11}X_1 + \gamma_{12}X_2 + \delta_1\varepsilon_1$$

$$Y_2 = \alpha_{21}Y_1 + \delta_2\varepsilon_2.$$

The disturbances are presumed uncorrelated with the X variables and with each other. The two X variables may be correlated.

The general formula for the variance of a linear combination

$$X = a_1X_1 + a_2X_2 + \cdots + a_nX_n$$

is given by

$$\sigma^2(X) = \sum_{i=1}^{n} a_i^2\sigma^2(X_i) + \sum_{\substack{i=1 \\ i\neq j}}^{n}\sum_{j=1}^{n} a_ia_j\sigma(X_i, X_j).$$

In other words, the variance of the linear combination is equal to the sum of the squared weight times the variance of each component plus twice the sum of the covariances between each component pair, each multiplied by the product of their weights.

Given two linear combinations X and Y,

$$X = a_1X_1 + a_2X_2 + \cdots + a_nX_n \quad \text{and} \quad Y = b_1Y_1 + b_2Y_2 + \cdots + b_mY_m,$$

the covariance between these two linear combinations is then given by

$$\sigma(X, Y) = \sum_{i=1}^{n}\sum_{j=1}^{m} a_i b_j \sigma(X_i, Y_j)$$

or the sum of the products of their respective weights times the covariance of each pair of component variables, one from each linear combination.

So, now we are able to compute the variances of the respective endogenous variables of our simple structural model. Because Y_2 is a function of Y_1, we have several ways of proceeding. Note that the covariance

$$\sigma(Y_2, Y_1) = \sigma[(\alpha_{21}Y_1 + \delta_2\varepsilon_2), Y_1)] = \alpha_{21}\sigma(Y_1, Y_1) + \delta_2\sigma(\varepsilon_2, Y_1) = \alpha_{21}\sigma^2(Y_1)$$
(5.2)

because the covariance of a variable with itself equals its variance and the covariance between Y_1 and ε_2 is zero by hypothesis. So, all we would need to do is find an expression for the variance of Y_1 and insert that in Equation 5.1. The variance of Y_1 is given as

$$\sigma^2(Y_1) = \gamma_{11}^2\sigma^2(X_1) + \gamma_{12}^2\sigma^2(X_2) + \delta_1^2\sigma^2(\varepsilon_1) + 2\gamma_{11}\gamma_{12}\sigma(X_1, X_2).$$
(5.3)

Hence

$$\sigma(Y_1, Y_2) = \alpha_{21}[\gamma_{11}^2\sigma^2(X_1) + \gamma_{12}^2\sigma^2(X_2) + \delta_1^2\sigma^2(\varepsilon_1) + 2\gamma_{11}\gamma_{12}\sigma(X_1, X_2)].$$
(5.4a)

An alternative approach is simply to substitute the full expression for Y_1 in the first right-hand expression of Equation 5.1, expand, and then collect similar terms. The result will be the same as multiplying α_{21} times each expression within the brackets of Equation 5.1:

$$\sigma(Y_1, Y_2) = \alpha_{21}\gamma_{11}^2\sigma^2(X_1) + \alpha_{21}\gamma_{12}^2\sigma^2(X_2) + \alpha_{21}\delta_1^2\sigma^2(\varepsilon_1)$$
$$+ 2\alpha_{21}\gamma_{11}\gamma_{12}\sigma(X_1, X_2).$$
(5.4b)

The variance of Y_2 is similarly a function of the variance of Y_1:

$$\sigma^2(Y_2) = \alpha_{21}^2\sigma^2(Y_1) + \delta_2^2\sigma^2(\varepsilon_2).$$

If we substitute the right-hand side of Equation 5.2 for the variance of Y_1 in this expression, we obtain

$$\sigma^2(Y_2) = \alpha_{21}^2\gamma_{11}^2\sigma^2(X_1) + \alpha_{21}^2\gamma_{12}^2\sigma^2(X_2) + \alpha_{21}^2\delta_1^2\sigma^2(\varepsilon_1)$$
$$+ 2\alpha_{21}^2\gamma_{11}\gamma_{12}\sigma(X_1, X_2) + \delta_2^2\sigma^2(\varepsilon_2)$$
(5.5)

Path Tracing Rules

A number of authors (Heise, 1975; Duncan, 1975) have described a method whereby an inspection of the path diagram for a structural model will yield expressions for the variances and covariances among variables determining the variables in question. In some respects Heise's discussion is easy to follow because he takes things step-by-step. But Duncan's use of the distinction between direct and indirect effects is useful. These rules will often come in handy as you seek to formulate equations involving variances and covariances among the variables and their relations to the model parameters when considering identification of a model and its parameters. They come in handy when we seek to estimate the overall effects of a causal variable on an effect variable that may be linked by several paths, or by paths with intermediate variables along the path.

Suppose that X_i is a causal variable that immediately effects a variable Y_j, which is illustrated by $X_i \xrightarrow{\gamma_{ji}} Y_j$, that is, by an arrow directly connecting X_i to Y_j. The structural coefficient γ_{ji} associated with the arrow is the *direct effect* of X on Y. The structural coefficient represents how much a unit change in X_i produces a change in Y_j. Now, suppose that there are intervening variables between X_i and Y_j:

$$X_i \xrightarrow{\gamma_{gi}} Y_g \xrightarrow{\alpha_{hg}} Y_h \xrightarrow{\alpha_{jh}} Y_j.$$

The effect of X_i on Y_j is now an *indirect effect* because the effect has to be mediated by the intervening variables Y_g and Y_h. The value of the indirect effect of X_i on Y_j is given by the product of the structural coefficients on the paths connecting X_i to Y_j: $\gamma_{gi} \cdot \alpha_{hg} \cdot \alpha_{jh}$. This allows the reduced expression $Y_j = \gamma_{ji} X_i = \gamma_{gi} \alpha_{hg} \alpha_{jh} X_i$. We may note here that if the absolute magnitudes of the structural coefficients are each less than 1.00, then the indirect effect of X_i on Y_j is less than any intervening direct effect between intervening variables on the open path from X_i to Y_j. The path is said to be *open* because it does not work its way back to the first variable X_i in the path. All paths in DAGS are open. This is the same as saying that all paths in a recursive model are open. On the other hand, if the path works its way back and connects with its starting variable, the path is known as a *loop*. A graph with at least one loop is a directed cyclic graph, corresponding to a *nonrecursive* model. For the present we only concern ourselves with DAGS and/or recursive models.

A causal variable may have more than one indirect effect on an effect variable. There may be several paths through different intervening variables from the cause to the effect variable. Each distinct path of intervening variables has an indirect effect on the effect variable. A distinct path of intervening variables is any sequence of variables traversed in the direction of the effect variable that passes through a distinct sequence of intervening variables. Two sequences are distinct if there is at least one variable in the sequence not found in the other. The *total effect* of a causal variable on an effect variable is the sum of its direct effect and all its distinct indirect effects on the effect variable.

We first consider the covariance between two variables. As an overview, consider that there are three fundamentally different ways in which two variables X and Y may covary. But all of these ways involve shared variance. (1) $X \to Y$: X may be a cause of Y. (2) $X \leftarrow Y$: Y may be a cause of X. (3) $X \leftarrow Z \to Y$: X and Y are not causes of each other but are effects of a common third variable Z. We may expand these cases by replacing any arrow by a sequence of intervening variables with arrows between them all pointing in the same direction. We may also consider that two variables may covary not only because one variable is the cause of the other, but additionally because there is some third variable that is a cause of both.

Given any two variables in a path diagram with the aim of finding the covariance between them, we first must look at the path diagram to identify the direct paths between them and the indirect paths between them. For instance, take the following simple path model with two manifest variables given in Figure 5.2. Note that the variance of the exogenous variable and the disturbance is indicated by an arch. [I am indebted to Professor Jack McArdle (*cf.* McArdle and Boker, 1990) for the use of arches (McArdle used arrows at the ends of his arches, but arches are easier to set up and draw) to represent variances in path diagrams. They are not always necessary, but do facilitate the use of path tracing rules.] If we wish to find the covariance between X_1 and Y_1, we see that there is only one path between them, shown by an arrow, and this is a direct path, with X_1 at the start and Y_1 at the end of the path. The rule is to start at the variable that is the origin of the direct path, here it is X_1, go back to the variance arch and go around it, picking up the variance expression as we go, then go back to the variable (X_1) and then go along the path from it to the other variable (Y_1), picking up the path coefficient along the way. Stop at Y_1. Multiply all coefficients passed through along the way. We should then have the covariance as $\sigma(X_1, Y_1) = \alpha_{11}\sigma^2(X_1)$. By tracing the path through the coefficients, we find the coefficients that must be multiplied together to obtain the covariance. In other words, the covariance is the variance times the effect of the causal variable on the effect variable.

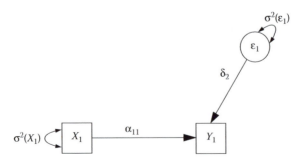

FIGURE 5.2 A simple path model with two manifest variables.

Consider now the model of four manifest variables that we used to illustrate the algebraic method. We now represent it by the path diagram in Figure 5.3. In this diagram Y_1 corresponds to X_1 and Y_2 corresponds to Y_1 in Figure 5.1. What is now a bit more complicated is that we do not know directly the variance of Y_1. This, however, is a function of the two exogenous variables X_1 and X_2, as well as the disturbance ε_1, their variances, and their covariances. Now, while we might preoccupy ourselves with finding the variance of Y_1, there is really no need to do so. The path tracing we do will accumulate all the expressions we need to obtain the covariance we seek. But while we are at it, notice that the covariance between the two exogenous variables is shown by a double-headed curved line with the covariance given by $\sigma(X_1, X_2)$.

We call a curve with double arrows a "bridge." Some of our traced paths will cross that bridge. So, let us begin with the causal variable Y_1. Note the three arrows coming into it. We start tracing paths up each one of these arrows. Let us begin with the arrow from X_1 to Y_1. From Y_1 go up along this arrow to X_1, picking up the path coefficient γ_{11} as we go, until we get to the exogenous variable X_1. Go now around the variance loop on X_1 and pick up the variance $\sigma^2(X_1)$ also. Now when you get back to X_1, go straight back home to Y_1. As we go back down the arrow from X_1 to Y_1, pick up the coefficient γ_{11}. When we arrive at Y_1, then take the path from it to Y_2, picking up the coefficient α_{21} also. Now, multiply all accumulated variances and coefficients together and write the result down somewhere. Algebraically we should now have $\gamma_{11}\sigma^2(X_1)\gamma_{11}\alpha_{21}$.

Next, let us take another path. This time we go down the path from Y_1 to X_2, picking up the path coefficient γ_{12} as we go; then we pick up the variance $\sigma^2(X_2)$; then come back to Y_1 and from there go directly to Y_2, picking up again the path coefficient α_{21}. We write the product of the coefficients collected along the tracing of this path. We should now have the second product $\gamma_{12}\sigma^2(X_2)\gamma_{12}\alpha_{21}$. We now have two product expressions. We save them for

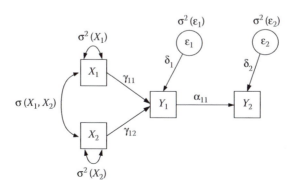

FIGURE 5.3 A simple graph model with four manifest variables.

later, when we add them, and additional expressions we are about to get, together.

Now, there are three other paths we must trace. The idea is that you go from the first variable of the covariance to any exogenous or disturbance variable connected to it. You then must seek a way to return to the first variable. You may (1) return by the way you came to the causal variable, after going around the variance loop for the exogenous variable, and then from there go to the effect variable of the covariance, or (2) take a bridge from the exogenous variable to another exogenous variable, that also is a cause of the first variable, and return to it from there, and then on to the second variable of the covariance. No two paths traced should pass through the same arrows, arches, and bridges *in the same order and direction.*

The first of these three additional paths to trace is to go from Y_1 to X_1, picking up γ_{11}, and instead of going around the variance arch, go across the bridge, picking up $\sigma(X_1, X_2)$, then directly from X_2 back to Y_1, picking up γ_{12}, and finally from Y_1 to Y_2, picking up α_{21}. So, multiplying the accumulated coefficients together, we obtain $\gamma_{11}\sigma(X_1, X_2)\gamma_{12}$.

The second of the remaining three is obtained by going in the opposite direction from Y_1 to X_2, across the bridge to X_1, back to Y_1, and then to Y_2. This produces $\gamma_{12}\sigma(X_1, X_2)\gamma_{11}$, which is equal to the value we obtained when we traced this path in the opposite direction.

The final path to trace is to go from Y_1 to ε_1, picking up δ_1, then around the variance loop, picking up $\sigma^2(\varepsilon_1)$, and then, because there is no other way to get back to Y_1, go back the way we came, picking up δ_1 a second time and from there to Y_2, picking up α_{21} once more. The result is $\delta_1\sigma^2(\varepsilon_1)\delta_1\alpha_{21}$.

We have now all of the component expressions for the covariance between Y_1 and Y_2. After we add them together and rearrange, we obtain

$$\sigma(Y_1, Y_2) = \gamma_{11}^2\sigma^2(X_1)\alpha_{21} + \gamma_{12}^2\sigma^2(X_2)\alpha_{21} + 2\gamma_{11}\gamma_{12}\sigma(X_1, X_2)\alpha_{21}$$
$$+ \delta_1^2\sigma^2(\varepsilon_1)\alpha_{21},$$

which we see is the same as the result shown in Equation 5.2 obtained algebraically.

Now, let us examine a case, illustrated by two models in Figure 5.4 with both a direct path and a covariance path between two variables for which we wish to determine their covariance.

In Figure 5.4a and b, we seek $\sigma(Y_1, Y_2)$. The direct path shown in Figure 5.4a goes from Y_1 to Y_2 with path coefficient α_{21}. We first obtain expressions for the direct path. Begin as before with the causal variable Y_1 and note that there are two arrows coming to it, one from ε_1 and the other from X_1. We are to trace a path back to either a disturbance or an exogenous variable that is a cause of Y_1, pick up any path coefficients on the way, then the variance term, and finally retrace our steps to Y_1 picking up the path coefficients again, and from there go to Y_2, picking up α_{21}. (There are no bridges to cross in this example.) So, let us begin with the path from ε_1. Go up this path, picking up δ_1, then around

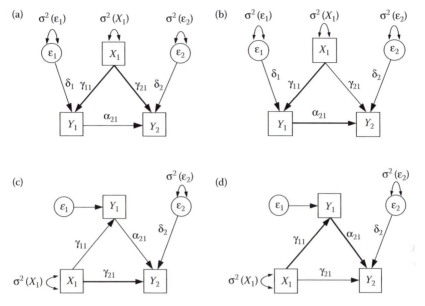

FIGURE 5.4 Two three-variable path models. At the top we seek $\sigma(Y_1, Y_2)$ composed of (a) direct and (b) covariance paths between variables Y_1 and Y_2 shown, respectively, in bold arrows. At the bottom we seek $\sigma(X_1, Y_1)$ via (a) a direct effect between X_1 and Y_2 shown in bold in (c) and an indirect effect between them in bold in (d).

the variance loop of ε_1, picking up $\sigma^2(\varepsilon_1)$, then back to Y_1, picking up δ_1 once again, and then from Y_1 to Y_2, picking up α_{21}. We should have $\delta_1 \sigma^2(\varepsilon_1) \delta_1 \alpha_{21}$. Let us now go up the path to X_1, picking up γ_{11}, then go around the variance loop of X_1, picking up $\sigma^2(X_1)$, then back to Y_1 again, picking up γ_{11} again, then finally from Y_1 to Y_2, picking up α_{21}. We should have $\gamma_{11} \sigma^2(X_1) \gamma_{11} \alpha_{21}$. So, now we have completed the direct terms.

The covariance path is due to the common cause X_1 of both Y_1 and Y_2. In this case we go from Y_1 to X_1, picking up γ_{11}, then around its variance loop, picking up $\sigma^2(X_1)$, and then, instead of returning to Y_1, we go from X_1 to Y_2, picking up γ_{21}. We should now have $\gamma_{11} \sigma^2(X_1) \gamma_{21}$. We do not go over this same path a second time in the reverse direction, because we are seeking a covariance, and only go in one direction in that case. (We only go a second time in the reverse direction when we find variances of the causal variable.) So, now the covariance between Y_1 and Y_2 is given by

$$\sigma(Y_1, Y_2) = \delta_1^2 \sigma^2(\varepsilon_1)\alpha_{21} + \gamma_{11}^2 \sigma^2(X_1)\alpha_{21} + \gamma_{11}\sigma^2(X_1)\gamma_{21}$$

or

$$\sigma(Y_1, Y_2) = \left[\delta_1^2 \sigma^2(\varepsilon_1) + \gamma_{11}^2 \sigma^2(X_1)\right]\alpha_{21} + \gamma_{11}\sigma^2(X_1)\gamma_{21}.$$

Note that $\sigma^2(Y_1) = \left[\gamma_{11}^2 \sigma^2(X_1) + \delta_1^2 \sigma^2(\varepsilon_1)\right]$.

In Figure 5.4c and d, we see another model with a direct and an indirect effect. We seek $\sigma(X_1, Y_2)$. The direct effect is shown in Figure 5.4c in bold arrows. As before, we first obtain the variance on X_1, then we go from X_1 to Y_2, picking up α_{21}. The covariance due to the direct effect is then $\sigma^2(X_1) \cdot \alpha_{21}$. Now, in Figure 5.4d, we see the path of the indirect effect in bold arrows. Again we begin with the variance $\sigma^2(X_1)$, then go from X_1 to Y_1, then from Y_1 to Y_2, picking up γ_{11} and α_{21}. The indirect contribution to the covariance is $\sigma^2(X_1) \cdot \gamma_{11} \cdot \alpha_{21}$. Now, add both the covariance due to the direct effect and the covariance due to the indirect effect and we obtain (after simplifying)

$$\sigma(X_1, Y_2) = \sigma^2(X_1)(\gamma_{21} + \gamma_{11} \cdot \alpha_{21}).$$

Next, let us consider Figure 5.5, which is the model with 16 manifest variables that we considered earlier. Suppose we wish to find the variance of variable Y_{18}. This is a more complex problem because the exogenous variables are at least once removed from variable Y_{18}. But as before, to find the variance, we must work our way backward from Y_{18} to various exogenous variable to find their variances, picking up path coefficients to multiply together as we go along our way. We presume that all exogenous latent variables and disturbance variables have unit variances. The key is to remember that the variance of a weighted linear combination is equal to the sum of the respective squared weights times the variance of each component variable plus the sum of twice the products of the respective weights times the covariances between each pair of component variables. This explain why we trace the paths that we do.

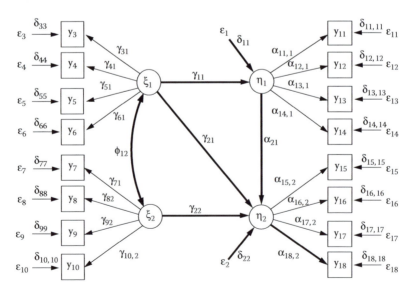

FIGURE 5.5 Model in which we find the variance of variable Y_{18} using path tracing rules. In the figure, bolder path lines in the diagram contribute to the variance of Y_{18}.

Let us first begin at Y_{18} and go to η_2, which is the immediate cause of Y_{18}. We see that we could write the variance of Y_{18} as $\sigma^2(Y_{18}) = \alpha_{18,2}^2 \sigma^2(\eta_2)$. But we do not know directly $\sigma^2(\eta_2)$, and so, we have to build it indirectly as we trace back to the ultimate exogenous variables and disturbances from which η_2 is derived as a weighted linear combination. So, from η_2 we go to ξ_1 and then back the way we came. The variance of ξ_1 is 1.00. Along the way we pick up $\alpha_{18,2}$ and γ_{21}, and then these again as we return. We end up with the product $\alpha_{18,2}^2 \gamma_{21}^2$. Next let us again go through η_2 to ξ_1, but this time we return to Y_{18} via η_1 and η_2. So the coefficients picked up along the way yield the product $\alpha_{18,2}^2 \gamma_{21} \gamma_{11} \alpha_{21}$. Similarly, we can go the same path, this time in reverse, to yield another $\alpha_{18,2}^2 \gamma_{21} \gamma_{11} \alpha_{21}$. We can also go to ε_1 by way of η_1 and return the same way, picking up the matrix product $\alpha_{18,2}^2 \alpha_{21}^2 \delta_{11}^2$. Next, we go to η_2, then to ξ_1, but this time we take the bridge to ξ_2, then go to η_2, and finally back to Y_{18}. Multiplying the coefficients we passed on the way, we obtain $\alpha_{18,2}^2 \gamma_{21} \phi_{12} \gamma_{22}$. We can go back over the same path in the reverse direction and again obtain $\alpha_{18,2}^2 \gamma_{21} \phi_{12} \gamma_{22}$. Next we go to η_2, then to ξ_2, and then back in the reverse direction. The coefficients we pick up on this run are $\alpha_{18,2}^2 \gamma_{22}^2$. Next, go to η_2, then η_1, then ξ_1, then across the bridge to ξ_2, then back to η_2, and finally to Y_{18}. The coefficients we gather yield the product $\alpha_{18,2}^2 \alpha_{21} \gamma_{11} \phi_{12} \gamma_{22}$. And we must go back over this same path in reverse to again yield $\alpha_{18,2}^2 \alpha_{21} \gamma_{11} \phi_{12} \gamma_{22}$. Finally, we go to η_2, then to ε_2, and return the way we came, yielding $\alpha_{18,2}^2 \delta_{22}^2$. The variance of Y_{18} is the sum of all these products:

$$\sigma^2(Y_{18}) = \alpha_{18,2}^2 \gamma_{21}^2 + \alpha_{18,2}^2 \gamma_{22}^2 + \alpha_{18,2}^2 \alpha_{21}^2 \delta_{11}^2 + \alpha_{18,2}^2 \delta_{22}^2 + 2\alpha_{18,2}^2 \gamma_{21} \gamma_{11} \alpha_{21}$$

$$+ 2\alpha_{18,2}^2 \gamma_{21} \gamma_{11} \alpha_{21} + 2\alpha_{18,2}^2 \gamma_{21} \phi_{12} \gamma_{22} + 2\alpha_{18,2}^2 \alpha_{21} \gamma_{11} \phi_{12} \gamma_{22}. \quad (5.6)$$

Now, if we wish to find the proportion of the variance of Y_{18} that is due to the variance of ξ_1, we collect all terms that involve a path through ξ_1 and divide these by the total variance given in Equation 5.6:

$$\text{variance due to } \xi_1 = \frac{\alpha_{18,2}^2 \gamma_{21}^2 + \alpha_{18,2}^2 \gamma_{11}^2 + 2\alpha_{18,2}^2 \gamma_{21} \gamma_{11} \alpha_{21} + 2\alpha_{18,2}^2 \gamma_{21} \phi_{12} \gamma_{22} + 2\alpha_{18,2}^2 \alpha_{21} \gamma_{11} \phi_{12} \gamma_{22}}{\sigma^2(Y_{18})}.$$

Next, the covariance between, say, Y_3 and Y_{18} involves finding all paths from Y_3 to Y_{18}. These are shown in Figure 5.6.

Begin with Y_3, go to ξ_1, then to η_2, and finally to Y_{18}. Multiply together all coefficients passed along the way: $\gamma_{31} \gamma_{21} \alpha_{18,2}$. Next from Y_3, again go to ξ_1, then to η_1, then to η_2, and finally to Y_{18}. Multiply together all coefficients passed along the way: $\gamma_{31} \gamma_{11} \alpha_{21} \alpha_{18,2}$. Finally, from Y_3, go to ξ_1, then cross the bridge to ξ_2, then to η_2, and finally to Y_{18}. Multiply coefficients passed on the way: $\gamma_{31} \phi_{12} \gamma_{22} \alpha_{18,2}$. Now, add all three products to yield

$$\sigma(Y_3, Y_{18}) = \gamma_{31} \gamma_{21} \alpha_{18,2} + \gamma_{31} \gamma_{11} \alpha_{21} \alpha_{18,2} + \gamma_{31} \phi_{12} \gamma_{22} \alpha_{18,2}.$$

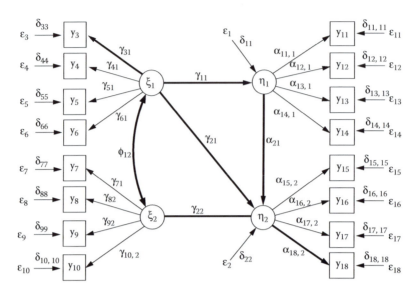

FIGURE 5.6 Paths to follow to yield the covariance between Y_3 and Y_{18} are in heavy lines. Latent variables in this case are assumed to have unit variance.

From these results we can see how the covariance between variables may be due not just to direct causal effects between variables, but to indirect paths and covariances between ancestor variables. Path tracing can make these associations much more perspicuous.

Matrix Equations

While ordinary algebraic equations are convenient for entering models into computer programs, they are less convenient for theoretical purposes. We now need to consider matrix equations for representing structural equation models and to derive from these equations the analogue of the fundamental theorem of factor analysis, that is, an equation that states how the variance-covariance matrix of the observed variables is a matrix function of model parameter matrices.

There are two ways by which we might express the model equation of the structural equation model with latent variables in matrix form. The first treats disturbances distinctly from exogenous variables:

$$\begin{bmatrix} \boldsymbol{\eta} \\ \mathbf{y} \end{bmatrix} = \mathbf{A} \begin{bmatrix} \boldsymbol{\eta} \\ \mathbf{y} \end{bmatrix} + \boldsymbol{\Gamma} \begin{bmatrix} \boldsymbol{\xi} \\ \mathbf{x} \end{bmatrix} + \begin{bmatrix} \boldsymbol{\Delta} & \mathbf{0} \\ \mathbf{0} & \boldsymbol{\Psi} \end{bmatrix} \begin{bmatrix} \boldsymbol{\zeta} \\ \boldsymbol{\varepsilon} \end{bmatrix}. \tag{5.7a}$$

or

$$\boldsymbol{\eta}^* = \mathbf{A}\boldsymbol{\eta}^* + \boldsymbol{\Gamma}^*\boldsymbol{\xi}^* + \boldsymbol{\Delta}^*\boldsymbol{\varepsilon}^*,$$

where $\eta^* = \begin{bmatrix} \eta \\ y \end{bmatrix}$ is the partitioned $(m + p) \times 1$ random vector of endogenous variables, with η the $m \times 1$ vector of latent endogenous variables and y the $p \times 1$ vector of manifest endogenous variables. A is the $(m + p) \times (m + p)$ matrix of structural coefficients relating endogenous variables to other endogenous variables. Note: the principal diagonal of A ordinarily contains zeros, because no endogenous variable can be a cause of itself. (When all elements above the principal diagonal of A are zeros, so that the only nonzero elements in A are below the principal diagonal, we have a *recursive model*, whose endogenous variables are only effects of variables "upstream" from them and never of variables "downstream" from them. Otherwise, when there are nonzero elements both below and above the principal diagonal of A, the model is a *nonrecursive model*. Nonrecursive models contain feedback loops and things like reciprocal causation and must be understood in ways different from the way we understand recursive models.)

Γ^* is the $(m + p) \times (n + q)$ matrix of structural coefficients relating endogenous variables to exogenous variables. $\xi^* = \begin{bmatrix} \xi \\ x \end{bmatrix}$ is the partitioned $(n + q) \times 1$ random vector of exogenous variables, with ξ the $n \times 1$ vector of latent exogenous variables and x the $q \times 1$ vector of q manifest exogenous variables.

Ψ is a $p \times p$ diagonal matrix of structural coefficients relating manifest endogenous variables to exogenous disturbance variables. ε is the $p \times 1$ vector of disturbance random variables on the manifest variables. In some treatments Ψ is regarded as an identity matrix and variances of the disturbances can differ from unity. In programs based on this latter treatment, when estimating variances of the disturbances, the program may wander into the negative region for these variances, which yields an inadmissible solution. In the development here, on the other hand, the variances of the disturbances can be fixed to unity, and variance relative to the endogenous variables will then be determined by the square of the ψ structural parameter for the corresponding disturbance. The ψ's can be positive or negative without implying inadmissible solutions.

Δ is an $m \times m$ diagonal matrix of structural coefficients relating m latent endogenous variables η to m disturbances ζ. A particular diagonal element of Δ is δ_{ii}. ε^* is the $(m + p) \times 1$ partitioned vector $\begin{bmatrix} \zeta \\ \varepsilon \end{bmatrix}$ of disturbances on latent and manifest endogenous variables collectively,

$$\Delta^* = \begin{bmatrix} \Delta & 0 \\ 0 & \Psi \end{bmatrix}.$$

An alternative way of expressing the fundamental model equation of structural equation modeling, which we use, is to treat both explicit exogenous variables and disturbances as exogenous variables. This leads to the following form for the equation:

$$\begin{bmatrix} \eta \\ y \end{bmatrix} = \mathbf{A} \begin{bmatrix} \eta \\ y \end{bmatrix} + \begin{bmatrix} \Gamma_\xi & \Gamma_X & \Gamma_\varepsilon \end{bmatrix} \begin{bmatrix} \xi \\ x \\ \varepsilon \end{bmatrix}$$

or

$$\eta^* = \mathbf{A}\eta^* + \Gamma^*\xi^*. \tag{5.7b}$$

ξ^* is an $(n+q) \times 1$ vector of exogenous and disturbance variables. ε is $(m+p) \times 1$. No distinction in notation is made between disturbances on latent endogenous or manifest endogenous variables, so there is no ζ. The submatrix Γ_ε equals Δ^*. I use this form because it allows us to treat γ's and δ's similarly when obtaining derivatives of the fit functions with respect to their elements. This form was used previously by James, Mulaik, and Brett (1982) and was inspired by Bentler and Weeks' (1982) simplification of the structural equation model, which in turn has affinities to McArdle's RAM model (McArdle, 1979, 1980; McArdle and McDonald, 1984). The LISREL© notation developed by Jöreskog (1973), although widely popularized, is now regarded as unnecessarily complex.

At this point we need to deal with a question that the reader may have: how is it that η^* appears on both the left and right sides of the model equation? Actually, if one returns to the algebraic equations developed for the path model previously, one will see that no variable occurs simultaneously on both sides of an equation. It only looks that way in the matrix notation. But recall that the diagonal elements of the $\mathbf{A}$ matrix are zeros, which means that no endogenous variable can occur on both sides of the equation. Nevertheless, to tidy things up a bit, structural equation modelers have resorted to the following device: subtract $\mathbf{A}\eta^*$ from both sides of Equation 5.7b to obtain

$$\eta^* - \mathbf{A}\eta^* = \Gamma^*\xi^*,$$

which can be simplified by factoring η^* out of each expression on the left to yield

$$(\mathbf{I} - \mathbf{A})\eta^* = \Gamma^*\xi^*. \tag{5.8}$$

Let $\mathbf{B} = (\mathbf{I} - \mathbf{A})$; then rewrite Equation 5.8 as

$$\mathbf{B}\eta^* = \Gamma^*\xi^*. \tag{5.9}$$

Finally, multiply both sides by $\mathbf{B}^{-1}$:

$$\eta^* = \mathbf{B}^{-1}\Gamma^*\xi^*. \tag{5.10}$$

This is known as the *reduced form* model equation. In this equation endogenous variables are matrix functions of the exogenous variables only.

Our next task is to derive variances and covariances for the manifest variables of the system. But first we need expressions for covariances among the exogenous variables. Let $\Phi = \mathrm{var}(\boldsymbol{\xi}^*)$ or

$$\Phi = \begin{bmatrix} \Phi_{\xi\xi} & \Phi_{\xi X} & 0 \\ \Phi_{X\xi} & \Phi_{XX} & 0 \\ 0 & 0 & \Phi_{\varepsilon\varepsilon} \end{bmatrix}$$

where $\Phi_{\xi\xi}$ is the $n \times n$ variance–covariance matrix for the latent exogenous variables. Φ_{XX} is the $q \times q$ variance–covariance matrix for the manifest exogenous variables. $\Phi_{\xi X}$ is the $n \times q$ matrix of covariances between the n latent exogenous and q manifest exogenous variables. $\Phi_{\varepsilon\varepsilon}$ is the $(m + p) \times (m + p)$ variance–covariance matrix for the $m + p$ disturbance variables. Finally, we indicate by null matrices that the covariances between exogenous variables and disturbances are zero (by hypothesis).

We now need to introduce a device for selecting the manifest variables from the vectors $\boldsymbol{\eta}^*$ and $\boldsymbol{\xi}^*$, respectively. What we use is known as a *selection matrix* (Bentler and Weeks, 1982). Let $\mathbf{G_y}$ be a $p \times (m + p)$ matrix, such that $\mathbf{G_y} = [0, \mathbf{I}]$, where the null matrix (used to ignore the latent endogenous variables) is a $p \times m$ null matrix and $\mathbf{I}$ is a $p \times p$ identity matrix for selecting the manifest endogenous variables. Then

$$y = \mathbf{G_y}\boldsymbol{\eta}^* = \begin{bmatrix} 0 & \mathbf{I} \end{bmatrix} \begin{bmatrix} \boldsymbol{\eta} \\ y \end{bmatrix}.$$

Similarly, let $\mathbf{G_x}$ be an $q \times (n + q + (m + p))$ matrix, such that $\mathbf{G_x} = [0, \mathbf{I}, 0]$, where the first null matrix is an $q \times n$ null matrix that leads one to ignore the latent exogenous variables, $\mathbf{I}$ is a $q \times q$ identity matrix used to select the manifest exogenous variables, and the second null matrix is a $q \times (m + p)$ null matrix used to ignore the disturbance variables. Thus

$$x = \mathbf{G_x}\boldsymbol{\xi}^* = \begin{bmatrix} 0 & \mathbf{I} & 0 \end{bmatrix} \begin{bmatrix} \boldsymbol{\xi} \\ x \\ \varepsilon \end{bmatrix}.$$

Bentler and Weeks (1982) suggested defining supervectors

$$Z = \begin{bmatrix} y \\ x \end{bmatrix} \tag{5.11}$$

and

$$\boldsymbol{v} = \begin{bmatrix} \boldsymbol{\eta}^* \\ \boldsymbol{\xi}^* \end{bmatrix} = \begin{bmatrix} \mathbf{B}^{-1}\boldsymbol{\Gamma}^*\boldsymbol{\xi}^* \\ \boldsymbol{\xi}^* \end{bmatrix} = \begin{bmatrix} \mathbf{B}^{-1} & 0 \\ 0 & \mathbf{I} \end{bmatrix} \begin{bmatrix} \boldsymbol{\Gamma} \\ \mathbf{I} \end{bmatrix} \boldsymbol{\xi}^*. \tag{5.12}$$

We may rewrite this result as $\boldsymbol{v} = \mathbf{B}^{*-1}\boldsymbol{\Gamma}^*\boldsymbol{\xi}^*$.

Next, define the supermatrix

$$G = \begin{bmatrix} G_y & 0 \\ 0 & G_x \end{bmatrix}. \tag{5.13}$$

Then $Z = Gv = GB^{*-1}\Gamma^*\xi^*$ or

$$Z = \begin{bmatrix} y \\ x \end{bmatrix} = \begin{bmatrix} G_y & 0 \\ 0 & G_x \end{bmatrix} v = \begin{bmatrix} G_y & 0 \\ 0 & G_x \end{bmatrix} \begin{bmatrix} \eta^* \\ \xi^* \end{bmatrix} = \begin{bmatrix} G_y & 0 \\ 0 & G_x \end{bmatrix} \begin{bmatrix} B^{-1}\Gamma\xi^* \\ \xi^* \end{bmatrix}$$

$$= \begin{bmatrix} G_y & 0 \\ 0 & G_x \end{bmatrix} \begin{bmatrix} B^{-1} & 0 \\ 0 & I \end{bmatrix} \begin{bmatrix} \Gamma \\ I \end{bmatrix} \xi^* \tag{5.14}$$

is the vector of observed random variables derived via the "grand" selection matrix G from the supervector v of endogenous and exogenous variables (including disturbances). The variance–covariance matrix of the observed variables (the fundamental theorem) is then

$$\Sigma = GB^{*-1}\Gamma^*\Phi\Gamma^{*\prime}B^{*\prime-1}G'. \tag{5.15}$$

It is worth noting that this form for the equation for the observed variables' variance–covariance matrix is a special case of McDonald's (1979, 1980) corresponding equation in the Covariance Structure Analysis (COSAN) model for analysis of covariance structures. From Equations 5.11 through 5.13 and Equation 5.15 we obtain

$$\Sigma_{yy} = G_y B^{-1}\Gamma\Phi\Gamma'B'^{-1}G'_y. \tag{5.16a}$$

$$\Sigma_{xy} = G_x\Phi\Gamma'B'^{-1}G_y. \tag{5.16b}$$

$$\Sigma_{xx} = G_x\Phi G'_x = \Phi_{xx}. \tag{5.16c}$$

Equations 5.14 and 5.16a, b, and c constitute, respectively, the model equation for the observed variables and the equation for the variance–covariance matrix of the observed variables as a matrix function of the parameters of the structural equation model. These equations correspond to the fundamental equation and the fundamental theorem of factor analysis, and represent the fundamental equation and theorem of the structural equation model.

6

Identification

Incompletely Specified Models

Before considering the problems of estimating the parameters in structural equation models, we must consider an early development of these models due to the Swedish mathematical statistician, Karl Jöreskog (1969), which was focused on developing confirmatory factor analysis, and which turns out to be a special case of structural equation modeling. Because the common factor analysis model is likely already familiar to the reader and because it is simpler to work with than the full structural equation model, we will discuss identification in the case of a common factor model. But the principles generalize immediately to structural equation models, of which the common factor model is a special case.

By way of review, the confirmatory factor analysis model has the model equation

$$\mathbf{Y} = \mathbf{\Lambda}\boldsymbol{\xi} + \mathbf{\Psi}\boldsymbol{\varepsilon}, \tag{6.1}$$

where $\mathbf{Y}$ is a $p \times 1$ random vector of observed variables, $\mathbf{\Lambda}$ is a $p \times n$ factor pattern matrix relating the p observed variables to the n common factors, $\boldsymbol{\xi}$ is an $n \times 1$ random vector of n common factor variables, $\mathbf{\Psi}$ is a $p \times p$ diagonal matrix of unique factor pattern coefficients, and $\boldsymbol{\varepsilon}$ is a $p \times 1$ vector of latent unique factor variables. Furthermore, assuming variables have zero means, the model usually assumes that $E(\boldsymbol{\xi}\boldsymbol{\varepsilon}) = \mathbf{0}$ and $E(\boldsymbol{\varepsilon}\boldsymbol{\varepsilon}') = \mathbf{I}$, implying that the unique factors have unit variances and are mutually uncorrelated. From the fundamental equation and the assumptions, we can derive the variances and covariances among the observed variables as

$$\Sigma = \Lambda \Phi_{\varepsilon\varepsilon} \Lambda' + \Psi^2 \qquad (6.2)$$

Jöreskog's major contribution in his (1969) paper arises from the fact that he appears to have realized that most researchers who use factor analysis cannot specify exactly all the loadings of the pattern matrix (the matrix Λ of coefficients indicating how much a unit change in a factor produces a change in a manifest variable) nor even the correlations $\Phi_{\varepsilon\varepsilon}$ among the factors (the latent variables of the model) when they generate hypotheses about the expected factors. However, frequently they can specify the zero elements of the pattern matrix, and sometimes, on the basis of past analyses, they can specify the exact values of a few of the nonzero loadings of the pattern matrix. Consequently, Jöreskog (1969) sought to formulate the factor analysis model in such a way that the researcher would have complete freedom in specifying or leaving unspecified various parameters (i.e., elements of the matrices Λ, $\Phi_{\varepsilon\varepsilon}$, and Ψ^2) of the model. Specified parameters would be *fixed* by hypothesis. Unspecified (free) parameters would be estimated by maximum-likelihood estimation, conditional on any fixed parameters. He also realized in later work (Jöreskog, 1974) that one could easily impose equality constraints between certain parameters, which would be useful in testing models where parameters are constrained to be equal against models where the parameters in question are free to take on distinct values during estimation. In summary, Jöreskog conceived that a researcher could treat each of the parameters of a model as one of the following:

1. As a *fixed* parameter, that is, a parameter that has a prespecified value by hypothesis that remains unchanged during the parameter estimation process.

2. As a *free* parameter, that is, a parameter that is free to vary during the iterations of the parameter estimation process until it attains a value that optimizes the fit function conditional on values of fixed parameters.

3. As a *yoked* (my term) parameter, that is, a parameter that is free to vary during parameter estimation but under the constraint that its value must *equal* the value of some other parameter (or parameters) at every iteration of the parameter estimation process. Because only one value must be determined for each group of yoked parameters, only one parameter from this group is counted when counting the number of distinct estimated parameters.

4. Later programs for structural equation modeling allowed one to place linear or nonlinear constraints on individual or groups of parameters.
   In recent years, methods have been developed by others wherein the researcher can constrain a parameter to be positive or negative, or

greater or less than a specific value. So, we might add a fifth category, which treats a parameter as the following:

5. As satisfying an *inequality constraint*, that is, that it is greater or less than some constant value. Methods for achieving inequality constraints on single parameters, however, do not represent a distinct technique for dealing with parameters apart from the first three ways of treating parameters. Imposing inequality constraints on a parameter usually involves clever uses of extra paths, "phantom variables," and/or duplicated parameters yoked under equality constraints combined with the fixing of certain other parameters (*cf.* Rindskopf, 1983, 1984). Like freeing a parameter, imposing an inequality constraint on a parameter also yields a net gain of an additional parameter to estimate and, unlike imposing equality constraints on several parameters, does not increase one's degrees of freedom.

The discussion that follows will illustrate these concepts with the simpler confirmatory common factor model, which at this point the student should be more familiar with, but they apply to the more general structural equation model that has more kinds of parameters.

Freeing Parameters Implies No Added Constraints on Their Value

In an incompletely specified model a free parameter, being an unspecified parameter, is a parameter about which the researcher asserts nothing constraining the value of the parameter beyond the constraints on it arising indirectly from explicitly specified constraints on other parameters on which the parameter is conditioned. Freeing a parameter is equivalent to expressing ignorance about its value.

A common misunderstanding has developed that freeing a parameter is equivalent to implicitly *specifying* or asserting a nonzero value and even a sign for the parameter. Jöreskog (1974) perhaps fostered this misunderstanding by recommending that if a researcher believes a test *is dependent on* a certain factor, but has no value for it, then in specifying his model he should treat the loading of that test on the factor as a free parameter. What he should have said was that if the researcher cannot assert that a test is independent of a certain latent variable by fixing the respective loading to zero, but has an unknown value, he should free the respective parameter: this does not imply that the test has a nonzero loading on the factor. Certain programs such as Bentler's EQS© program or the SIMPLIS language of LISREL© 8 also foster this misunderstanding by requiring the researcher to specify, for each dependent variable, an equation that asserts that the dependent variable is a linear function of certain independent variables. Leaving out an independent variable in an equation implies that the structural parameter relating that independent variable to the dependent variable is fixed to zero. The parameters associated with independent variables explicitly included in this equation, on the other hand, may be free parameters, fixed, yoked-equal, or inequality-constrained. There is nothing intrinsically wrong with freeing these parameters. But the researcher must recognize them for what they are—ways of simplifying the model specification process for researchers. But asserting that a parameter in these equations is a free parameter does not impose any additional constraint on the parameter during the parameter estimation process beyond those constraints on other parameters on which the parameter estimate is conditioned. The estimated value of the parameter may turn out to be negative, positive, or zero!

Where this misunderstanding leads to confusions is when researchers speak about "misspecifying a model" by freeing up a zero parameter that actually is zero in the population. This is not a misspecification, for nothing is specified by freeing the parameter. (To *specify* is to assert something specific, and freeing a parameter does not assert anything specific about it.) Furthermore, if the parameter is really zero, then the expected estimate will be zero if constraints on other parameters permit. Actually less is specified. Furthermore, in very large samples the estimate of the parameter will converge in probability to its population value of zero. The overall hypothesis involving all fixed and constrained parameters also suffers because by freeing a parameter it loses an independent condition by which it could confirm or disconfirm the model's validity.

If you feel strongly enough about it, you should impose inequality constraints, even though that will not increase your degrees of freedom. On the other hand, instead of simply freeing a parameter that one expects to be nonzero and imposing an inequality constraint on it, specify a fixed value within the range of values expected as a way of committing oneself to a stronger statement of one's hypothesis. Such an approach gains degrees of freedom at the possible risk of greater lack of fit, meaning you test something. If you obtain a good approximate fit, it may mean that your specification was close to the mark. Hence something is gained, although the specific value specified is still in doubt. We will take up inequality constraints in greater detail in the next chapter.

The specified portion of the incompletely specified hypothesis, on the other hand, concerns explicitly specified fixed and constrained parameters of the model.

Identification

Even though no values are asserted for the free parameters, values must nevertheless be obtained for them to allow one to reproduce the variance–covariance matrix $\hat{\Sigma}_0$ for the hypothesized model so that it may be compared with (usually) an unrestricted sample covariance matrix $\mathbf{S}$ taken from the population. The *identification problem* concerns whether or not one can determine unique values for the unknown parameters using the observed data and constraints placed on other parameters. Here let us consider the simpler and possibly more familiar special case of the common factor model. The observed data in this case are the $p(p+1)/2$ independent elements of the variance–covariance matrix $\mathbf{S}$, given by adding the elements along the principal diagonal and, because of symmetry, off one side of the matrix. The potential number of parameters of the common factor model can exceed the number of independent observed elements of the covariance matrix. Consider that the fundamental theorem of the common factor model can be written in the following way:

$$\Sigma_{YY} = \mathbf{\Lambda}^* \mathbf{\Phi}^* \mathbf{\Lambda}^{*'} = \begin{bmatrix} \mathbf{\Lambda} & | & \mathbf{\Psi} \end{bmatrix} \begin{bmatrix} \mathbf{\Phi}_{\varepsilon\varepsilon} & 0 \\ 0 & \mathbf{I} \end{bmatrix} \begin{bmatrix} \mathbf{\Lambda}' \\ - \\ \mathbf{\Psi}' \end{bmatrix}$$

Here $\mathbf{\Lambda}^*$ is a partitioned $p \times (n+p)$ factor pattern matrix and $\mathbf{\Phi}^*$ is an $(n+p) \times (n+p)$ factor covariance matrix. Potentially there could be $p \times (n+p) + (n+p+1) \times (n+p)/2$ free, unknown parameters to determine in $\mathbf{\Lambda}^*$ and $\mathbf{\Phi}^*$ from the $n(n+1)/2$ independent elements of $\mathbf{\Phi}_{\varepsilon\varepsilon}$. In this case we would have far more unknowns than knowns with which to determine them. Some of the indeterminacy is reduced by requiring that $\mathbf{\Psi}$ be a diagonal matrix, then requiring $\mathbf{\Sigma}_{\xi\varepsilon} = 0$ and $\mathbf{\Sigma}_{\varepsilon\varepsilon} = \mathbf{I}$, so that we have only $p \times (n+p) + n(n-1)/2$ unknown parameters to determine, which, depending on n, could be less than $p(p+1)/2$. To minimally identify the solution, we have to impose an additional n^2 constraints on $\mathbf{\Lambda}$ and $\mathbf{\Phi}_{\varepsilon\varepsilon}$, usually by imposing that $[\text{diag}\,\mathbf{\Phi}_{\varepsilon\varepsilon}] = \mathbf{I}$ and $n(n-1)$ on elements of $\mathbf{\Lambda}$; however, this will have to be done in a proper way. A necessary but not sufficient condition for identification of parameters is that the number of unknown, free parameters be less than or equal to the number of independent elements of the observed variables' covariance matrix, that is $k \leq p(p+1)/2$.

Let us now arrange all of the parameters of $\mathbf{\Lambda}^*$ and $\mathbf{\Phi}^*$ above into a single vector $\boldsymbol{\theta}$ of $p \times (n+p) + (n+p+1) \times (n+p)/2$ parameters. We further partition $\boldsymbol{\theta}$ as $\boldsymbol{\theta} = (\hat{\boldsymbol{\theta}}, \boldsymbol{\theta}^*)$, with $\hat{\boldsymbol{\theta}}$ being the independent free and yoked parameters and $\boldsymbol{\theta}^*$ the fixed parameters of the matrices $\mathbf{\Lambda}^*$ and $\mathbf{\Phi}^*$ discussed previously. Suppose the vector $\boldsymbol{\theta}$ of model parameters is allowed to take on arbitrary values. Each $\boldsymbol{\theta}$ generates a specific reproduced covariance matrix $\boldsymbol{\Sigma}(\boldsymbol{\theta})$. This implies that the covariance matrix $\boldsymbol{\Sigma}$ is a matrix function $\boldsymbol{\Sigma} = \boldsymbol{\Sigma}(\boldsymbol{\theta})$ of the parameters in $\boldsymbol{\theta}$. Because of the symmetry of $\boldsymbol{\Sigma}$, this matrix function is equivalent to a system of $p(p+1)/2$ distinct simultaneous nonlinear equations $\sigma_{ij} = \sigma_{ij}(\boldsymbol{\theta}), i = 1, \ldots, p; j = 1, \ldots, i$.

The question now before us is, given some select subset of the parameters of $\boldsymbol{\theta}$ placed in the subvector $\hat{\boldsymbol{\theta}}$, what constraints must we place on the remaining parameters of this system of equations so that we might in turn solve for the subset of free parameters in the subvector $\hat{\boldsymbol{\theta}}$ using the elements of $\boldsymbol{\Sigma}$? We say that a parameter $\hat{\theta}_k$ in $\hat{\boldsymbol{\theta}}$ is *identified* if there exists a subset of equations $\sigma_{ij} = \sigma_{ij}(\boldsymbol{\theta})$ that is uniquely solvable for $\hat{\theta}_k$ given $\boldsymbol{\Sigma}$, fixed parameters $\boldsymbol{\theta}^*$ and constraints. We say that $\hat{\boldsymbol{\theta}}$ is identified if every element in $\hat{\boldsymbol{\theta}}$ is identified. If every element in $\hat{\boldsymbol{\theta}}$ is identified, then we say that the model is identified. On the other hand, we say that a parameter is underidentified if it is not identified. A model is underidentified if at least one of its parameters is underidentified. We say that a parameter $\hat{\theta}_k$ is *overidentified* if more than one distinct subset of equations $\sigma_{ij} = \sigma_{ij}(\boldsymbol{\theta})$ may be found that is solvable for $\hat{\theta}_k$. If at least one free parameter of a model is overidentified, we say the model is *overidentified*. If each free parameter is identified, but none is overidentified, we say the model is *just-identified*.

Just-identified models cannot themselves be tested, because they will always perfectly reproduce the data. (A test must have a logical possibility of failing what is tested.) They can only be used in the testing of more constrained models as a basis for comparison. Based on a system of as many equations as there are unknowns that is uniquely solvable for the unknowns, there is one and only one way to solve for the unknowns of a just-identified model. When solved for in terms of the empirical covariance matrix and the constraints, the estimated parameters in turn will perfectly reproduce the empirical covariance matrix. But this is a mathematical necessity, not an empirical finding.

We cannot call a situation a test if there is no logical possibility in the situation of failing the test. Thus evaluating a just-identified model by how well it reproduces the empirical covariance matrix is not a test of the model because it is logically impossible to fail to fit the covariance matrix other than perfectly. This might not disturb us if there was only one just-identified model that could be formulated for a given covariance matrix. But actually there are infinitely many such models. A principal component analysis model $\boldsymbol{\Sigma}_{YY} = \boldsymbol{\Lambda}\boldsymbol{\Lambda}'$ with as many principal component factors as there are observed variables, and the constraints $\mathbf{A}'\mathbf{A} = \mathbf{I}$ and $\mathbf{A}'\boldsymbol{\Sigma}_{YY}\mathbf{A} = \mathbf{D}$, with $\boldsymbol{\Lambda} = \mathbf{A}\mathbf{D}^{1/2}$, where $\mathbf{A}$ is the eigenvector matrix and $\mathbf{D}$ is the eigenvalue matrix of $\boldsymbol{\Sigma}_{YY}$, is a just-identified model. $\mathbf{A}'\mathbf{A} = \mathbf{I}$ specifies $p(p+1)/2$ constraints (the p diagonal and $p(p-1)/2$ off-diagonal elements of $\mathbf{I}$). $\mathbf{A}'\boldsymbol{\Sigma}_{YY}\mathbf{A} = \mathbf{D}$ specifies the $p(p-1)/2$ distinct off-diagonal elements of $\mathbf{D}$. Hence, there are $p(p+1)/2 + p(p-1)/2 = p^2$ constraints. All models are just-identified when based on nonsingular linear transforms of the components $\boldsymbol{\Sigma}_{YY} = (\boldsymbol{\Lambda}\mathbf{T})(\mathbf{T}^{-1}\mathbf{T}'^{-1})(\mathbf{T}'\boldsymbol{\Lambda}')$, where $\mathbf{T}$ is any nonsingular $p \times p$ matrix. The constraints imposed to achieve a just-identified condition may reflect simply a subjective view of the researcher.

This is especially evident when $\boldsymbol{\Lambda}_c = \boldsymbol{\Lambda}\mathbf{T}$ is a Cholesky factor of $\boldsymbol{\Sigma}_{YY}$. In this case $\boldsymbol{\Sigma}_{YY} = \boldsymbol{\Lambda}_c\mathbf{I}\boldsymbol{\Lambda}_c'$. The matrix $\boldsymbol{\Lambda}_c$ is lower triangular with $p(p+1)/2$ nonzero elements on and below the diagonal. These are the elements that are estimated, while the zero elements above the diagonal of $\boldsymbol{\Lambda}_c$ and the off-diagonal 0's and the diagonal 1's of the identity matrix $\mathbf{I}$ are fixed. Hence, there are as many distinct elements to estimate in Λ as there are distinct elements in Σ_{YY}. This implies as many knowns as there are unknowns (estimated parameters) to solve for.

Because of the relation $\Sigma = \Sigma(\theta)$, all solutions for an overidentified parameter $\hat{\theta}_k$ will be consistent across these different subsets of equations that yield solutions for $\hat{\theta}_k$. But that might not be the case if we use an empirical covariance matrix such as S (a sample covariance matrix)—or Σ (the corresponding population covariance matrix)—in place of $\hat{C}_0$ (the reproduced covariance matrix) and seek variant solutions for an overidentified parameter with the corresponding subsets of the elements of the empirical covariance matrix. If we obtain inconsistent (i.e., different) values for the overidentified parameter from different subsets of equations, this alerts us to the fact that the empirical covariance matrix is not consistent with the model. Thus the goal is to overidentify as many of the free parameters as possible in order to test the model, where testing a model implies possibly disconfirming it by showing that it is inconsistent with the model. Now, we do not in actual practice evaluate models this way. The inconsistency in solutions for values of overidentified parameters translates itself into lack of fit between the reproduced model covariance matrix $\hat{\Sigma}_0$ that minimizes a discrepancy function in the parameter estimation process and the sample covariance matrix S (that we work with in lieu of the population covariance matrix Σ). We use this lack of fit as evidence for the misspecification of a model.

Because the estimated values of the free and yoked parameters $\hat{\theta}$ are those values that minimize the discrepancy function conditional on the fixed parameters θ^* and the equality and inequality constraints, we require the free parameters to all be at least identified. Otherwise, we will be unable to distinguish between distinct models having the same fixed parameters and parameter constraints, but different values for the free and yoked parameters that nevertheless reproduce the same $\hat{\Sigma}_0$ and consequently have the same fit to S. By having identified models, we put the onus of any lack of fit squarely onto the fixed parameters and constraints specified by hypothesis.

So far, we have only defined the meaning of identified parameters. We have not discussed how to determine whether a given model is identified. In theory one can establish that a parameter is identified by taking a hypothetical covariance matrix $\Sigma = \Sigma(\theta)$, the constraints on the model parameters, the individual model equations for determining each element of Σ, and then finding some subset of these equations by which one can solve uniquely for a parameter in question. Except for very simple cases, such as establishing that the parameters of a single common factor model for three variables are identified, this approach to identification can become horrendously complex and burdensome.

Nevertheless, because it is an important concept, we will illustrate this brute force approach to identification: suppose we have two variables, Y_1 and Y_2, and hypothesize that they have a single common factor. Going to the level of individual model equations for each variable, we obtain

$$Y_1 = \lambda_1 \xi + \psi_1 \varepsilon_1$$
$$Y_2 = \lambda_2 \xi + \psi_2 \varepsilon_2 \tag{6.3}$$

where ξ is the common factor, ε_1 and ε_2 are unique factors, λ_1 and λ_2 are common factor pattern coefficients, and ψ_1 and ψ_2 are unique factor pattern coefficients. We presume (without loss of generality) that all variables have zero means, which implies, for example, that $\sigma^2(Y) = E(Y^2)$. We also presume that $\sigma^2(\xi) = 1$ and $\sigma^2(\varepsilon_1) = \sigma^2(\varepsilon_2) = 1$. Before going further, we need to consider simplifying our notation: by σ_i^2 we mean $\sigma^2(Y_i)$; by σ_{ij} we mean $\sigma(Y_i, Y_j)$. From equations in Equation 6.3 and the usual assumptions of no correlation between common and unique factors or between distinct unique factors, we derive the variances and covariances for these two variables:

$$\sigma_1^2 = E(Y_1^2) = E(\lambda_1 \xi + \psi \varepsilon_1)^2 = \lambda_1^2 + \psi_1^2.$$

By a similar reasoning we obtain

$$\sigma_2^2 = \lambda_1^2 + \psi_2^2$$
$$\sigma_{12} = \lambda_1 \lambda_2.$$

This represents a system of three equations in four unknowns, $\lambda_1, \lambda_2, \psi_1$, and ψ_2. There is no way we can solve uniquely for $\lambda_1, \lambda_2, \psi_1$, and ψ_2. These parameters are *underidentified* and the model is consequently *underidentified*. On the other hand, if we fix $\lambda_2 = 1$ in addition to other constraints, then a solution for λ_1 is possible, namely $\lambda_1 = \sigma_{12}$. But in some cases the solution for ψ_2^2, namely $\psi_2^2 = \sigma_2^2 - 1$, may be inadmissible when $\sigma_2^2 < 1$, implying a negative variance for the unique-variance component of Y_2. Fixing *both* $\sigma^2(\xi) = 1$ and $\lambda_2 = 1$ also represents a fairly strong hypothesis, in effect, that when the variance of the common factor is unity, a unit change in ξ leads to a unit change in Y_2. This may not be a hypothesis one wishes to test. One can also achieve identification by constraining $\lambda_1 = \lambda_2$. In that case $\lambda_1 = \lambda_2 = \sqrt{\sigma_{12}}$. This again is a strong hypothesis that may not represent a hypothesis one wishes to test. Furthermore, if $\sigma_{12} < 0$, then λ_1 and λ_2 will take on inadmissible, imaginary values.

Let us next consider the case where we have not two but three variables, Y_1, Y_2, and Y_3. We try to fit a single common factor model to these three variables. We similarly assume zero means for variables, that variances of the common factor and the unique variances are unity. Thus we have the following model equations:

$$Y_1 = \lambda_1 \xi + \psi_1 \varepsilon_1$$
$$Y_2 = \lambda_2 \xi + \psi_2 \varepsilon_2 \tag{6.4}$$
$$Y_3 = \lambda_3 \xi + \psi_3 \varepsilon_3$$

from which we can derive, using the assumption of unit variances for the common and unique factors and the usual assumptions of the common factor model concerning correlations between common and unique factors, expressions for the variances and covariances among these observed variables as functions of the model parameters:

$$\sigma_1^2 = \lambda_1^2 + \psi_1^2$$

$$\sigma_2^2 = \lambda_2^2 + \psi_2^2$$

$$\sigma_3^2 = \lambda_3^2 + \psi_3^2$$

$$\sigma_{12} = \lambda_1 \lambda_2$$

$$\sigma_{13} = \lambda_1 \lambda_3$$

$$\sigma_{23} = \lambda_2 \lambda_3.$$

Here we have six equations in six unknowns. We can solve for each of the unknowns in terms of the observed variances and covariances: from the equation $\sigma_{12} = \lambda_1 \lambda_2$ we can obtain $\lambda_2 = \sigma_{12}/\lambda_1$, and from $\sigma_{13} = \lambda_1 \lambda_3$, we can obtain $\lambda_3 = \sigma_{13}/\lambda_1$. Substituting these two expressions into the equation $\sigma_{23} = \lambda_2 \lambda_3$, we obtain

$$\sigma_{23} = \frac{\sigma_{13}\sigma_{12}}{\lambda_1^2}$$

or

$$\lambda_1^2 = \frac{\sigma_{13}\sigma_{12}}{\sigma_{23}},$$

hence

$$\lambda_1 = \sqrt{\frac{\sigma_{13}\sigma_{12}}{\sigma_{23}}}.$$

By similar arguments we obtain

$$\lambda_2 = \sqrt{\frac{\sigma_{23}\sigma_{12}}{\sigma_{13}}}$$

and

$$\lambda_3 = \sqrt{\frac{\sigma_{23}\sigma_{13}}{\sigma_{12}}}.$$

Once we have solutions for the common factor pattern coefficients, the unique variances are readily obtained from the equations for the variances. For example, from

$$\sigma_1^2 = \lambda_1^2 + \psi_1^2$$

and the equation

$$\lambda_1^2 = \frac{\sigma_{13}\sigma_{12}}{\sigma_{23}},$$

we obtain

$$\sigma_1^2 = \frac{\sigma_{13}\sigma_{12}}{\sigma_{23}} + \psi_1^2$$

or

$$\psi_1^2 = \sigma_1^2 - \frac{\sigma_{13}\sigma_{12}}{\sigma_{23}}.$$

Similarly,

$$\psi_2^2 = \sigma_2^2 - \frac{\sigma_{23}\sigma_{12}}{\sigma_{13}}$$

$$\psi_3^2 = \sigma_3^2 - \frac{\sigma_{23}\sigma_{13}}{\sigma_{12}}.$$

We now illustrate an overidentified model. Given four observed variables, $Y_1, \ldots, Y_4$, and a model in which each shares a single common factor, along with the usual assumptions of unit variances for common and unique factors and zero correlations between common and unique factors and among unique factors,

$$Y_1 = \lambda_1 \xi + \psi_1 E_1,$$
$$Y_2 = \lambda_2 \xi + \psi_2 E_2,$$
$$Y_3 = \lambda_3 \xi + \psi_3 E_3,$$
$$Y_4 = \lambda_4 \xi + \psi_4 E_4,$$

we can obtain expressions for the variances and covariances among these variables in terms of the model's parameters:

$$\sigma_1^2 = \lambda_1^2 + \psi_1^2 \tag{6.5a}$$
$$\sigma_2^2 = \lambda_2^2 + \psi_2^2 \tag{6.5b}$$
$$\sigma_3^2 = \lambda_3^2 + \psi_3^2 \tag{6.5c}$$
$$\sigma_4^2 = \lambda_4^2 + \psi_4^2 \tag{6.5d}$$

$$\sigma_{12} = \lambda_1 \lambda_2 \tag{6.5e}$$

$$\sigma_{13} = \lambda_1 \lambda_3 \tag{6.5f}$$

$$\sigma_{14} = \lambda_1 \lambda_4 \tag{6.5g}$$

$$\sigma_{23} = \lambda_2 \lambda_3 \tag{6.5h}$$

$$\sigma_{24} = \lambda_2 \lambda_4 \tag{6.5i}$$

$$\sigma_{34} = \lambda_3 \lambda_4. \tag{6.5j}$$

Here we have $10 = 4(4+1)/2$ equations (the number of distinct elements in the observed variables' covariance matrix) in eight unknowns. We have more equations than unknowns, which is a necessary but not sufficient condition that some parameters will be overidentified.

We know from our work with the previous three-variable case that there is a solution for λ_1: using Equations 6.5e through 6.5h, we obtain

$$\lambda_1 = \sqrt{\frac{\sigma_{13}\sigma_{12}}{\sigma_{23}}}.$$

But we can also obtain an alternative solution using a different subset of the equations—Equations 6.5f through 6.5j:

$$\lambda_1 = \sqrt{\frac{\sigma_{13}\sigma_{14}}{\sigma_{34}}}.$$

A third solution for λ_1 is obtained using Equations 6.5e through 6.5i:

$$\lambda_1 = \sqrt{\frac{\sigma_{12}\sigma_{14}}{\sigma_{24}}}.$$

Thus λ_1 is overidentified.

In similar ways, we can find three distinct solutions for each of the other three common factor pattern coefficients, implying that each of them is also overidentified. Consequently, the one-factor model for four indicator variables is overidentified. It can be disconfirmed when applied to four-indicator covariance matrices that are not generated by this model.

Having only three indicators of a common factor will lead to a just-identified single-factor model, having as many unknowns to solve for as there are independent data points to determine them with. In such circumstances, as long as the covariances are all nonzero, one can always solve for the parameters, which in turn will perfectly reproduce the covariances. In these cases there is no logical possibility of disconfirming the single-factor hypothesis, so no test of it is possible. Four indicators, on the other hand, will yield an overidentified single-factor model, with more than one subset of equations available to solve for a parameter, the solutions to which may be inconsistent when the single common factor model is not appropriate for the empirical covariances. Consequently, researchers are encouraged, in formulating confirmatory factor analysis models, to include at least four indicators of each factor in the model. This allows one to test the hypothesis that a single common factor underlies each of the indicators of it. This is also indirectly a provisional test of the assumption that the model is closed and self-contained against the influences of any extraneous, unspecified variables. With only three indicators, one would always be able to fit a single-factor model to them (as long as they have nonzero correlations). If one can have more than four indicators, that is even better, because that creates more overidentified conditions by which one's hypothesis of a single common factor may be tested.

Working out identification for these simple two-, three-, and four-variable cases with single common factors allows us to generalize to other situations. The following model is underidentified if $c_{12} = 0$:

$$\Lambda = \begin{bmatrix} \lambda_{11} & 0 \\ \lambda_{21} & 0 \\ 0 & \lambda_{32} \\ 0 & \lambda_{42} \end{bmatrix}, \quad \Phi_{XX} = \begin{bmatrix} 1 & \phi_{12} \\ \phi_{21} & 1 \end{bmatrix},$$

$$\Psi^2 = \begin{bmatrix} \psi_1^2 & 0 & 0 & 0 \\ 0 & \psi_2^2 & 0 & 0 \\ 0 & 0 & \psi_3^2 & 0 \\ 0 & 0 & 0 & \psi_4^2 \end{bmatrix}.$$

If $\phi_{12} = 0, \sigma_{13} = \sigma_{14} = \sigma_{23} = \sigma_{24} = 0$ and the estimation of λ_{11} and λ_{21} breaks down because we have only the block of variances and covariances among Y_1 and Y_2 to work with, which we know from analysis of the two indicator model is an underidentified case. A parallel breakdown in the estimation of λ_{32} and λ_{42} occurs as well. On the other hand, when $\phi_{12} \neq 0$, all of the free parameters are overidentified, with estimates of pattern coefficients λ_{11} and λ_{21} and λ_{32} and λ_{42}, respectively, made possible by covariation of variables Y_1 and Y_2 with Y_3 and Y_4. For example,

$$\lambda_{11} = \sqrt[+]{\frac{\sigma_{41}\sigma_{21}}{\sigma_{42}}} = \sqrt[+]{\frac{\sigma_{31}\sigma_{21}}{\sigma_{32}}}, \quad \psi_1^2 = \sigma_1^2 - \frac{\sigma_{41}\sigma_{21}}{\sigma_{42}}$$

$$\lambda_{21} = \sqrt[+]{\frac{\sigma_{42}\sigma_{21}}{\sigma_{41}}} = \sqrt[+]{\frac{\sigma_{32}\sigma_{21}}{\sigma_{31}}}, \quad \psi_2^2 = \sigma_2^2 - \frac{\sigma_{42}\sigma_{21}}{\sigma_{41}}$$

$$\lambda_{32} = \sqrt[+]{\frac{\sigma_{31}\sigma_{43}}{\sigma_{41}}} = \sqrt[+]{\frac{\sigma_{32}\sigma_{43}}{\sigma_{42}}}, \quad \psi_3^2 = \sigma_3^2 - \frac{\sigma_{31}\sigma_{43}}{\sigma_{41}}$$

$$\lambda_{42} = \sqrt[+]{\frac{\sigma_{41}\sigma_{43}}{\sigma_{31}}} = \sqrt[+]{\frac{\sigma_{42}\sigma_{43}}{\sigma_{32}}}, \quad \psi_4^2 = \sigma_4^2 - \frac{\sigma_{41}\sigma_{43}}{\sigma_{31}}.$$

$$\phi_{12} = \frac{\sigma_{31}}{\sqrt{\sigma_{41}\sigma_{21}/\sigma_{42}}\sqrt{\sigma_{31}\sigma_{43}/\sigma_{41}}} = \frac{\sigma_{41}}{\sqrt{\sigma_{31}\sigma_{21}/\sigma_{32}}\sqrt{\sigma_{42}\sigma_{43}/\sigma_{32}}}$$

$$\phi_{12} = \frac{\sigma_{32}}{\sqrt{\sigma_{42}\sigma_{21}/\sigma_{41}}\sqrt{\sigma_{31}\sigma_{43}/\sigma_{41}}} = \frac{\sigma_{42}}{\sqrt{\sigma_{32}\sigma_{21}/\sigma_{31}}\sqrt{\sigma_{41}\sigma_{43}/\sigma_{31}}}.$$

The following model is identified whether ϕ_{12} is zero or not:

$$\Lambda = \begin{bmatrix} \lambda_{11} & 0 \\ \lambda_{21} & 0 \\ \lambda_{31} & 0 \\ 0 & \lambda_{42} \\ 0 & \lambda_{52} \\ 0 & \lambda_{62} \end{bmatrix}, \quad \Phi_{XX} = \begin{bmatrix} 1 & \phi_{12} \\ \phi_{21} & 1 \end{bmatrix},$$

$$\Psi^2 = \begin{bmatrix} \psi_1^2 & 0 & 0 & 0 & 0 & 0 \\ 0 & \psi_2^2 & 0 & 0 & 0 & 0 \\ 0 & 0 & \psi_3^2 & 0 & 0 & 0 \\ 0 & 0 & 0 & \psi_4^2 & 0 & 0 \\ 0 & 0 & 0 & 0 & \psi_5^2 & 0 \\ 0 & 0 & 0 & 0 & 0 & \psi_6^2 \end{bmatrix}.$$

With three indicators per factor there is sufficient information within the block of variances and covariances among the indicators of a factor to find solutions for the factor pattern coefficients. The solution does not depend on using information about relations with the second set of indicators.

Restricted versus Unrestricted Factor Analysis Models

In Jöreskog (1969), Jöreskog was concerned to make a distinction between specifying a restricted solution and specifying an unrestricted solution when one fixes or frees certain parameters of the confirmatory factor analysis model. Unrestricted models often are found nested or embedded within the more constrained "measurement models" and/or structural equation models. On the one hand, for a fixed value of the number of common factors, n, an unrestricted solution imposes no restrictions on the solution for $\Lambda\Phi_{\varepsilon\varepsilon}\Lambda'$ other than those equivalent to what is needed to obtain a principal axes solution for n common factors. In that case we require $\Phi_{\varepsilon\varepsilon} = I$, which fixes n diagonal elements to unity and $n(n-1)/2$ distinct off-diagonal elements to zero, and requires further that $\Lambda'\Psi^{-2}\Lambda$ is diagonal, placing $n(n-1)/2$ additional constraints on the columns of $\Psi^{-1}\Lambda$, to make them mutually orthogonal. The diagonal elements of Ψ^2 are free parameters. In all, n^2 constraints are placed on the parameters of the model to yield a principal axes solution for n common factors. But there can be more than one unrestricted model. Suppose the principal axis solution is $\Sigma_{YY} = \Lambda_0 \Lambda_0' + \Psi^2$. Then an equivalent model is

$$\Sigma_{YY} = (\Lambda_0 T)(T^{-1}T'^{-1})(T'\Lambda_0') + \Psi^2 = \Lambda_0 \Lambda_0' + \Psi^2$$

where T is any nonsingular $n \times n$ transformation matrix. Because T contains n^2 elements, this suggested to Jöreskog that a necessary but not sufficient condition that one has an unrestricted solution is that exactly n^2 constraints are distributed across the Λ and $\Phi_{\varepsilon\varepsilon}$ matrices (Jöreskog, 1979). The significance of unrestricted solutions is that they correspond to exploratory factor analysis solutions where only minimal constraints are placed on the model to achieve a solution for n common factors. An unrestricted model for n common factors will yield the best fit of any model with n common factors. Jöreskog (1979; personal communication) suggested the following as one way of specifying an unrestricted solution: (1) Fix the n diagonal elements of $\Phi_{\varepsilon\varepsilon}$ to unity. This fixes the metric for the solution. The remaining off-diagonal elements of $\Phi_{\varepsilon\varepsilon}$ are left free. (2) In each column of Λ free up one coefficient corresponding to an expected high loading; make sure that each freed parameter is in a different row. (3) In each row with a freed high loading, fix the remaining $n - 1$ other coefficients to zero. (4) Free all other parameters in Λ. There should now be $n - 1$ zeros in each column. In contrast, a restricted solution restricts the solution for $\Lambda \Phi_{\varepsilon\varepsilon} \Lambda'$ and in turn for Ψ^2. This will occur whenever more than r^2 independent parameters of Λ and $\Phi_{\varepsilon\varepsilon}$ are specified or when some of the diagonal elements of Ψ^2 are specified. The fit of an unrestricted solution to a given sample variance–covariance matrix S will usually be better than a restricted solution.

Identification of Metric

So far we have considered identification in the context of assuming that the common factors have unit variances. The effect of fixing the variances of latent variables to unity is to fix their "metric," that is, their units of measurement. Because a factor pattern coefficient indicates how many units of change will occur in an indicator variable given a unit change in the value of the common factor, the value of a factor pattern coefficient depends on the units of measurement of both the indicator variable and the common factor. In most single-sample studies, fixing the variance of the common factor arbitrarily to unity is sufficient to establish a metric for it, and this aids the identification of other parameters involving the factor as well. But it is quite possible to fix the metric of the common factor in another way: by fixing the pattern loading of one of its indicators while freeing the variance of the common factor. (When fixing a metric for the common factor, one should fix only one parameter—a pattern loading on that factor or the variance of the factor, *but not both*. Otherwise one specifies a rather restricted model that may not be of interest at all. For example, if we set both the pattern loading and the factor variance to unity, this says that when, in the metric that gives it unit variance, the common factor changes one unit, the respective indicator variable changes by one unit. That is a very strong hypothesis, particularly in connection with the variance of the observed indicator.)

There are occasions when one especially wishes to fix the pattern coefficient and not the factor variance. Suppose in a previous study one has obtained an exploratory factor analytic solution and wishes to see if the same solution will be found in a new setting. One will fix the pattern loadings on the factors to the values found in the previous study, while freeing the factor variances. According to Meredith (1964a,b), factor pattern coefficients are invariant under selection effects on the factors while the variances and covariances among the factors are not. So, we might expect to find changes in the factor variances and covariances in the new setting. Leaving these parameters free while fixing the pattern coefficients to values obtained from the previous study allows us to see if there are any effects on the factor variances and covariances. We may also wish to test the hypothesis of whether there has been any change from the previous to the new setting. To do this, we can free up all the pattern loadings except r of them on each factor, with the fixed values of the pattern coefficients set to their values obtained in the previous study. Choosing the largest loading and $n - 1$ lowest loadings on indicators of other factors to fix in each column while freeing the rest would be my recommendation (see also Jöreskog, 1979), although further study of this problem is required. Note that this fixes n^2 parameters. This fixes both the factor metric and the solution for the pattern loadings to what it would be in an unrestricted model rotated provisionally to the position of the previous study as defined by the fixed parameters. We leave the factor variances and covariances free. We then compare this unrestricted solution against the more restricted solution to see where any changes may have come into play. Changes may come in the form of different pattern loadings and different factor variances and covariances.

Many models studied by researchers are themselves formed by joining together several simpler models. Frequently when it is known that the simpler models are all identified by themselves, it is likely then that the parameters of the more complex model formed from them are identified. But one would like more assurances than that. Most contemporary programs for confirmatory factor analysis and linear structural equations modeling (a generalization of factor analysis) check for identification by examining the *information matrix* to see if it is positive definite.

According to Jöreskog and Sörbom (1989), "the information matrix is the probability limit of the matrix of second order derivatives of the fit function used to estimate the model" (p. 17). Another way of putting it is that the information matrix contains estimates of the variances and covariances among the parameter estimates, and is derived from the matrix of second derivatives of the fit function with respect to the free parameters evaluated at the values for the parameters that minimize the fit function. (More about this matrix will be given in a later section.) This matrix is of the order of the number of free and distinct yoked parameters. Jöreskog and Sörbom (1989) then say, "If the model is identified, the information matrix is almost certainly positive definite. If the information matrix is singular, the model is not identified, and *the rank of the information matrix* indicates how many parameters are identified" (p. 17).

Another way of evaluating whether a model is identified, according to Jöreskog and Sörbom (1989), is to arbitrarily assign reasonable values to the free parameters and then, using both fixed and assigned free parameters, generate a reproduced covariance matrix for the raw variables presumed to be dependent on these parameters. Next, take this reproduced covariance matrix and subject it to one's structural model, with free parameters now unknowns to be determined. If the solution does not yield the same parameters that generated the covariance matrix in the first place, then the model is likely not identified.

7

Estimation of Parameters

Discrepancy Functions

When measuring a model's fit to the matrix $\mathbf{S}$, we want to regard lack of fit as due exclusively to misspecifying certain constraints on certain parameters asserted in the hypothesis. Thus the values for the free parameters, which are not given by hypothesis but are needed to complete the model, are required to be those values for these parameters that uniquely minimize the discrepancy between the model's reproduced covariance matrix $\hat{\mathbf{\Sigma}}_0$ and the sample covariance matrix $\mathbf{S}$ *conditional* on the explicitly constrained parameters of the model. The reason for conditioning on the constraints is that we want the estimates to be dependent on the constraints. Then any discrepancy will be due to the constraints. In other words, if there were no constraints other than those to minimally just-identify the model, then the model would fit the data perfectly. But, of course, without an over-identifying set of constraints, there would be no test of a hypothesis. The discrepancy is measured with a discrepancy function, $F[\hat{\mathbf{\Sigma}}_0(\mathbf{\theta}), \mathbf{S}]$, where $\mathbf{\theta}$ is a vector $\mathbf{\theta} = (\hat{\mathbf{\theta}}, \mathbf{\theta}^*)$, with $\hat{\mathbf{\theta}}$ the independent free and yoked parameters and $\mathbf{\theta}^*$ the fixed parameters of the structural equation model.

The most frequently used discrepancy functions are the following:
Ordinary least squares

$$L = \frac{1}{2}\text{tr}[(\mathbf{S} - \hat{\mathbf{\Sigma}}_0(\mathbf{\theta}))'(\mathbf{S} - \hat{\mathbf{\Sigma}}_0(\mathbf{\theta}))]$$

Maximum likelihood

$$F = \ln\left|\hat{\boldsymbol{\Sigma}}_0(\boldsymbol{\theta})\right| + \text{tr}(\mathbf{S}\hat{\boldsymbol{\Sigma}}_0(\boldsymbol{\theta})^{-1}) - \ln|\mathbf{S}| - p$$

Generalized least squares

$$G = \frac{1}{2}\text{tr}[\mathbf{S}^{-1/2}(\mathbf{S} - \hat{\boldsymbol{\Sigma}}_0(\boldsymbol{\theta}))'\mathbf{S}^{-1/2}\mathbf{S}^{-1/2}(\mathbf{S} - \hat{\boldsymbol{\Sigma}}_0(\boldsymbol{\theta}))\mathbf{S}^{-1/2}]$$

$$= \frac{1}{2}\text{tr}[(\hat{\boldsymbol{\Sigma}}_0(\boldsymbol{\theta}) - \mathbf{S})\mathbf{S}^{-1}]^2 = \frac{1}{2}\text{tr}[(\hat{\boldsymbol{\Sigma}}_0(\boldsymbol{\theta})\mathbf{S}^{-1} - \mathbf{I})]^2,$$

where $\mathbf{S}$ is the sample estimate of the unrestricted variance–covariance matrix and $\hat{\boldsymbol{\Sigma}}_0(\boldsymbol{\theta})$ is the reproduced variance–covariance matrix under a hypothesis, generated as a multidimensional function of the free parameters of the structural model arranged for convenience in a single vector $\boldsymbol{\theta}$, with fixed parameters and constraints on parameters implicitly in the function. $\boldsymbol{\Sigma}_{YY} = \boldsymbol{\Sigma}_0(\boldsymbol{\theta})$ means that each element of the variance–covariance matrix, for example, σ_{ij}, is a certain function $\sigma_{ij} = \sigma_{ij}(\boldsymbol{\theta})$ of the model parameters. When the values of the free parameters are chosen to minimize a discrepancy function conditional on the explicit constraints on the parameters of the model, we say that the free parameters are *estimated* according to that discrepancy function.

Ordinary least squares seeks to minimize the sum of squared residuals remaining in the observed covariance matrix $\mathbf{S}$ after subtracting from it the reproduced model covariance matrix $\hat{\boldsymbol{\Sigma}}_0$. This method of estimation has the advantage of requiring no distributional assumptions and it is usually easier to work out the estimating equations and their implementation. Its main disadvantages are that when multivariate normality cannot be presumed and the matrix analyzed is not a covariance matrix, it lacks a distributional theory within which one can make probabilistic inferences about fit. In those cases there is no significance test for fit. The assessment of fit is also affected by arbitrary choices for the units of measurement of the variables.

Maximum-likelihood estimation seeks simultaneously to minimize the difference between $\ln|\mathbf{S}|$ and $\ln|\hat{\boldsymbol{\Sigma}}_0|$ and between $\text{tr}(\mathbf{S}\hat{\boldsymbol{\Sigma}}_0^{-1})$ and $\text{tr}(\mathbf{I}) = n$. These differences will be zero if $\mathbf{S} = \hat{\boldsymbol{\Sigma}}_0$. Maximum-likelihood estimation requires the making of distributional assumptions and is usually more difficult to work with in terms of developing estimating equations and algorithms. But if one can make distributional assumptions, then one gains in the ability to perform probabilistic inferences about the degree of fit, although this may require fairly large samples. Measures of fit based on the likelihood ratio are also metric invariant.

Generalized least squares is a variant of ordinary least squares. As in ordinary least squares, one seeks to minimize the sum of squared residuals. But in generalized least squares the residuals are transformed by pre- and postmultiplying the residual matrix by the inverse of the square root of the sample covariance matrix before evaluating their sum of squares. Generalized least squares has the advantage of not requiring distributional assumptions but still allowing for probabilistic inference about model fit. The complexity of its estimating equations is usually intermediate between those of ordinary least squares and maximum-likelihood estimation. Probabilistic inference, however, may require larger samples than needed for maximum likelihood.

The principles of identification that we discussed in connection with factor analysis transfer themselves readily to structural equation modeling. Although special situations arise in structural equations modeling that require specific analysis to establish the identification of a model, the general principles are the same as those in confirmatory factor analysis. The principles of parameter estimation are also essentially the same in structural equation modeling as in confirmatory factor analysis (which has to be seen as a special case of the structural equation model). The very same general algorithms used to estimate parameters of the structural equation model are also used to estimate parameters of the confirmatory factor analysis model. However, because the structural equation model has in general more sets of parameters than the common factor model, the derivation of the specifics of these algorithms involving the finding of derivatives of a discrepancy function to minimize will be unique to the structural equation model. We will thus now look at this problem in connection with maximum-likelihood, least-squares, and generalized least-squares estimation.

Maximum-likelihood estimation proceeds by seeking to minimize the following discrepancy function:

$$F_{\mathrm{ML}}(\hat{\boldsymbol{\Sigma}}_{ZZ}) = \ln \left| \hat{\boldsymbol{\Sigma}}_{ZZ} \right| + \mathrm{tr}\left(\hat{\boldsymbol{\Sigma}}_{ZZ}^{-1} \mathbf{S} \right) - \ln |\mathbf{S}| - (k + h), \qquad (7.1)$$

where $\hat{\boldsymbol{\Sigma}}_{ZZ}$ is the estimated model variance–covariance matrix among the observed variables, $\mathbf{S}$ is the sample unrestricted estimate of the population covariance matrix for the observed variables, p is the number of manifest endogenous variables, and q the number of manifest exogenous variables. The derivative of the maximum-likelihood discrepancy function F with respect to an arbitrary parameter q of the model is given by

$$\frac{\partial F}{\partial \theta_i} = \mathrm{tr}\left[\left(\boldsymbol{\Sigma}_{zz}^{-1} - \boldsymbol{\Sigma}_{zz}^{-1} \mathbf{S} \boldsymbol{\Sigma}_{zz}^{-1} \right) \frac{\partial \boldsymbol{\Sigma}_{zz}}{\partial \theta_i} \right] = \mathrm{tr}\left[\mathbf{Q} \frac{\partial \boldsymbol{\Sigma}_{zz}}{\partial \theta_i} \right]. \qquad (7.2)$$

Derivatives of Elements of Matrices

In obtaining algorithms for the minimization of discrepancy functions over variations in the values of estimated parameters, we need to know how to obtain the derivatives of the discrepancy function with respect to the free parameters of the model. To find the first derivatives of a discrepancy function

F, we need to develop rules and notation for the taking of partial derivatives of matrices with respect to scalar values. In this regard, we will follow fairly closely the notation used by Bock and Bargmann (1966, pp. 514–515). Their rules and notation are as follows:

In the following development, let the elements of the matrices involved be differentiable functions of a scalar x. Now consider the $n \times r$ matrix $\mathbf{A}$.

$$\frac{\partial \mathbf{A}}{\partial x} = \left[\frac{\partial a_{ij}}{\partial x} \right]_{(n \times r)}, \quad i = 1, \ldots, n, \ j = 1, \ldots, r.$$

In other words, we take the partial derivative of each element in turn and place it in its respective place in a similarly sized matrix.

Let $\mathbf{A}$ be an $n \times r$ matrix and $\mathbf{B}$ a $p \times q$ matrix. Then

$$\frac{\partial (\mathbf{A} + \mathbf{B})}{\partial x} = \frac{\partial \mathbf{A}}{\partial x} + \frac{\partial \mathbf{B}}{\partial x} \quad n = p, \quad r = q, \tag{7.3a}$$

$$\frac{\partial \mathbf{AB}}{\partial x} = \mathbf{A} \frac{\partial \mathbf{B}}{\partial x} + \frac{\partial \mathbf{A}}{\partial x} \mathbf{B} \quad r = p, \tag{7.3b}$$

$$\frac{\partial \mathbf{A}^{-1}}{\partial x} = -\mathbf{A}^{-1} \frac{\partial \mathbf{A}}{\partial x} \mathbf{A}^{-1} \quad n = r, \quad |\mathbf{A}| \neq 0. \tag{7.3c}$$

Let $\mathbf{C}$ be a constant matrix; then

$$\frac{\partial \mathrm{tr}(\mathbf{AC})}{\partial x} = \mathrm{tr} \frac{\partial \mathbf{A}}{\partial x} \mathbf{C}. \tag{7.3d}$$

Given two $n \times n$ square matrices $\mathbf{A}$ and $\mathbf{C}$ with $\mathbf{A}$ nonsingular and $\mathbf{C}$ a constant matrix, the following is a useful consequence of the relationships cited above:

$$\frac{\partial \mathrm{tr}(\mathbf{A}^{-1}\mathbf{C})}{\partial x} = -\mathrm{tr}\left(\mathbf{A}^{-1} \frac{\partial \mathbf{A}}{\partial x} \mathbf{A}^{-1} \mathbf{C} \right) = -\mathrm{tr}\left(\frac{\partial \mathbf{A}}{\partial x} \mathbf{A}^{-1} \mathbf{C} \mathbf{A}^{-1} \right) \tag{7.4a}$$

with the latter expression on the right resulting from the invariance of the trace under cyclic permutation of the matrices. Finally [with proof given by Bock and Bargmann (1966, pp. 514–515)]

$$\frac{\partial \ln |\mathbf{A}|}{\partial x} = -\mathrm{tr} \mathbf{A}^{-1} \frac{\partial \mathbf{A}'}{\partial x}. \tag{7.4b}$$

Next consider the derivative of an $n \times r$ matrix $\mathbf{A}$ with respect to one of its elements a_{ij}. If each of the elements of $\mathbf{A}$ is independent of the other elements,

$$\frac{\partial \mathbf{A}_{ij}}{a_{ij}} = \mathbf{1}_{ij} \quad \text{and} \quad \frac{\partial \mathbf{A}'_{ij}}{a_{ij}} = \mathbf{1}_{ji}, \tag{7.5}$$

where $\mathbf{1}_{ij}$ denotes an $n \times r$ matrix with zeros in every position except the i, j position, which contains a 1. However, if $\mathbf{A}$ is square $n \times n$ symmetric,

$$\frac{\partial \mathbf{A}}{\partial a_{ij}} = \mathbf{1}_{ij} + \mathbf{1}_{ji} - \mathbf{1}_{ij}\mathbf{1}_{ij}, \tag{7.6}$$

where $\mathbf{1}_{ij}\mathbf{1}_{ij} = \mathbf{1}_{ij}$ when $i = j$ and $\mathbf{1}_{ij}\mathbf{1}_{ij} = \mathbf{0}$ when $i \neq j$, $i,j = 1,\ldots,n$.

We will also need the following result: let $\mathbf{A}$ be an $n \times r$ matrix and $\mathbf{B}$ a $p \times q$ matrix, $\mathbf{1}_{ji}$ an $r \times p$ matrix, and $\mathbf{1}_{ts}$ a $q \times n$ matrix. Then

$$\mathrm{tr}(\mathbf{A}\mathbf{1}_{ji}) = [\mathbf{A}]_{ij}. \tag{7.7a}$$

For example,

$$\mathrm{tr}(\mathbf{A}\mathbf{1}_{24}) = [\mathbf{A}]_{42},$$

that is,

$$\mathrm{tr}\begin{bmatrix} a_{11} & a_{12} & a_{13} \\ a_{21} & a_{22} & a_{23} \\ a_{31} & a_{32} & a_{33} \\ a_{41} & a_{42} & a_{43} \end{bmatrix} \begin{bmatrix} 0 & 0 & 0 & 0 \\ 0 & 0 & 0 & 1 \\ 0 & 0 & 0 & 0 \end{bmatrix}_{24} = \mathrm{tr}\begin{bmatrix} 0 & 0 & 0 & a_{12} \\ 0 & 0 & 0 & a_{22} \\ 0 & 0 & 0 & a_{32} \\ 0 & 0 & 0 & a_{42} \end{bmatrix} = [\mathbf{A}]_{42}.$$

Postmultiplying the matrix $\mathbf{A}$ by the matrix $\mathbf{1}_{ji}$ has the effect of extracting the jth column of $\mathbf{A}$ and inserting it into the ith column of an $n \times n$ null matrix. This leaves only the ith element of the jth column of $\mathbf{A}$ as a diagonal element in the resulting square matrix; hence the trace of this matrix is simply this one element, the ith element of the jth column of $\mathbf{A}$, that is, $[\mathbf{A}]_{ij}$.

Next,

$$\mathrm{tr}(\mathbf{A}\mathbf{1}_{ji}\mathbf{B}\mathbf{1}_{ts}) = \mathrm{tr}[\mathbf{A}]_{sj}[\mathbf{B}]_{it} = a_{sj}b_{it}. \tag{7.7b}$$

This result follows from the fact that $\mathbf{A}\mathbf{1}_{ji}$ is an $n \times p$ matrix containing zeros everywhere except in its ith column, which contains the jth column $[\mathbf{A}]_j$ of $\mathbf{A}$, and $\mathbf{B}\mathbf{1}_{ts}$ is a $p \times n$ matrix that contains zeros everywhere except in its sth column, which contains the tth column $[\mathbf{B}]_t$ of $\mathbf{B}$. The product of these two resulting matrices is an $n \times n$ matrix containing zeros everywhere except in its sth column, which contains the equivalent of $[\mathbf{A}]_j[\mathbf{B}]_{it}$. The only nonzero diagonal element of the resulting $n \times n$ matrix is in the sth row and sth column and is equal to $[\mathbf{A}]_{sj}[\mathbf{B}]_{it}$. For example,

$$\begin{bmatrix} a_{11} & a_{12} & a_{13} \\ a_{21} & a_{22} & a_{23} \\ a_{31} & a_{32} & a_{33} \\ a_{41} & a_{42} & a_{43} \end{bmatrix} \begin{bmatrix} 0 & 0 & 0 & 0 & 0 \\ 0 & 0 & 0 & 1 & 0 \\ 0 & 0 & 0 & 0 & 0 \end{bmatrix}_{24} = \begin{bmatrix} 0 & 0 & 0 & a_{12} & 0 \\ 0 & 0 & 0 & a_{22} & 0 \\ 0 & 0 & 0 & a_{32} & 0 \\ 0 & 0 & 0 & a_{42} & 0 \end{bmatrix}$$

and

$$\begin{bmatrix} b_{11} & b_{12} & b_{13} \\ b_{21} & b_{22} & b_{23} \\ b_{31} & b_{32} & b_{33} \\ b_{41} & b_{42} & b_{43} \\ b_{51} & b_{52} & b_{53} \end{bmatrix} \begin{bmatrix} 0 & 1 & 0 & 0 \\ 0 & 0 & 0 & 0 \\ 0 & 0 & 0 & 0 \end{bmatrix}_{12} = \begin{bmatrix} 0 & b_{11} & 0 & 0 \\ 0 & b_{21} & 0 & 0 \\ 0 & b_{31} & 0 & 0 \\ 0 & b_{41} & 0 & 0 \\ 0 & b_{51} & 0 & 0 \end{bmatrix};$$

hence

$$\text{tr}(\mathbf{A1}_{24}\mathbf{B1}_{12}) = \text{tr}\left\{ \begin{bmatrix} 0 & 0 & 0 & a_{12} & 0 \\ 0 & 0 & 0 & a_{22} & 0 \\ 0 & 0 & 0 & a_{32} & 0 \\ 0 & 0 & 0 & a_{42} & 0 \end{bmatrix} \begin{bmatrix} 0 & b_{11} & 0 & 0 \\ 0 & b_{21} & 0 & 0 \\ 0 & b_{31} & 0 & 0 \\ 0 & b_{41} & 0 & 0 \\ 0 & b_{51} & 0 & 0 \end{bmatrix} \right\}$$

or

$$\text{tr}(\mathbf{A1}_{24}\mathbf{B1}_{12}) = \text{tr}\begin{bmatrix} 0 & a_{12}b_{41} & 0 & 0 \\ 0 & a_{22}b_{41} & 0 & 0 \\ 0 & a_{32}b_{41} & 0 & 0 \\ 0 & a_{42}b_{41} & 0 & 0 \end{bmatrix} = [\mathbf{A}]_{22}[\mathbf{B}]_{41}.$$

Equation 7.7 generalizes (Mulaik, 1971) to

$$\text{tr}(\mathbf{A1}_{ji}\mathbf{B1}_{ts}\mathbf{C}_{pq}\cdots\mathbf{1}_{mn}\mathbf{Z1}_{uv}) = [\mathbf{A}]_{vj}[\mathbf{B}]_{it}[\mathbf{C}]_{sp}\cdots[\mathbf{Z}]_{nu}. \qquad (7.8)$$

Note that the rightmost subscript on the rightmost **1** matrix "cycles" around to the left and the subscripts are then assigned pairwise from the left to the matrices **A**, **B**, **C**, and so on.

The traces of the matrix expressions for the derivatives may be obtained by rearranging these expressions (taking advantage of the invariance of the trace under cyclic permutations) into the form of Equations 7.7 or Equation 7.8 and simplifying the result.

There has been considerable work on the topic of working out first and second derivatives of fit functions in connection with the development of algorithms for nonlinear optimization in the realm of multivariate statistics. McDonald and Swaminathan (1973) produced one of the major advances in this area, directly in the service of developing algorithms for analysis of covariance structures. These developments have since been expanded and systematized by mathematicians, and readers may consult Rogers (1980) and Magnus and Neudecker (1988) for integrated treatments of this topic.

Derivatives of Discrepancy Functions in Structural Equation Modeling

Thus it remains to work out the partial derivatives of the observed variables' variance–covariance matrix with respect to the respective parameters of the structural equation model. To do this, we need first to express the variance–covariance matrix among the observed variables as a matrix function of the parameters of the structural equation model in a form suitable for this derivation. This is given in the following equation:

$$
\begin{bmatrix} \Sigma_{YY} & \Sigma_{YX} \\ \Sigma_{XY} & \Sigma_{XX} \end{bmatrix} = \begin{bmatrix} G_y & 0 \\ 0 & G_x \end{bmatrix} \begin{bmatrix} B^{-1} & 0 \\ 0 & I \end{bmatrix} \begin{bmatrix} \Gamma \\ I \end{bmatrix}
$$
$$
\times \Phi \begin{bmatrix} \Gamma' & I \end{bmatrix} \begin{bmatrix} B'^{-1} & 0 \\ 0 & I \end{bmatrix} \begin{bmatrix} G'_y & 0 \\ 0 & G'_x \end{bmatrix}. \tag{7.9}
$$

The elements of the **G** matrix, the null matrix, and the identity matrices may be regarded as fixed parameters in the parameter matrices of Equation 6.10. The **B** and **Γ** matrices, respectively, contain both free and fixed parameters. But individual free parameters in these matrices may be regarded as simply free parameters of the whole supermatrix in which they are embedded. So, we can drop the asterisks on the matrices in Equation 5.15 and write Equation 7.9 (recalling that $\mathbf{Z} = \begin{bmatrix} Y \\ X \end{bmatrix}$) as simply

$$
\Sigma_{ZZ} = GB^{-1}\Gamma\Phi\Gamma'B'^{-1}G. \tag{7.10}
$$

[This equation is essentially the same as Bentler and Weeks' (1982) Equation (2.6).]

Drawing upon the formulas for obtaining derivatives of matrices with respect to individual parameters given in Equations 7.3 through 7.6, we obtain the following partial derivatives:

$$
\frac{\partial \Sigma_{ZZ}}{\partial \phi_{gh}} = GB^{-1}G(1_{gh} + 1_{hg} - 1_{gh}1_{gh})\Gamma'B'^{-1}G', \tag{7.11}
$$

$$
\frac{\partial \Sigma_{ZZ}}{\gamma_{ij}} = GB^{-1}\Gamma\Phi 1_{ji}B'^{-1}G' + GB^{-1}1_{ij}\Phi\Gamma'B'^{-1}G', \tag{7.12}
$$

$$
\frac{\partial \Sigma_{ZZ}}{\partial b_{st}} = GB^{-1}\Gamma\Phi\Gamma'\frac{\partial B'^{-1}}{\partial b_{st}}G' + G\frac{\partial B'^{-1}}{\partial b_{st}}\Gamma\Phi\Gamma'B'^{-1}G'
$$
$$
= GB^{-1}\Gamma\Phi\Gamma'\left(-B'^{-1}\frac{\partial B'}{\partial b_{st}}B'^{-1}\right)G' + G\left(-B^{-1}\frac{\partial B}{\partial b_{st}}B^{-1}\right)\Gamma\Phi\Gamma'B'^{-1}G'
$$
$$
= GB^{-1}\Gamma\Phi\Gamma'(-B'^{-1}1_{ts}B'^{-1})G' + G(-B^{-1}1_{st}B^{-1})\Gamma\Phi\Gamma'B'^{-1}G'. \tag{7.13}
$$

Letting $\mathbf{Q} = (\Sigma_{ZZ}^{-1} - \Sigma_{ZZ}^{-1}S\Sigma_{ZZ}^{-1})$ and substituting, respectively, Equations 7.12 and 7.13 into the formula for the partial derivatives of the maximum-likelihood discrepancy function in Equation 7.8, and simplifying by means of Equation 7.7 and 7.8 and the invariant properties of traces (*cf.* Chapter 2), we then obtain

$$\frac{\partial F}{\partial \phi_{gh}} = (2 - [\mathbf{I}]_{gh})[\mathbf{\Gamma}'\mathbf{B}'^{-1}\mathbf{G}'\mathbf{Q}\mathbf{G}\mathbf{B}^{-1}\mathbf{\Gamma}]_{gh} \tag{7.14}$$

$$\frac{\partial F}{\partial \gamma_{ij}} = 2[\mathbf{B}'^{-1}\mathbf{G}'\mathbf{Q}\mathbf{G}\mathbf{B}^{-1}\mathbf{\Gamma}\mathbf{\Phi}]_{ij} \tag{7.15}$$

$$\frac{\partial F}{\partial b_{st}} = -2[\mathbf{B}'^{-1}\mathbf{G}\mathbf{Q}\mathbf{G}\mathbf{B}^{-1}\mathbf{\Gamma}\mathbf{\Phi}\mathbf{\Gamma}'\mathbf{B}'^{-1}]_{st} \tag{7.16}$$

The partial derivatives for least-squares estimation are given by

$$\frac{\partial L}{\partial \theta_i} = \mathrm{tr}\left[(\mathbf{\Sigma}_{ZZ} - \mathbf{S})\frac{\partial \mathbf{\Sigma}_{ZZ}}{\partial \theta_i}\right] = \mathrm{tr}\left[\mathbf{Q}\frac{\partial \mathbf{\Sigma}_{ZZ}}{\partial \theta_i}\right]. \tag{7.17}$$

If we define $\mathbf{Q} = (\mathbf{\Sigma}_{ZZ} - \mathbf{S})$ then we should get comparable forms for the partial derivatives of the least-squares criterion as found for maximum likelihood:

$$\frac{\partial G}{\partial \phi_{gh}} = (2 - [\mathbf{I}]_{gh})[\mathbf{G}'\mathbf{B}'^{-1}\mathbf{G}'\mathbf{Q}\mathbf{G}\mathbf{B}^{-1}\mathbf{\Gamma}]_{gh} \tag{7.18}$$

$$\frac{\partial G}{\partial \gamma_{ij}} = 2[\mathbf{B}'^{-1}\mathbf{G}\mathbf{Q}\mathbf{G}\mathbf{B}^{-1}\mathbf{\Gamma}\mathbf{\Phi}]_{ij} \tag{7.19}$$

$$\frac{\partial G}{\partial b_{st}} = -2[\mathbf{B}'^{-1}\mathbf{G}'\mathbf{Q}\mathbf{G}\mathbf{B}^{-1}\mathbf{\Gamma}\mathbf{\Phi}\mathbf{\Gamma}'\mathbf{B}'^{-1}]_{st} \tag{7.20}$$

As for *generalized least-squares estimation*, the partial derivatives with respect to the parameters of the structural equation model are of the same form as those for maximum-likelihood estimation, with $\mathbf{Q} = (\mathbf{S}^{-1}\mathbf{\Sigma}_{ZZ}\mathbf{S}^{-1} - \mathbf{S}^{-1})$.

From this point on we may implement one of the quasi-Newton algorithms to be discussed in a subsequent section.

Estimation with Nonnormally Distributed Variables

The techniques described so far presume that the joint distribution of the observed variables is multivariate normal, or nearly so. But many problems in the social sciences involve cases where this assumption is questionable. For instance, the variables may be categorical, as in scores from Likert scale ratings, where raters rate persons or preferences for things on scales that range from 1 to 5 or 1 to 7. In other situations the distributions of the variables are highly skewed or platykurtotic. To deal with these cases, several researches have proposed alternative methods of estimation.

Asymptotic Distribution Free Estimation

Michael Browne (1974, 1977, 1984) investigated extensions of generalized least-squares estimation to the analysis of the covariance structures case, which readily generalizes to structural equation modeling. He suggested thinking of the parameter estimation problem as a kind of regression problem. Let $\Sigma(\theta)$ be the population variance–covariance matrix for p observed is variates in the random vector $\mathbf{Y}$ under a linear model with latent variates involving the $m \times 1$ parameter vector θ. When the model is *true*, because the sample variance–covariance matrix $\mathbf{S}$ is an unbiased estimator, $\mathbf{E} = [\mathbf{S} - \Sigma(\theta)]$ is a $p \times p$ symmetric matrix of errors of sampling having expectation equal to the null matrix. We may express the sample variance–covariance matrix as a matrix function of the parameters and errors of sampling:

$$\mathbf{S} = \Sigma(\theta) + \mathbf{E}.$$

We may transform this equation into a form resembling a typical regression problem by obtaining

$$\text{vecs}(\mathbf{S}) = \text{vecs}[\Sigma(\theta)] + \text{vecs}(\mathbf{E}), \tag{7.21}$$

where the notation vecs($\mathbf{M}$) applied to a symmetric matrix $\mathbf{M}$ means the extraction and reordering of the $p(p+1)/2$ nonredundant elements of the symmetric matrix $\mathbf{M}$ into a $p(p+1)/2 \times 1$ column vector. Elements are extracted by rows, up to the diagonal element, and inserted in a column vector; thus

$$\text{vecs}\left(\begin{bmatrix} 1 & 2 & 4 \\ 2 & 3 & 5 \\ 4 & 5 & 6 \end{bmatrix}\right) = \begin{bmatrix} 1 \\ 2 \\ 3 \\ 4 \\ 5 \\ 6 \end{bmatrix}.$$

Thus m_{ij}, $i \geq j$, is stored in the $g = i(i-1)/2 + j$ position of vecs($\mathbf{M}$). (While we are at it, if $\mathbf{M}$ is nonsymmetric, then the analogue operator vecs($\mathbf{M}$) rearranges the elements of $\mathbf{M}$ by stacking its columns into a single column supervector.) Equation 7.12 may be expressed then as

$$\mathbf{s}^* = \sigma^*(\theta) + \mathbf{e}^*$$

because $E(\mathbf{E}) = \mathbf{0}, E(\mathbf{e}^*) = \mathbf{0}$, where $\mathbf{s}^* = \text{vecs}(\mathbf{S})$, $\sigma^*(\theta) = \text{vecs}[\Sigma(\theta)]$, and $\mathbf{e}^* = \text{vecs}(\mathbf{E})$. We let $p^* = p(p+1)/2$. Note that $\mathbf{e}^* = \mathbf{s}^* - \sigma^*(\theta)$.
  Because $E(\mathbf{E}) = \mathbf{0}, E(\mathbf{e}^*) = \mathbf{0}$. But

$$E(\mathbf{e}^*\mathbf{e}^{*\prime}) = \Psi^*$$

is the variance–covariance matrix of the errors. Unlike ordinary regression, as the sample size increases, and the model is a true model, the size p^* of the vectors $\mathbf{s}^*$, $\mathbf{\sigma}^*(\mathbf{\theta})$, and $\mathbf{e}^*$ remains constant, while the population variance–covariance matrix $\mathbf{\Psi}f^*$ for samples of size $N = n + 1$ vanishes in the limit, that is,

$$\lim_{N \to \infty} (\mathbf{\Psi}^*) = \mathbf{0}$$

rather than being some positive definite matrix. This reflects the vanishing of sampling error variance as sample size increases indefinitely.

Browne (1982, 1984) cites Kendall and Stuart (1969, Section 13.16) for the following:

$$[n\mathbf{\Psi}^*]_{ij,kl} = \sigma_{il}\sigma_{jl} + \sigma_{il}\sigma_{jk} + (n/N)\kappa_{ij,kl}, \tag{7.22}$$

where $[n\mathbf{\Psi}^*]_{ij,kl} = n \operatorname{cov}(s_{ij}, s_{kl})$, σ_{il}, σ_{jl}, σ_{il}, and σ_{jk} are population covariances between the variables whose subscripts are shown, and $\kappa_{ij,kl}$ is a fourth-order cumulant given by

$$\kappa_{ij,kl} = \sigma_{ij,kl} - \sigma_{ij}\sigma_{kl} - \sigma_{ik}\sigma_{jl} - \sigma_{il}\sigma_{jk},$$

where

$$\sigma_{ij,kl} = E[(Y_i - \mu_i)(Y_j - \mu_j)(Y_k - \mu_k)(Y_l - \mu_l)].$$

In the limit, as n increases without bound, the covariance for $n\mathbf{\Psi}^*$ thus equals in general

$$\lim_{n \to \infty} [n\mathbf{\Psi}^*]_{ij,kl} = \sigma_{ij}\sigma_{jl} + \sigma_{il}\sigma_{jk} + \kappa_{ij,kl} = \sigma_{ij,kl} - \sigma_{ij}\sigma_{kl}. \tag{7.23}$$

The asymptotic distribution for $n^{1/2}(\mathbf{s}^* - \mathbf{\sigma}^*)$ is furthermore known to be normal (Muirhead, 1982, Theorems 1.2.16 through 1.2.18) with mean vector $\mathbf{0}$ and variance–covariance matrix $\mathbf{V}$ *under fairly general conditions*.

Browne (1984) notes that when we do not know the actual sampling distribution for $n^{1/2}\mathbf{e}^*$ in some finite (but large) sample case, we may use the asymptotic distribution as an approximation and proceed to estimate model parameters under the presumption of that distribution.

By $\hat{\mathbf{\theta}}$ we mean an estimate of the parameter vector $\mathbf{\theta}$, while $\mathbf{\Sigma}(\hat{\mathbf{\theta}})$ is the reproduced variance–covariance matrix of the observed variables based on the parameter estimates. Browne notes that the aim of estimation is to minimize a "discrepancy function" $F(\mathbf{S}, \mathbf{\Sigma}(\hat{\mathbf{\theta}}))$ of the observed variance–covariance matrix $\mathbf{S}$ and the reproduced variance–covariance matrix $\mathbf{\Sigma}(\hat{\mathbf{\theta}})$. Discrepancy functions have the property

(i) $F(\mathbf{S}, \mathbf{\Sigma}(\hat{\mathbf{\theta}})) \geq 0$,

(ii) $F(\mathbf{S}, \mathbf{\Sigma}(\hat{\mathbf{\theta}})) = 0$ iff $\mathbf{\Sigma}(\hat{\mathbf{\theta}}) = \mathbf{S}$,

(iii) $F(\mathbf{S}, \mathbf{\Sigma}(\hat{\mathbf{\theta}}))$ is twice continuously differentiable.

Furthermore, all estimators estimated by minimizing a discrepancy function as defined here are consistent estimators (Browne, 1984). We have already discussed the three major discrepancy functions for ordinary least squares, generalized least squares, and maximum likelihood. We will focus in particular on discrepancy functions in the class of *generalized least squares*. These, Browne (1982) indicates, are of the form

$$F(\mathbf{S}, \mathbf{\Sigma}(\hat{\mathbf{\theta}})|\mathbf{U}) = \frac{1}{2}(\mathbf{s}^* - \sigma^*(\mathbf{\theta}))'\mathbf{U}^{-1}(\mathbf{s}^* - \sigma^*(\mathbf{\theta})), \qquad (7.24)$$

where $\sigma^*(\mathbf{\theta}) = \text{vecs}(\mathbf{\Sigma}(\mathbf{\theta}))$ and $\mathbf{U}$ is a $p^* \times p^*$ positive definite matrix. In some situations $\mathbf{U}$ might be simply the identity matrix $\mathbf{I}$. The estimator that minimizes Equation 7.24 will be the "best" generalized least-squares (BGLS) estimator if it produces parameter estimates with the smallest asymptotic variances in this class. This will happen, Browne (1982) indicates, if (but not only if) the matrix $\mathbf{U}$, selected usually on the basis of sample information, converges in probability to any constant multiple of $\lim_{n\to\infty} n\mathbf{\Psi}^*$. For convenience, we let $\mathbf{U}$ be chosen so that it converges in probability to $\lim_{n\to\infty} n\mathbf{\Psi}^*$, that is,

$$\operatorname*{p\,lim}_{n\to\infty} \mathbf{U} = \lim_{n\to\infty} n\mathbf{\Psi}^*.$$

According to Browne (1982) the elements of $n\mathbf{\Psi}^*$ depend on the kurtosis of the distribution of the observed variables $\mathbf{Z}$. He argues that when the multivariate distribution is neither leptokurtotic with positive excess kurtosis nor platykurtotic with negative excess kurtosis—as is the case if the variables have a multivariate normal distribution—then the situation can be simplified. We do not need then to work with the $p^{*^2} \times p^{*^2}$ matrix $\mathbf{U}$, but can do our estimation with the maximum-likelihood and/or generalized least-squares estimators described previously that work with $p \times p$ matrices. Both the generalized least-squares and the maximum-likelihood estimators will converge to the same solution in the asymptotic case. More serious difficulties arise when there is an excess kurtosis. In the case of excess kurtosis the use of these usual estimators involving $p \times p$ matrices can produce incorrect test statistics and incorrect standard errors for the estimated parameters.

Browne (1984) then proposed that we obtain estimators by choosing $\mathbf{U}$ to be a matrix that is a consistent estimator of $\lim_{n\to\infty} n\mathbf{\Psi}^*$. We may obtain such a matrix if we substitute sample moments for population moments in Equation 7.24. Thus Browne (1984) defines the following sample quantities,

$$\dot{z}_i = N^{-1} \sum_{r=1}^{N} z_{ir},$$

$$w_{ij} = N^{-1} \sum_{r=1}^{N} (z_{ir} - \dot{z}_i)(z_{jr} - \dot{z}_j) = \frac{n}{N}[\mathbf{S}]_{ij}, \quad i \geq j,$$

$$w_{ijkl} = N^{-1} \sum_{r=1}^{N} (z_{ir} - \dot{z}_i)(z_{jr} - \dot{z}_j)(z_{kr} - \dot{z}_k)(z_{lr} - \dot{z}_l), \quad i \geq j, \, k \geq l,$$

and lets

$$[\mathbf{U}]_{ij,kl} = w_{ijkl} - w_{ij}w_{kl}, \quad i \geq k, \, i \geq j, \, k \geq l.$$

$\mathbf{U}$ is a $p^* \times p^*$ symmetric matrix. To be in conformity with the elements in $\mathbf{s}^* = \text{vecs}(\mathbf{S})$ and $\sigma^*(\boldsymbol{\theta}) = \text{vecs}(\boldsymbol{\Sigma}(\boldsymbol{\theta}))$, an element $u_{gh} = [\mathbf{U}]_{ij,kl}$ must be positioned in row $g = i(i-1)/2 + j$ and column $h = k(k-1)/2 + l$.

The BGLS estimate $\hat{\boldsymbol{\theta}}$ of $\boldsymbol{\theta}$ is then obtained by minimizing

$$F[\mathbf{S}, \boldsymbol{\Sigma}(\boldsymbol{\theta})|\mathbf{U}] = \frac{1}{2}(\mathbf{s}^* - \sigma^*(\boldsymbol{\theta}))'\mathbf{U}^{-1}(\mathbf{s}^* - \sigma^*(\boldsymbol{\theta})). \tag{7.25}$$

We will not go further into the details of how to perform this minimization. It would be along the lines of those already described.

Browne (1984) notes, however, that the asymptotic distribution free (ADF) method of estimating parameters is not likely to have a practical usefulness for models in which there are many variables. For example, if the number of variables p is 40, $\mathbf{s}^*$ contains $p^* = 820$ elements and the number of non-redundant elements of the $p^* \times p^*$ matrix $\mathbf{U}$ equals 336,610. Several matrices of each size would be needed in computer memory. But the number of calculations would also become enormous and time-consuming to complete. Clever programming might be able to push upward the upper limit for the size of problems that could be managed with this method of estimation.

But there are other limitations for the ADF method of estimation. Bentler (1994), reported Monte Carlo studies that reveal that the ADF method of estimation does not attain its promised superiority in typical research size samples (250–500). It does indeed display correct results asymptotically, but this usually means in samples of 5000 or larger. This has been a big disappointment to theorists who had hoped ADF estimation would give practical solutions to the problem of nonnormal data generally.

Maximum-Likelihood Estimation with Elliptical Distribution

Browne (1982, 1984) argued that in these strongly nonnormal cases there are ways of proceeding that involve working theoretically with distributions similar to but more general than the normal distribution, specifically, the family of elliptical distributions. Members of this family have distributions of the form

$$c_{\mathrm{m}} = |\boldsymbol{\Sigma}|^{-1/2} \, h\left[(\mathbf{y} - \boldsymbol{\mu})'\boldsymbol{\Sigma}^{-1}(\mathbf{y} - \boldsymbol{\mu})\right]$$

where c_m is a normalizing constant designed to make the volume under the function equal to unity, and $h(\ )$ is a function. The multivariate normal distribution, the multivariate t-distribution, and certain other "contaminated" normal distributions are members of this family, but there are many others that have excess kurtosis.

Using elliptical theory, Browne (1982, 1984) found that when the covariance structure $\Sigma(\theta)$ is invariant under a constant scaling factor, that is, there exists another value for the parameter vector, θ^*, such that $\Sigma(\theta^*) = a^2\Sigma(\theta)$, then we can use $\bar{U}$ in Equation 7.24 where a typical element of this matrix is

$$\bar{u}_{ij,kl} = w_{ijkl} - w_{ij}w_{kl} \tag{7.26}$$

with

$$w_{ijkl} = N^{-1}\sum_{r=1}^{N}(z_{ir} - \dot{z}_i)(z_{jr} - \dot{z}_j)(z_{kr} - \dot{z}_k)(z_{lr} - \dot{z}_l), \quad i \ge j, \, k \ge l$$

$$w_{ij} = N^{-1}\sum_{r=1}^{N}(z_{ir} - \dot{z}_i)(z_{jr} - \dot{z}_j) = \frac{n}{N}[\mathbf{S}]_{ij}, \quad i \ge j.$$

The drawbacks of estimation using Equations 7.23 and 7.24, according to Browne (1982), are that (1) the matrix $\bar{U}$ may be very large and thus demand considerable computer memory. West, Finch, and Curran (1995) review further the topic of estimation with nonnormal variables.

Parameter Estimation Algorithms

General Principles

We have already seen that to find estimates of unspecified, free parameters, we must find those values for the free parameters that, together with the fixed and constrained parameters, minimize a discrepancy function, such as the ordinary least-squares, maximum-likelihood, or generalized least-squares function. Jöreskog made a significant breakthrough using this function-minimizing methodology in obtaining maximum-likelihood solutions for these parameters in exploratory common factor analysis. He made a similar breakthrough using the same methodology to implement the development of confirmatory factor analysis and analysis of covariance structures. Following this he developed his LISREL© program for analyzing structural equation models based on the same algorithm. With function-minimizing techniques at the heart of structural equation modeling, we need to understand their general principles. Because the discrepancy functions for the confirmatory factor analysis model involve many parameters, it is not convenient to illustrate the function-minimizing methods with these discrepancy functions. Rather we will consider how we might go about minimizing a simple function with two free variables.

The minimization problem is analogous to finding oneself on a mountain side in some mountains surrounding a deep valley. It is foggy, so visibility is poor. One's task is to find a house one knows is at the lowest place in the valley below. One cannot see directly to where the house is located, but can see a few yards ahead and knows that the pull of gravity will give some guidance as to which direction to go. So, one starts off, taking steps in a downhill direction, changing direction toward a new downhill heading, if to step further in the current direction would lead one back uphill. Eventually, after many steps one eventually finds one's way to the house at the bottom of the valley. With that scenario in mind, we will now consider how mathematicians solve analogous problems in parameter space.

The principle underlying algorithms for the minimization of nonlinear functions is that beginning at some initial point in parameter space, where the value of each free parameter is a coordinate of the point, one will move in successive steps from that point to other points, choosing at each step points to move to that are ever lower and so closer to the minimum point. Along the way, at each intermediate point one will determine a direction and a distance to move to get to the next point.

Consider the problem of minimizing the function [suggested by Bazaraa and Shetty (1979)]:

$$z = (x_1 - 2)^4 + (x_1 - 2x_2)^2.$$

The calculus states that a necessary (but not sufficient) condition for a function to be at a (local) minimum at a point (x_1, x_2) is that the partial derivatives of the function with respect to x_1 and x_2, respectively, evaluated at the point (x_1, x_2), be each equal to zero, that is,

$$\frac{\partial}{\partial x_1}\left[(x_1 - 2)^4 + (x_1 - 2x_2)^2\right] = 4(x_1 - 2)^3 + 2(x_1 - 2x_2) = 0,$$

$$\frac{\partial}{\partial x_2}\left[(x_1 - 2)^4 + (x_2 - 2x_2)^2\right] = -4(x_1 - 2x_2) = 0.$$

The partial derivative of a function with respect to one of the independent variables evaluated at a given point represents the slope of a line drawn tangent to the curve at that point in the direction parallel to the axis of the independent variable in question. We show in Figure 7.1 on the graph of our function the tangent lines in the direction of the variables x_1 and x_2, respectively, at the point (2, 1), which happens to be the minimum point for this curve. The slope of each of these lines should be zero, if (2, 1) is the minimum.

In Figure 7.2, we show a contour plot of the same function. The lines represent loci of equal altitude of the function. The same principle is used in map

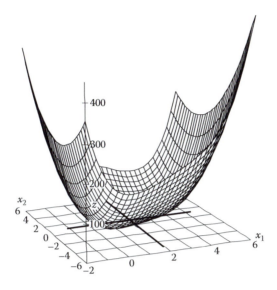

FIGURE 7.1 Graph of function $z = (x_1 - 2)^4 + (x_1 - 2x_2)^2$ showing tangent lines corresponding to directional partial derivatives of the function at the function's minimum. The graph has been cut to show only values within a rectangular interval around the minimum.

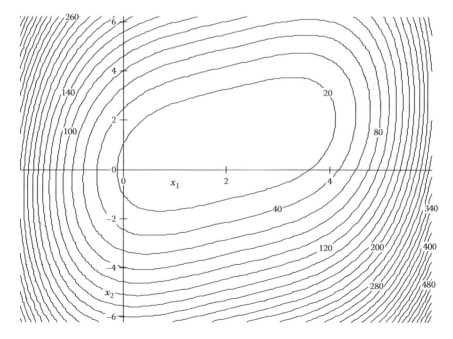

FIGURE 7.2 Contour plot of function $z = (x_1 - 2)^4 + (x_1 - 2x_2)^2$ showing loci of points (contours) with equal value for z. Outer contours represent higher values of the function. (The scale of the x_1 axis has been stretched so that it is consistent with Figure 7.1.)

making to represent altitudes of mountains and valleys. Contour plots are better most of the time for discussing our next topic, gradient vectors.

The *gradient vector* at a point $\mathbf{x}' = (x_1, x_2)$ is the vector $\nabla z(\mathbf{x})$ whose coordinates are the values of the partial derivatives of the function $z = z(\mathbf{x})$ evaluated at that point with respect to the independent variables:

$$\nabla z(\mathbf{x}) = \begin{bmatrix} \dfrac{\partial z}{\partial x_1} \\ \dfrac{\partial z}{\partial x_2} \end{bmatrix}$$

For example, suppose we evaluate the gradient vector at the point $(1.17, 1.5)$. At this point

$$\nabla z(\mathbf{x}) = \begin{bmatrix} 4(x_1 - 2)^3 + 2(x_1 - 2x_1) \\ -4(x_1 - 2x_2) \end{bmatrix} = \begin{bmatrix} -5.9471 \\ 7.32 \end{bmatrix}.$$

We illustrate how the gradient vector, when added to the point $(1.17, 1.5)$, has the coordinates $(-4.7771, 8.42)$, which is shown in the following contour plot (Figure 7.3).

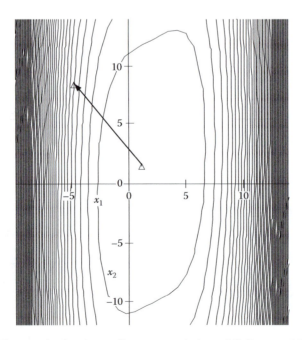

FIGURE 7.3 Contour plot showing gradient vector pointing uphill. Because of the length of the gradient vector, we have expanded the view of the graph.

The gradient vector points "uphill" from the point (1.17, 1.5). In fact the gradient vector is orthogonal to the line drawn tangent to the contour of equal functional values passing through the point (1.17, 1.5). So, if we take the negative of the gradient vector,

$$-\nabla z(\mathbf{x}) = \begin{bmatrix} -\dfrac{\partial z}{\partial x_1} \\ -\dfrac{\partial z}{\partial x_2} \end{bmatrix},$$

and add this to the point $\mathbf{x}' = (x_1, x_2)$, that is, $\mathbf{x} - \nabla z(\mathbf{x})$, we have an arrow pointing in the most direct *downhill* direction from that point, a direction that is also orthogonal to the line drawn tangent to the contour passing through the point. This direction is called *the direction of steepest descent*. We can use the direction of the negative gradient vector to suggest a direction to move from the initial point $\mathbf{x} = (x_1, x_2)$ toward a region much closer to the true minimum point. Frequently we will use a normalized negative gradient vector with unit length. We will call it $\mathbf{d}$ (Figure 7.4).

Actually, what we want to do at this point is to search along a line $\mathbf{x} - \beta \nabla z(\mathbf{x})$ to find the value b_{min} of β $(0 \leq \beta)$ where the function z is a minimum along the line. Then the point $\mathbf{x} - \beta_{min} \nabla z(\mathbf{x})$ is an excellent place to evaluate the gradient vector for a new provisional direction heading downhill toward either the global minimum or (in some situations) a local minimum of the function.

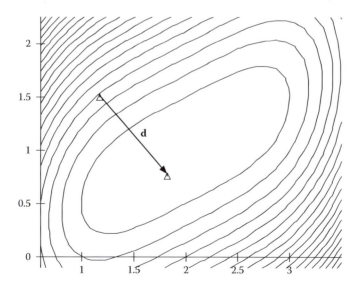

FIGURE 7.4 Magnified contour graph of function with normalized negative gradient **d** pointing in the direction of steepest descent.

Method of Steepest Descent

A method suggested by this analysis of the problem is known as the method of steepest descent. Given initial values for the independent variables x_1 and x_2 in a two-dimensional vector $\mathbf{x}_0$, the algorithm constructs iteratively successively better and better choices for the vector of minimizing values of the independent variables by iterating with the equation

$$\mathbf{x}_{k+1} = \mathbf{x}_k - \alpha_{k+1} \nabla z(\mathbf{x}_k)$$

where $\mathbf{x}_k$ is the value of the vector of independent variables after k iterations, $\nabla z(\mathbf{x}_k)$ is the value of the gradient vector computed using values of the parameters in $\mathbf{x}_k$, and α_{k+1} is a step-size value determined at each iteration of the algorithm that establishes how far to move in the direction of the negative gradient evaluated at the kth iteration to find the next, improved solution $\mathbf{x}_{k+1}$ for the point $\mathbf{x}$ at which the function is minimized. At each iteration, the value of α_{k+1} is determined either by a line search in the direction of the negative gradient (also known as the *direction of steepest descent*) to find the minimum of the function *along that line* or by some heuristic single variable minimization method that yields in most cases a close approximation to the minimum of the function along the line.

Jöreskog (1967) described and used, in his programs for factor analysis and structural equation modeling, a heuristic method that was originally suggested by Fletcher and Powell (1963). This method quickly finds a near minimum for the function along the line of steepest descent. We will repeat it here in a little more detail: let $\mathbf{d} = -\nabla z(\mathbf{x}_k)/\|\nabla z(\mathbf{x}_k)\|$ be a unit length vector in the direction of the negative gradient vector. $\|\nabla z(\mathbf{x}_k)\| = \left(\nabla z(\mathbf{x}_k)'\nabla z(\mathbf{x}_k)\right)^{1/2}$, the length of the gradient vector (evaluated at $\mathbf{x}_k$). Then any point on the line drawn through $\mathbf{x}_k$ in the direction of $\mathbf{d}$ is given by $\mathbf{x}_k + \beta\mathbf{d}$. Suppose we select two points. Let the first point be simply $\mathbf{x}_k = \mathbf{x}_k + \beta_0\mathbf{d}$, where $\beta_0 = 0$, which is just the point $\mathbf{x}_k$. Let the second point be, according to Fletcher and Powell (1963), $\mathbf{x}_k^* = \mathbf{x}_k + \beta_1\mathbf{d}$, where

$$\beta_1 = \text{minimum of} \left[1, \frac{-2z(\mathbf{x}_k)}{\nabla z(\mathbf{x}_k)'\mathbf{d}}\right].$$

The denominator of the expression on the right, inside the brackets, is the directional derivative of the function $z(\mathbf{x})$ in the direction of the vector $\mathbf{d}$ from the point $\mathbf{x}_i$. So, if the slope of the function $z(\mathbf{x})$ at the point $\mathbf{x}_k$ is small and negative relative to the magnitude of the function $z(\mathbf{x})$, this may make the term on the right larger than unity, and so, when that happens, β_1 can be set to unity. Now let us evaluate the function $z(\mathbf{x})$ at the points $\mathbf{x}_k = \mathbf{x}_k + \beta_0\mathbf{d}$ and $\mathbf{x}_k^* = \mathbf{x}_k + \beta_1\mathbf{d}$, and designate these values as $z(\mathbf{x}_k + \beta_0\mathbf{d})$ and $z(\mathbf{x}_k + \beta_1\mathbf{d})$, respectively. Let us further evaluate the slope of the function $z(\mathbf{x})$ in the direction of the vector $\mathbf{d}$ at these same two points. The slope of a multidimensional function $z(\mathbf{x})$ in the direction of a unit length vector $\mathbf{d}$ at the point $\mathbf{x}$ is given by $\nabla z(\mathbf{x})'\mathbf{d}$.

Let us designate these two slopes as

$$z'(\mathbf{x}_k + \beta_0\mathbf{d}) = \nabla z(\mathbf{x}_k)'\mathbf{d}$$

and

$$z'(\mathbf{x}_k + \beta_1\mathbf{d}) = \nabla z(\mathbf{x}_k^*)'\mathbf{d},$$

respectively. Note that $\nabla z(\mathbf{x}^*)'\mathbf{d}$ is evaluated at the second temporary point $\mathbf{x}_k^* = \mathbf{x}_k + \beta_1\mathbf{d}$.

Now, what we are trying to do is best understood in terms of the two graphs in Figure 7.5.

The top graph shows the value of the function $z(\mathbf{x}_k + \beta\mathbf{d})$ along the line $\mathbf{x}_k + \beta\mathbf{d}$. We seek the point $\mathbf{x}_k + \beta_{\min}\mathbf{d}$ at which this function is a minimum along the line. The bottom graph is the corresponding derivative of the function $z(\mathbf{x}_k + \beta\mathbf{d})$ taken with respect to β. By the properties of similar triangles, if $z'(\mathbf{x}_k + \beta_1\mathbf{d})$ is also negative we can extend a line connecting the two points $[z'(\mathbf{x}_k + \beta_0\mathbf{d}), \beta_0]$ and $[z'(\mathbf{x}_k + \beta_0\mathbf{d}), \beta_1]$ in the bottom graph to extrapolate to the point β_h where this line intersects with the horizontal axis by noting that (recalling that $\beta_0 = 0$)

$$\frac{\beta_h}{z'(\beta_0)} = \frac{\beta_1}{z'(\beta_1) - z'(\beta_0)},$$

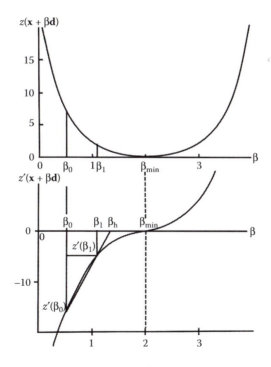

FIGURE 7.5 Graphs of function and derivative of function in the direction of steepest descent.

which we easily solve for β_h by

$$\beta_h = \frac{z'(\beta_0)\beta_1}{z'(\beta_1) - z'(\beta_0)}.$$

The value of β_h should be a good approximation to β_{min}. When the derivative $z'(x_k + \beta_1 d)$ is positive, this same formula may be used to interpolate to a point near the minimum along the line. However if $|z'(x_k + \beta_1 d)|$ is more than twice $|z'(x_k + \beta_0 d)|$, then choosing a new $\tilde{\beta}_1 = 0.6081\beta_1$ might lead to a better point for interpolating to a β_h close to β_{min}. But in this situation, Fletcher and Powell (1963) suggest another approach: construct a cubic polynomial function of β with the same altitude and slope as the empirical function at $\beta_0 = 0$ and β_1 along the line. Find the value of β where this cubic function has its minimum. The value of β that minimizes the cubic function should be close to the point along the line where z has its minimum along the line.

Cubic Approximation to Find Approximate Minimum

Let $p(\beta) = c_0 + c_1\beta + c_2\beta^2 + c_3\beta^3$ be a cubic polynomial function of β, and let $p'(\beta) = c_1 + 2c_2\beta + 3c_3\beta^2$ be the derivative of the cubic function. Let $p(0) = z(x_k + \beta_0 d) = z(x_k)$ be the value of the function z at the point along the line $x_k + \beta d$ where $\beta_0 = 0$, and let $p(\beta_1)$ be the value of the function z at the point along the same line where $\beta = \beta_1$. Correspondingly, let $p'(0) = z'(x_k) = \nabla z(x)'d$ be the derivative of the function z at the point along the line where $\beta_0 = 0$, and let $p'(\beta_1) = z'(x_k + \beta_1 d) = \nabla z(x_k + \beta_1 d)'d$ be the derivative of the function z at the point along the line where $\beta = \beta_1$.

In theory the coefficients c_0, c_1, c_2, and c_3 of the polynomial $p(\beta)$ may be obtained by solving a system of four simultaneous equations representing constraints on the coefficients of the polynomial, that is, in matrix terms

$$\begin{bmatrix} 1 & \beta_0 & \beta_0^2 & \beta_0^3 \\ 1 & \beta_1 & \beta_1^2 & \beta_1^3 \\ 0 & 1 & 2\beta & 3\beta \\ 0 & 1 & 2\beta & 3 \end{bmatrix} \begin{bmatrix} c_0 \\ c_1 \\ c_2 \\ c_3 \end{bmatrix} = \begin{bmatrix} p(\beta_0) \\ p(\beta_1) \\ p'(\beta_0) \\ p'(\beta_1) \end{bmatrix} \quad \text{or} \quad \mathbf{Ac = P,}$$

which may be further simplified because $\beta_0 = 0$ to

$$\begin{bmatrix} 1 & 0 & 0 & 0 \\ 1 & \beta_1 & \beta_1^2 & \beta_1^3 \\ 0 & 1 & 2\beta & 3\beta \\ 0 & 1 & 2\beta & 3 \end{bmatrix} \begin{bmatrix} c_0 \\ c_1 \\ c_2 \\ c_3 \end{bmatrix} = \begin{bmatrix} p(\beta_0) \\ p(\beta_1) \\ p'(\beta_0) \\ p'(\beta_1) \end{bmatrix}.$$

Then the solution for $\mathbf{c}$ is given by $\mathbf{c = A^{-1}P}$.

Once the solution for the coefficients of the cubic polynomial is obtained, then one can solve for the required (local) minimum point of the cubic polynomial by setting the derivative of this function equal to zero, that is, $c_1 + 2c_2\beta + 3c_3\beta^2 = 0$. Since this is a quadratic equation, we may solve for its roots by the equation

$$\beta_h = \frac{-2c_2 \pm \sqrt{(2c_2)^2 - 4(3c_3)c_1}}{2(3c_3)} = \frac{-2c_2 \pm \sqrt{4c_2^2 - 12c_3c_1}}{6c_3}.$$

Since two roots are provided, one must choose the one between zero and β_1. We then set $\alpha_{k+1} = \beta_h$.

To illustrate, suppose we are searching down a line collinear with the x_1 axis $z = (x_1 - 2)^4 + (x_1 - 2)^2$ where $x_2 = 1$. Our function then is simply $z = (x_1 - 2)^4 + (x_1 - 2)^2$. For convenience, let us refer to this as the line $x' = x + \beta d$, where $d = 1$, implying $x' = x + \beta$. Suppose further that $x = 1.5$ is our initial point where $\beta = \beta_0 = 0$ and $\beta_1 = 0.9$ leads us to the point $x' = x + \beta_1 = 1.5 + 0.9 = 2.4$, which we believe is beyond the minimum point along the line. The z function along this line is given in Figure 7.6.

The column vector $\mathbf{P}$ is then given as

$$\mathbf{P} = \begin{bmatrix} p(\beta_0) \\ p(\beta_1) \\ p'(\beta_0) \\ p'(\beta_1) \end{bmatrix} = \begin{pmatrix} 0.3125 \\ 0.1856 \\ -1.5 \\ 1.056 \end{pmatrix},$$

and

$$\mathbf{A} = \begin{pmatrix} 1 & 0 & 0 & 0 \\ 1 & \beta & \beta^2 & \beta^3 \\ 0 & 1 & 0 & 0 \\ 0 & 1 & 2\beta & \beta^2 \end{pmatrix} = \begin{pmatrix} 1 & 0 & 0 & 0 \\ 1 & 0.9 & 0.81 & 0.729 \\ 0 & 1 & 0 & 0 \\ 0 & 1 & 1.8 & 0.81 \end{pmatrix},$$

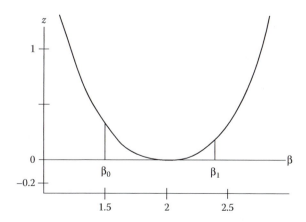

FIGURE 7.6 Graph of z function along a line.

and we have as a result

$$\mathbf{c} = \mathbf{A}^{-1}\mathbf{P} = \begin{pmatrix} 0.3125 \\ -1.5 \\ 1.33 \\ 0.2 \end{pmatrix}.$$

The cubic polynomial function sought is then

$$y = 0.3125 - 1.5x + 1.33x^2 + 0.2x^3$$

and has the graph shown in Figure 7.7, which appears shifted to the left by 1.5 units from where it ought to be. But it is the correct function, since it will have the same height and slope at zero and 0.9 as the function z has at 1.5 and 2.4, respectively. We can see how this function aligns itself with the z function when we change the variable from x to $x' = (x - 1.5)$ and merge the graph of the resulting cubic function

$$t = 0.3125 - 1.5(x - 1.5) + 1.33(x - 15)^2 + 0.2(x - 1.5)^3$$

with the previous graph of the z function, as shown in Figure 7.8.

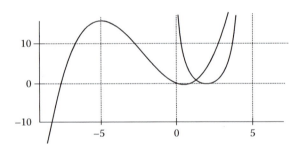

FIGURE 7.7 Cubic polynomial function (dark line) that has comparable values and derivatives as an empirical function.

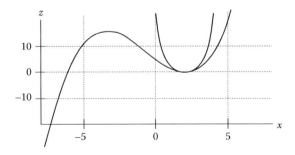

FIGURE 7.8 Approximation of cubic function to empirical function.

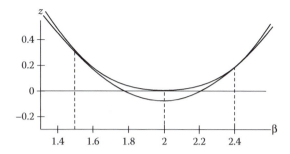

FIGURE 7.9 Magnified view showing that the cubic function (dark line) has a local minimum at nearly the same value as does the empirical function (fine line).

When magnified at the minimum, as in Figure 7.9, we see the match in heights at 1.5 and 2.4, respectively. The (local) minimum of the cubic curve also occurs at about the same point as does the original quartic function.

Returning now to the untransformed cubic function, the solution for the value of β that minimizes the cubic polynomial along the line is given by

$$\beta_h = \frac{-2(1.33) + \sqrt{(2(1.33))^2 - 12(0.2)(-1.5)}}{6(0.2)} = 0.50613.$$

We see that if this value is added to the original value of x at 1.5, it yields 2.00613, which is very close to the minimum of the quartic function at 2.00. Thus we would set α to 0.50613 in this step of an iterative algorithm.

Figure 7.10 illustrates how the method of steepest descent typically makes bold steps at the outset, but then slows down to a crawl and sometimes gets stuck as it nears the minimum point. After 11 iterations it was at the point (1.93114, 0.965492) and moving slower and slower to the minimum (2, 1). Frequently near the minimum, when the minimum is located in a shallow valley that is longer than it is wide, the method of steepest descent tends to show a zig-zag, back-and-forth movement across the long axis of the valley and slows down and fails to converge to the true minimum. This tendency is shown in Figure 7.11 in the magnification of the central region of the preceding graph, which shows the steps taken near the minimum of the quartic function in greater detail. So the method of steepest descent is usually not the most efficient method for minimizing a function in the region of the sought minimum.

It must be understood that generally these algorithms never reach the minimum in a finite number of steps. Most computer programs stop the iterations when the difference between the parameter estimates on two successive iterations becomes less in absolute magnitude than some very small quantity, for example, 0.0001, in the case of each parameter. Or one may stop the iterations when the last discrepancy function value F_k and the previous F_{k-1} differ by some very small amount in absolute value,

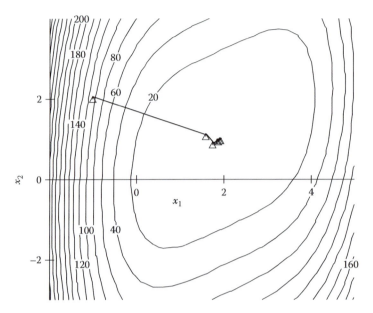

FIGURE 7.10   A graphical illustration of the method of steepest descent with the line search methods just described to find the minimum of the equation is shown where the starting point is $(-1, 2)$.

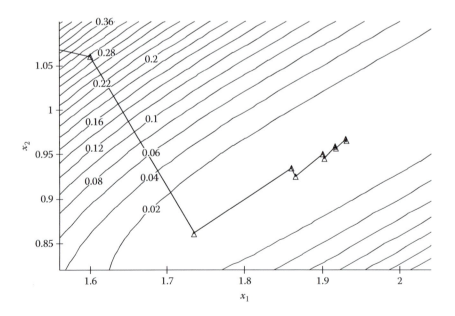

FIGURE 7.11   Zig-zag tendency of the method of steepest descent when approaching a minimum along a valley that is longer than wide.

say, $\left|(F_k - F_{k-1})/(1 + |F_k|)\right| < 0.001$ (Steiger, 1994). With inefficient algorithms such as the method of steepest descent, many iterations may be required to satisfy this convergence criterion.

The Newton–Raphson Algorithm

According to Bazarraa and Shetty (1979), a very efficient algorithm for finding the minimum of a function is the Newton–Raphson method. The idea underlying this algorithm is that one can form a quadratic approximation $q(\mathbf{x})$ to the function $f(\mathbf{x})$ to be minimized at a given point $\mathbf{x}_k$ by using the first and second derivative terms of a Taylor's series expansion of the function at that point:

$$q(\mathbf{x}) = f(\mathbf{x}) + \nabla f(\mathbf{x})'(\mathbf{x} - \mathbf{x}_k) + \frac{1}{2}(\mathbf{x} - \mathbf{x}_k)'\mathbf{H}(\mathbf{x} - \mathbf{x}_k),$$

where $\nabla f(\mathbf{x}_k)$ is the vector of partial derivatives of $f(\mathbf{x})$ evaluated at the point $\mathbf{x}_k$,

$$\nabla f(\mathbf{x}_k) = \begin{bmatrix} \dfrac{\partial f(\mathbf{x})}{\partial x_1} \\[2mm] \dfrac{\partial f(\mathbf{x})}{\partial x_2} \\[1mm] \vdots \\[1mm] \dfrac{\partial f(\mathbf{x})}{\partial x_n} \end{bmatrix}_{\mathbf{x}_k},$$

and

$$\mathbf{H}(\mathbf{x}_k) = \begin{bmatrix} \dfrac{\partial f(\mathbf{x})}{\partial x_1^2} & \dfrac{\partial f(\mathbf{x})}{\partial x_1 \partial x_2} & \cdots & \dfrac{\partial f(\mathbf{x})}{\partial x_1 \partial x_n} \\[3mm] \dfrac{\partial f(\mathbf{x})}{\partial x_2 \partial x_1} & \dfrac{\partial f(\mathbf{x})}{\partial x_2^2} & \cdots & \dfrac{\partial f(\mathbf{x})}{\partial x_2 \partial x_n} \\[2mm] \vdots & \vdots & \ddots & \vdots \\[2mm] \dfrac{\partial f(\mathbf{x})}{\partial x_n \partial x_1} & \dfrac{\partial f(\mathbf{x})}{\partial x_n \partial x_2} & \cdots & \dfrac{\partial f(\mathbf{x})}{\partial x_n^2} \end{bmatrix}_{\mathbf{x}_k}$$

is the Hessian matrix of second partial derivatives of the function with respect to its parameters evaluated at the point $\mathbf{x}_k$.

The reason for obtaining the quadratic function $q(\mathbf{x})$ to approximate the function $f(\mathbf{x})$ to minimize, is because it is very easy to minimize a quadratic function in a single step, once we know its partial derivatives with respect to the independent variables. If the equation to minimize $f(\mathbf{x})$ is a quadratic function, then we will be able to proceed directly to its minimum; and if $f(\mathbf{x})$ is not quadratic, the quadratic approximation to $f(\mathbf{x})$ may still have a minimum in the vicinity of the minimum of $f(\mathbf{x})$. In fact, in the region immediately around a minimum of $f(\mathbf{x})$, the minimum of a quadratic approximation to $f(\mathbf{x})$ at a point in that region will be very close to a minimum of $f(\mathbf{x})$.

In Figure 7.12, we show in three dimensions the function to be minimized (the net of the function to fit cut away in the front) and the surface of a quadratic function approximating the function to minimize, looking like a tornado.

Thus, think of the quadratic function as a tornado that forms itself at a point in such a way as to conform to the function to be minimized very well at that point. Then it skips to its current minimum point, reforming itself again to conform to the function to be minimized at this point, and then skips to its new minimum point, reforming again, and so on. As it approaches the

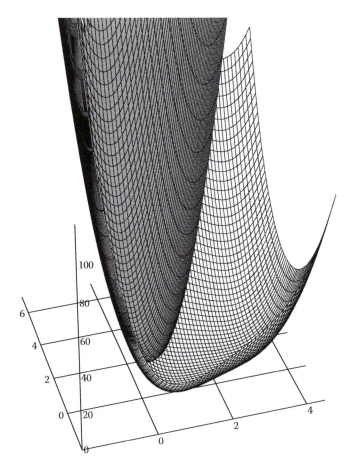

FIGURE 7.12 Quadratic function $q(\mathbf{x})$ ("dark tornado") is an approximation to function $z(\mathbf{x})$ (open mesh) and the minimum of $q(\mathbf{x})$ is an approximation to the minimum of $z(\mathbf{x})$. The near side of $z(\mathbf{x})$ is cut away to show $q(\mathbf{x})$.

minimum, the quadratic function makes a better and better approximation to the function to be minimized in the region of that function's minimum, and the quadratic function's minimum thus rapidly approaches ever more closely to the minimum of the function to be minimized.

In Figure 7.13 a contour plot of the approximating quadratic function $q(\mathbf{x})$ is superimposed on the contour plot of the function to minimize $f(\mathbf{x})$ so that the two functions coincide at the point $(-1, 2)$.

Thus to find the minimum of the function $q(\mathbf{x})$, we need to find its partial derivatives, set them equal to zero, and solve for the value of $\mathbf{x}$ at its minimum:

$$q(\mathbf{x}) = f(\mathbf{x}) + \nabla f(\mathbf{x})'(\mathbf{x} - \mathbf{x}_k) + \frac{1}{2}(\mathbf{x} - \mathbf{x}_k)'\mathbf{H}(\mathbf{x} - \mathbf{x}_k)$$

$$\frac{\partial q(\mathbf{x})}{\partial \mathbf{x}} = \frac{\partial f(\mathbf{x}_k)}{\partial \mathbf{x}} + \frac{\partial [\nabla f(\mathbf{x}_k)'(\mathbf{x} - \mathbf{x}_k)}{\partial \mathbf{x}} + \frac{\partial}{\partial \mathbf{x}}\frac{1}{2}[(\mathbf{x} - \mathbf{x}_k)'\mathbf{H}(\mathbf{x}_k)(\mathbf{x} - \mathbf{x}_k).$$

The expressions $\mathbf{x}_k, f(\mathbf{x}_k), f(\mathbf{x}_k)$, and $\mathbf{H}(\mathbf{x}_k)$ represent constant expressions at the point $\mathbf{x}_k$. We may now take advantage of certain rules for taking the partial derivatives of scalars and traces of matrices with respect to a vector: let c, $\mathbf{a}$, and $\mathbf{A}$ be a constant scalar, vector and matrix, respectively. Then

$$\frac{\partial c}{\partial \mathbf{x}} = \mathbf{0}$$

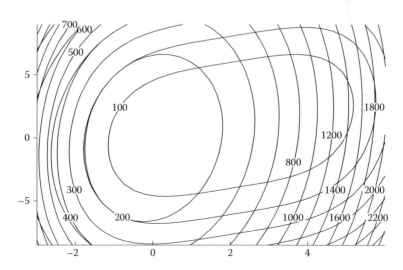

FIGURE 7.13 Contour plot of $q(\mathbf{x})$ fit to the point $(-1, 2)$ superimposed on the contour plot of $f(\mathbf{x})$.

$$\frac{\partial \mathbf{a}'\mathbf{x}}{\partial \mathbf{x}} = \mathbf{a}$$

$$\frac{\partial \text{tr}(\mathbf{x}'\mathbf{A}\mathbf{x})}{\partial \mathbf{x}} = 2\mathbf{A}\mathbf{x} \quad \text{for symmetric } \mathbf{A}.$$

Thus,

$$\frac{\partial q(\mathbf{x})}{\partial \mathbf{x}} = \nabla f(\mathbf{x}_k) + \mathbf{H}(\mathbf{x}_k)(\mathbf{x} - \mathbf{x}_k)$$

$$= \nabla f(\mathbf{x}_k) + \mathbf{H}(\mathbf{x}_k)\mathbf{x} - \mathbf{H}(\mathbf{x}_k)\mathbf{x}_k.$$

Since this represents the partial derivatives of the quadratic approximation with respect to $\mathbf{x}$, setting the expression equal to a null vector and solving for $\mathbf{x}$, we obtain the value of $\mathbf{x}$ at the minimum:

$$\nabla f(\mathbf{x}_k) + \mathbf{H}(\mathbf{x}_k)\mathbf{x} - \mathbf{H}(\mathbf{x}_k)\mathbf{x}_k = \mathbf{0}.$$

Moving the constant expressions to the right, we obtain

$$\mathbf{H}(\mathbf{x}_k)\mathbf{x} = \mathbf{H}(\mathbf{x}_k)\mathbf{x}_k - \nabla f(\mathbf{x}_k)$$

or, by multiplying both sides by the inverse $\mathbf{H}(\mathbf{x}_k)^{-1}$,

$$\mathbf{x} = \mathbf{x}_k - \mathbf{H}(\mathbf{x}_k)^{-1}\nabla f(\mathbf{x}_k).$$

This yields the basic equation of the Newton–Raphson algorithm:

$$\mathbf{x}_{k+1} = \mathbf{x}_k - \mathbf{H}(\mathbf{x}_k)^{-1}\nabla f(\mathbf{x}_k).$$

We illustrate in Figure 7.14 its implementation with the search for the minimum of $z = (x_1 - 2)^4 + (x_1 - 2x_2)^2$ starting at the point $(-1, 2)$.

First we obtain the gradient vector of first partial derivatives of z with respect to $\mathbf{x}$:

$$\nabla z(\mathbf{x}_k) = \begin{bmatrix} \dfrac{\partial z}{\partial x_1} \\ \dfrac{\partial z}{\partial x_2} \end{bmatrix} = \begin{bmatrix} 4(x_1 - 2)^3 + 2(x_1 - 2x_2) \\ -4(x - 2x_2) \end{bmatrix}.$$

Next

$$\mathbf{H}(\mathbf{x}) = \begin{bmatrix} \dfrac{\partial^2 z}{\partial x_1^2} & \dfrac{\partial^2 z}{\partial x_1 \partial x_2} \\ \dfrac{\partial^2 z}{\partial x_2 \partial x_1} & \dfrac{\partial^2 z}{\partial x_2^2} \end{bmatrix} = \begin{bmatrix} 12(x_1 - 2)^2 + 2 & -4 \\ -4 & 8 \end{bmatrix}.$$

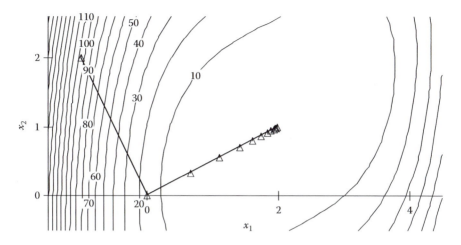

FIGURE 7.14 The first and subsequent steps toward the minimum made by the Newton–Raphson method.

Evaluating these expressions at the starting point of $\mathbf{x}_0 = (-1, 2)$, we obtain

$$\nabla z(\mathbf{x}_0) = \begin{bmatrix} -118 \\ 20 \end{bmatrix},$$

$$\mathbf{H}(\mathbf{x}_0) = \begin{bmatrix} 110 & -4 \\ -4 & 8 \end{bmatrix}.$$

So, the first step leads us from $(-1, 2)$ to

$$\begin{bmatrix} x_1 \\ x_2 \end{bmatrix}_1 = \begin{bmatrix} -1 \\ 2 \end{bmatrix}_0 - \begin{bmatrix} 110 & -4 \\ -4 & 8 \end{bmatrix}^{-1} \begin{bmatrix} -118 \\ 20 \end{bmatrix},$$

$$\begin{bmatrix} x_1 \\ x_2 \end{bmatrix}_1 = \begin{bmatrix} -1 \\ 2 \end{bmatrix}_0 - \begin{bmatrix} 0.0092593 & 0.0046296 \\ 0.0046296 & 0.12731 \end{bmatrix} \begin{bmatrix} -118 \\ 20 \end{bmatrix}$$

$$= \begin{bmatrix} 0 \\ 0 \end{bmatrix}.$$

The major drawback of the Newton–Raphson method is the need to have expressions for not only the first but also the second derivatives of the function to be minimized. In many applications, especially in factor analysis and structural equation modeling, the derivation of such expressions is very difficult. Furthermore, when the Hessian matrix is very large, because there are many free parameters to estimate, there is a heavy storage requirement, as well as a heavy price to pay in computing, since one has to compute the Hessian and then its inverse at each iteration. These difficulties and costs have spurred mathematicians to find alternative algorithms that keep many of the advantages of the Newton–Raphson method without its costs.

Quasi-Newton Methods

The quasi-Newton algorithms take the general form

$$\mathbf{x}_{k+1} = \mathbf{x}_k - \alpha_{k+1}\mathbf{H}(\mathbf{x}_k)\nabla f(\mathbf{x}_k).$$

These methods require only measurements of the gradient at each iteration and thus forego the need to derive expressions for second derivatives. The **H** matrix is not the Hessian of second derivatives, but as these procedures progress, the **H** matrix becomes an increasingly better approximation to the *inverse* of the Hessian at the minimum of the function $f(\mathbf{x})$. As in the method of steepest descent, line searches for a minimum are taken in the direction of a vector, but this time the direction is given by the gradient vector $\nabla f(\mathbf{x}_k)$, deflected by the matrix $\mathbf{H}_k = \mathbf{H}(\mathbf{x}_k)$. Initially the matrix $\mathbf{H}_0 = \mathbf{I}$, so the first iteration is a step to a minimum along the line in the direction of the gradient. At each iteration $\alpha_{k+1} = \beta_h$ computed by the method of cubic approximation described earlier. At the end of each iteration, a new $\mathbf{H}_{k+1}$ is constructed from the previous $\mathbf{H}_k$ and gradient $\nabla f(\mathbf{x}_k)$.

Why use some other matrix than the Hessian? Press et al. (1992) indicate that if the function to be minimized is not quadratic, starting at a point distant from the minimum sought will produce a Hessian that is not positive definite. The effect in that case will be that the point moved to for the next iteration may actually correspond to a $z(\mathbf{x}_{k+1})$ that is larger than $z(\mathbf{x}_k)$. In other words, we are not moving in a descending direction. So, in the early iterations it is actually better to use a $\mathbf{H}_k$ that is positive definite, and the identity matrix **I** at the start is positive definite. Subsequent constructions of the **H** matrix are made in a way that guarantees that they are positive definite so that the iterations are moving in a descending direction.

Citing his source as Fletcher (1981), Pike (1986) gives the following update equation for the Broyden, Fletcher, Goldfarb, and Shanno (BGFS) method, which is currently regarded as the most robust and efficient of the quasi-Newton methods: let

$$\delta_{k+1} = \mathbf{x}_{k+1} - \mathbf{x}_k$$

be the difference between the new and the old step point, and let

$$\gamma_{k+1} = \nabla(\mathbf{x}_{k+1}) - \nabla(\mathbf{x}_k)$$

be the difference between the gradients at the new and old step points. Then let

$$\mathbf{H}_{k+1} = \left[\frac{\mathbf{H}_k\gamma_k\delta'_k + \delta_k\gamma'_k\mathbf{H}_k}{\delta'_k\gamma_k}\right] + \left[\frac{\gamma'_k\mathbf{H}_k\gamma_k}{\delta'_k\gamma_k}\right]\left[\frac{\delta_k\delta'_k}{\delta'_k\gamma_k}\right].$$

Note that in this equation two matrices are constructed, and then added together. The matrix in the numerator of the left-hand expression on the right is constructed by adding the products of two square matrices. The matrix on the right $[\delta_k \delta_k']$ is a rank-one square matrix.

Sometimes the cubic approximation methods of line search used in connection with quasi-Newton algorithms still run into problems. Jöreskog and Sörbom (1989) describe alternative procedures that may be used in these cases to assure accuracy. Each program on the commercial market for performing confirmatory factor analysis uses methods that its programmer found efficient and accurate. But the methods we have described give one a description of generally how these programs go about finding parameter values that minimize the respective discrepancy functions.

8

Designing SEM Studies

Preliminary Considerations

Any study with a structural equation model (SEM) begins with a focus on the causal and effect constructs that one wishes to study. Each of these constructs must be a variable. In the framework of this book, a variable is a set of attribute states or qualities such that, to any object that may be described in terms of these states or qualities, one and only one member of the set may be assigned at any one time. In other words, there is an implicit schema that objects (e.g., *persons*) become bearing attributes, that attributes (e.g., *blonde, brunette, redhead, brownette*) are segregated into sets, each having some characteristic in common such as *hair color*, and no object may have more than one value of the same attribute at the same time (e.g., *Joe is blonde*). Frequently, these states or qualities are represented as a given quantity of some attribute and thus have numerical values. For example, *Joe weighs 174 pounds.*

Now, it is important as one proceeds to think about possible causal relations to be thinking of the specific variables that enter into these relations. Causal relations are functional relations between variables, that is, sets of attribute states of objects, not just between the objects themselves as objects. To illustrate how this is often misunderstood, in Mulaik (2004) I reported attending a graduate student's orals in which the student said, echoing the literature of the field, that one of the variables in the student's model was "leader–follower exchange." This does not tell us what specific attribute of the "leader–follower exchange" is the variable in question. It was supposed to be a quantity. But what quantity? I cited these possibilities: "... There are numerous variables one could focus on in exchanges between leaders and followers: *How many orders does the leader give to the follower? To what degree does the subordinate feel he/she has freedom to decide what to do? To what extent does the subordinate say negative things about his/her supervisor? To what extent does the subordinate say negative things about the boss's orders directly to the boss? To what extent does the follower like the leader?"* (pp. 428–430). By using phrases such as *how many . . . , to what extent . . . , to what degree . . . , number of . . . , rate of . . .*, and *how often . . .*, we are forced to complete the phrase by indicating what specific attribute the variable refers to. So, I teach students to always preface their descriptions of their variables by such phrases, so that they will be forced to focus on the specific attribute in question. If the variable in question is an effect variable, then this will clarify one's thinking about what the possible causes of it are. It is difficult to think of what the causes of "leader–follower exchange" are. It is easier to imagine what the causes are of "the degree to which the follower feels free to disagree with the leader." Variables in SEMs must be unidimensional quantities.

Because causal relations concern functional relations between variables, these variables involve the attributes of objects, and this constrains us to select only those objects (e.g., subjects to study) that are *homogeneous* in representing the same functional relations between the variables. We call this the *causal homogeneity condition* (Mulaik and James, 1995). This means that there must not be a subset of the population of subjects studied in which the functional relations between specified variables differ from those in other subsets of subjects of the same population. Now, this may not be initially easy to bring about in a study, because we may not yet have studied the functional relations in various subsets of the population, and are only beginning to select subjects for our study. So, it may take a series of studies before we begin to be certain of how to select subjects that are causally homogeneous. (It is unrealistic to suppose that one will always be able to perform a study with an SEM and get everything right the first time. Science progresses in graduated steps with series of studies.)

But one way to enhance the possibility of selecting subjects that are causally homogeneous is to pay attention to those characteristics of the subjects selected. They should be alike on any other variables not studied that may have some causal influence that will affect the causal relations hypothesized in the model. For example, subjects should be essentially equivalent in age, race, sex, and education, if it is believed these variables will have an effect on the causal relations studied.

Multiple Indicators

My philosophy in designing SEM studies is to include in the study as many indicators of each latent variable as is practically feasible, with a minimum of four indicators for each latent variable. The aim is to establish objective validation for the latent constructs by demonstrating that they are measured simultaneously from at least four distinct points of view by being common factors of these measures. Furthermore, this helps raise the squared multiple correlation for predicting the latent variable from the indicators, reducing indeterminacy for the latent. Multiple indicators demonstrate the researcher's capability to select or construct repeatedly variables he or she claims are independent measures of a given construct that is represented by a latent variable in the model. The researcher identifies the attributes that vary and constitute the cause and shows that the indicators all reflect the same causal variable by having a common factor. This lends objectivity to the construct by there being procedures or rules by which the researcher—or any other researcher—can at will repeatedly construct or select, or put in place measures of the same construct. Others can study the procedures used in the selection or construction process and repeat the study in other contexts. However, objective validity does not imply "absolute truth." Objective validity is provisional, meaning that a test of something that is independent of the observer and invariant from several points of view has been passed. The requirement for at least four indicators per latent variable is to make it possible to test the hypothesis that the indicators believed to have a factor in common, indeed, have a single common factor. The common factor represents an invariant form of variation among the indicators. Also, by testing whether a common factor model with a single common factor conforms to the indicators, we test assumptions of closure and self-containment, since by conditioning on the common factor, the residual variables should be uncorrelated by local independence, since they correspond to the unique factors of the model.

Multiple Indicators versus Single Indicators

There are others, such as Hayduk (1996), who advocate using models with single indicators. Hayduk recommends selecting a "single best indicator" of a latent variable, fixing its structural coefficient on the latent to unity, and, additionally, fixing the "error" variance of the variable to that value to indicate what proportion of the total variance of the indicator is not due to the hypothesized latent. Then by including the latent and its indicator with other latents and their single indicators in a model that is relatively sparse in the number of free structural coefficients between the latents, one is able to have a model with an ample number of degrees of freedom. In this case, Hayduk believes the latent is defined not just by being a cause of its indicator, but by its being a cause (sometimes indirectly) of certain other latents and their indicators while not being a cause, directly or indirectly, of certain other latents and their indicators. If your model with a complex but sparse causal network involving many variables fits, this supports your concepts of the latent variables in the model.

The contrast between the multiple indicator approach and the single indicator approach in the conceptualization of the latent variable is reminiscent of the contrast logicians and philosophers of language make between intensional and extensional meanings of words. According to Copi (1978), words applicable to more than one thing have both kinds of meanings. The intensional meaning of a word is a set of *attributes* common to all and only those objects to which the word refers. The extensional meaning is the set of *objects* to which the word may correctly refer. The attributes of the intensional meaning provide criteria for what objects the word can be extended to. But the set of objects of an extension does not determine the intension. Different sets of attributes may have the same extension. But words with different extensions cannot have the same intension (Copi, 1978).

The causal analogue with these two kinds of meaning resides in the fact that causes are variable attributes of things that determine certain other variable attributes of things. So, the *intensional* meaning of a cause is those and only those attributes of objects whose variation produces effects (covariation) in certain other attributes of things. The *extensional* meaning of a cause is the set of all those attributes that are effects by varying as a result of the variation of the causal attributes. The intensional meaning of a cause determines the extension of its effects to other attributes (variables). However, the set of attributes (variables) that are the effects of a given cause, and define its extension, does not determine uniquely its intension. *A cause cannot be uniquely determined by its effects.* More than one causal variable, say, can have the same set of attributes affected. Consider any variable that is a node in a causal network that has more than one causal input. Each causal input to the node will have the same downstream set of effect variables determined by the nodal variable. On the other hand, two different extensions cannot have the same cause.

What we have said so far applies only to absolute extensions and intensions, that is to say, extensions and intensions of variables from a God's-eye view in which all possible variables that belong to the extension and all possible attributes that define the intension of a given cause are known. With any finite model involving a finite number of variables, extensions and intensions may be relative to the variables involved and what is currently known about the individual variables.

Consider Figure 8.1a that shows two models with single indicators for each latent. The models are simplex models, which will be discussed further in Chapter 13 on longitudinal models. Models (1) and (2) are identical, with the exception that the model in Figure 8.1a(1) has a latent ξ_1 with single indicator Y_1 at the front end on the left, whereas the model in Figure 8.1a(2) has a latent ξ_0 with single indicator Y_0 in the same place at the front end on the left. Suppose researcher (1) has chosen Y_1 to be the indicator of a latent he believes is a direct cause of η_2, whereas researcher (2) has chosen Y_0 to be the indicator of a latent she believes is a direct cause of η_2.

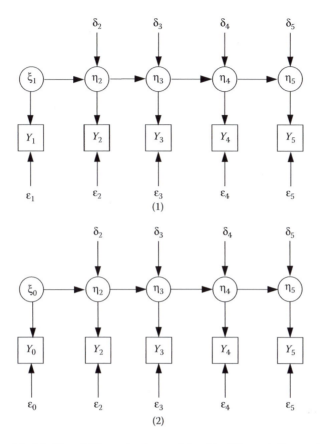

FIGURE 8.1a Models (1) and (2) are simplex models with different indicators Y_0 and Y_1 of corresponding latents ξ_0 and ξ_1, each, respectively, a cause of η_2.

If we try to determine the nature of the cause ξ_1 or ξ_0, respectively, of η_2 from the downstream extension of η_1, which includes all downstream variables that are direct or indirect effects of η_2, we might be inclined (wrongly) to infer that ξ_1 is the same cause as ξ_0. But they need not be. If we include both Y_1 and Y_0, along with their latents ξ_1 and ξ_0, each regarded as a cause of η_2, in the same model with $Y_2, \ldots, Y_5$ and $\eta_2, \ldots, \eta_5$ and their disturbances, we may discover a situation as in Figure 8.1b(1) in which ξ_1 and ξ_0 are completely distinct, uncorrelated variables. Or ξ_1 and ξ_0 may be actually the same variable ξ_1 as in Figure 8.1b(2).

Returning now to our discussion of the causal intension and extension of a variable, if we limit ourselves to a particular set of variables, a causal variable in that set may have the same extension as another and not be distinguished by its extension. One could substitute one for the other and have the same extension. And it would be a fallacy to think that by showing that a causal variable has a given extension, this determines the identity of the cause. This

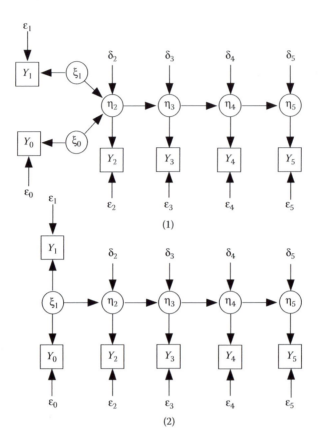

FIGURE 8.1b Models (1) and (2) are simplex models. In (1) latents ξ_1 and ξ_0 are distinct, uncorrelated variables. In (2) Y_0 and Y_1 are indicators of a single latent ξ_1.

suggests that the way to distinguish causes is to enlarge the set of variables in such a way as to show that one cause causes some different variables than does another cause with which it may be otherwise confounded in some of the variables. Distinct extensions have different intensions.

Indicators of causal variables should have some set of attributes that suggest the attributes that are varied of the cause indicated. Other attributes of the indicator are the effects of the cause.

Guttman (1965), for example, argued that analytic ability, which some equate with general intelligence, is the ability to correctly discern relations. He argued that measures of this ability show subjects one or several pairs in a relation, for example, "Dog is to puppy," and then present just one member of a new pair from the same relation, for example, Cat is to ___?, that is, either in the domain or the range of the relation. The subject must then demonstrate the "best" or the "correct" perception of the relation by either providing the corresponding other member of the pair (e.g., fill in the blank) or correctly picking the other member from a set of alternatives, for example, (a) cub, (b) kitten, (c) flea, (d) mouse. (b) is "correct" or the "best" according to the tester, whose criteria are what would be most salient to most members of the community. (With no further constraints, such as a limited set of alternatives and the implication that there is a "correct" or "best" answer, there is no unique response that one could make.) The relation sought by the tester is names of offspring of specified animals. Displaying one or several pairs in an unnamed relation and a limited number of choice responses to make is the same stimulus given to all subjects. The judgment in determining the relation is made in each subject's brain. The latent causal variable is the variation among subjects in the way whatever centers of their brains process the problem to yield the judgment of the relation and then the alternative chosen. The latent causal variable thus represents a capacity of individuals to properly detect relations. Next, the judgment has to be communicated to the tester. There are numerous media by which the judgment could be communicated; a few possibilities are by pointing to a given alternative, by saying in words what the alternative chosen is, and by making a mark on an answer sheet. This confounds the output from the center of the brain that made the judgment with the medium of communication conveying the judgment. Communicating may involve other centers of the brain involved in processing the overt response. And there could be individual variation in communicating in a given medium.

Other extraneous factors can also enter in, such as the input medium in which the problem is represented to the subject, for example, as a verbal problem, a figural problem, or a numerical problem. Guttman (1965) held that many of the intellectual factors obtained by others represented not pure analytic ability, but human variation in ability to detect relations in a given input medium. This would suggest that analytic ability is a second-order factor to be distinguished from first-order factors that confound analytic ability with the ability to process a problem in a given input medium and the ability to communicate the judgment in a given output medium. In other words, an indicator of the ability to grasp relations involves variation in processing certain types of inputs, central processing, and in processing certain types of outputs. But to isolate these factors we need multiple indicators of analytic ability varying in different input and output media.

Multiple indicators allow for testing whether those indicators believed from their content to represent a latent causal variable varying in certain specified attributes across subjects do indeed have a factor in common. It also allows for the detection of extraneous common factors as well as doublets not represented in the model. So, the focus is on the intensional nature of a factor as opposed to its extensions to other variables.

The use of models with single indicators of latent variables is possible. But in the author's opinion they are fraught with possible indeterminacies and ambiguities, especially at the outset of a research program where clear concepts of latent variables have not been formulated and an understanding of how their single indicators are effects of just the latent in question and not of other variables has not been worked out and tested. These models rely chiefly on the extension of causes to determine that the causes are what they are supposed to be. But we have shown that extensions cannot uniquely determine the causes. Merely saying "Choose the best indicator of the latent" does not direct the researcher to focus on the intensional meaning of the cause, to specify attributes that are varied as the cause. So the cause is not clearly specified. With multiple indicators one is compelled to select indicators that pertain to the same set of attributes that when varied constitute the cause. The causal extensions of causes to other variables, on the other hand, are more problematic than are the intensional attributes of the causes. One first seeks to establish clearly and distinctly the intensional attributes that constitute the causes, and only then does one focus on the extensional properties of the causes. Usually it is the causal extensions that are problematic and the reason for a study. But we first wish to establish that our choice of indicators has the proper intensional meaning. And this is difficult to demonstrate with a single indicator.

Tetrad Difference Tests with Multiple Indicators

As pointed out in Chapter 6, three indicators of a single common factor yield, for three positively correlated variables, a just-identified system, and no test of a single common factor may be performed in such a case, since mathematically it would be impossible to fail such a test. With four indicators per factor, a system with a single common factor is overidentified and a test may be performed. But more specifically Hart and Spearman (1913) showed that with four manifest variables Y_1, Y_2, Y_3, and Y_4, with a common factor, each of the following three "tetrad difference equations" should hold: $\rho_{12}\rho_{34} - \rho_{13}\rho_{24} = 0$, $\rho_{14}\rho_{23} - \rho_{13}\rho_{24} = 0$, and $\rho_{13}\rho_{24} - \rho_{12}\rho_{34} = 0$, where ρ_{ij} denotes the correlation between variables i and j. To see why this is so, consider now a path diagram for these four variables having a single common factor (Figure 8.2).

Assume that all variables have unit variances (for convenience). Applying path-tracing rules, the correlations between any pair of distinct observed variables are

$$
\begin{aligned}
\rho_{12} &= \gamma_{11}\gamma_{21}, \\
\rho_{13} &= \gamma_{11}\gamma_{31}, \\
\rho_{14} &= \gamma_{11}\gamma_{41}, \\
\rho_{23} &= \gamma_{21}\gamma_{31}, \\
\rho_{24} &= \gamma_{21}\gamma_{41}, \\
\rho_{34} &= \gamma_{31}\gamma_{41}.
\end{aligned}
\tag{8.1}
$$

Note in Equation 8.1 that if we multiply ρ_{12} with ρ_{34}, we obtain

$$\rho_{12}\rho_{34} = \gamma_{11}\gamma_{21}\gamma_{31}\gamma_{41}.$$

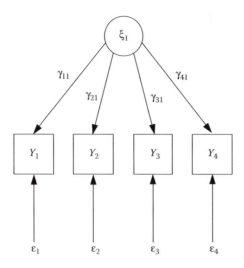

FIGURE 8.2   Four variables with a single common factor.

Similarly,

$$\rho_{13}\rho_{24} = \gamma_{11}\gamma_{31}\gamma_{21}\gamma_{41}.$$

But now, we can see that these two products yield the same value, so we may write

$$\rho_{12}\rho_{34} = \rho_{13}\rho_{24}$$

or

$$\rho_{12}\rho_{34} - \rho_{13}\rho_{24} = 0,$$

which is known as a *tetrad difference equation*.

Now, there are three distinct tetrad difference equations we can form from these six expressions in Equation 8.1. We have found the first equation, and the remaining two are

$$\rho_{13}\rho_{24} - \rho_{14}\rho_{23} = 0,$$

$$\rho_{14}\rho_{23} - \rho_{13}\rho_{24} = 0.$$

However, once two are given the third follows necessarily and is not an independent equation.

The question naturally arises at this point of a statistical test for each of the two independent tetrad difference equations. Spirtes et al. (2000) say that they use a test given by Wishart (1928), and we will show it here. The problem is to estimate the variance of the sampling distribution of a tetrad difference, and Wishart gave the following expression for this variance in terms of determinants of the covariance matrix among the four observed variables. The tetrad differences are computed from covariances instead of correlations. So, let $\hat{T} = s_{12}s_{34} - s_{13}s_{24}$ be a tetrad difference computed using the sample covariances s_{ij} between the respective pairs of variables. Given now the sample covariance matrix,

$$\mathbf{S} = \begin{bmatrix} s_{11} & s_{12} & s_{13} & s_{14} \\ s_{21} & s_{22} & s_{23} & s_{24} \\ s_{31} & s_{32} & s_{33} & s_{34} \\ s_{41} & s_{42} & s_{43} & s_{44} \end{bmatrix},$$

the estimated variance of the sampling distribution of a tetrad difference is given by Wishart (1928) as

$$\hat{V}(\hat{T}) = \frac{D_{12}D_{34}(N+1)}{(N-1)(N-2)} - D,$$

where $D_{12} = \begin{vmatrix} s_{11} & s_{12} \\ s_{21} & s_{22} \end{vmatrix}$ and $D_{34} = \begin{vmatrix} s_{33} & s_{34} \\ s_{43} & s_{44} \end{vmatrix}$ are determinants of submatrices of $\mathbf{S}$ in the upper-left quadrant and lower-right quadrants, respectively, of $\mathbf{S}$, whereas D is the determinant of the matrix $\mathbf{S}$ itself. The second-order determinants are not difficult to compute, but the fourth-order determinant can present a problem if computed by hand. It would be better to estimate D by finding eigenvalues of $\mathbf{S}$ and computing their product, which can be done with matrix manipulation routines in programs such as SPSS.

Now, a statistic to use would be a z statistic

$$z = \frac{\hat{T}}{\sqrt{\hat{V}(T)}}.$$

This presents a two-tail test of the null hypothesis that the tetrad difference equals zero. The null hypothesis is rejected if z falls in the regions of rejection at some level of significance.

This test is only asymptotically accurate in samples of 300 or more and greater observations and is more powerful when the correlations among the variables are large as opposed to small (e.g., 0.30) (Glymour et al., 1987).

This establishes a basis for requiring four indicators of each latent variable. But with more than four indicators, the calculation of tetrad differences is to be reserved for exploratory programs such as Tetrad II, which seeks to discover causal structures in correlation matrices and uses the brute force of the computer to compute all possible tetrad differences and various criteria to ascertain these structures. The usual researcher may find it more convenient simply to fit a single common factor confirmatory factor analysis model to the indicators of a latent variable and ascertain whether the fit is satisfactory.

There are, however, situations in which using single indicators is reasonable. Some researchers may regard "sex," "age," "weight," "height," "years of schooling," and so on to be sufficiently measured by single variables. It will be important in such cases to consider how these variables function in one's model. Is a variable of this class a cause of other variables, or is it an effect of some other causal variables? In what way? Perhaps the single indicator serves as an indicator of some other variable, which is really what the researcher is (or should be) interested in. For example, if sex is determined by the possession of certain kinds of genital organs, which vary in a population, is that what one wants to measure? Or is one measuring a variation of a certain social category commonly, but neither exclusively nor universally possessed by everyone with such organs? In such a case, one may need to treat this as a latent variable of interest and seek multiple indicators of it to pin down what it is and also to have some indicators that are *not* effects or causes of it. "Sex" would then be merely one of several indicators of the latent variable in question. Is "age" a cause or an effect of other variables? What is the mechanism by which these causal relations are effected? Age, technically, is merely the amount of time since birth. Variation in age is often correlated with other variables. It might be causally connected as a criterion for selection of individuals which are then exposed to other causal (e.g., educational) variables. Educational variables become more immediate causes of other variables. So, one should include measures of these educational variables to identify the causal pathway by which age is related to some educational outcome variables. In this case age may be used as an exogenous, manifest variable, but other aspects of the model may require latent variables with multiple indicators to determine the particular causal pathway by which age influences an educational outcome. Age may also be an exogenous moderator variable that interacts with other variables to influence the strength of the causal effects of these other variables, expressed in sizes of structural coefficients. In these cases multiple group studies or multilevel studies may involve the use of age, sex, group membership, and so on as exogenous classification variables defining groups or levels.

But some problems remain. Hayduk (1996) points out that merely finding that the four or more indicators have a single common factor does not establish in any absolute sense that the factor is indeed the latent variable hypothesized by the researcher. He illustrates this with the following two models.

Suppose in Figure 8.3 that four indicators $Y_2, \ldots, Y_5$ are generated according to model A, but a researcher hypothesizes according to model B. In the real world ξ_1 is a common cause but not a parent of $Y_2, \ldots, Y_5$. But the researcher thinks it is a common parent of $Y_2, \ldots, Y_5$. So, the researcher conducts a confirmatory factor analysis and determines that indeed $Y_2, \ldots, Y_5$ have a single common factor, and she assumes this supports her hypothesis that this is ξ_1. But in reality the common factor of $Y_2, \ldots, Y_5$ is $\eta_1 = \gamma_{11}^* \xi_1 + \delta_{11}\varepsilon_1$. ξ_1 is a disturbance and may, in addition to unsystematic errors of measurement, contribute other causes of η_1, such as method factors. We see then that the estimate of the effect of the hypothesized variable ξ_1 on, say, Y_2 according to the common factor model would be equal to $\gamma_{21} = \alpha_{21}$, when in fact the true

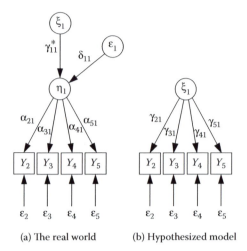

(a) The real world   (b) Hypothesized model

FIGURE 8.3   The world generates variables according to model A, but the researcher hypothesizes according to model B.

effect is $\gamma_{11}^* \alpha_{21}$. So, the estimate of its effect from the common factor model would be biased, and would likely overestimate the true effect if γ_{11}^* is less than unity.

Hayduk (1996) argues that "These observations carry the strong conclusion that even though a specific variable is named and justifiably identified as a common cause of a set of indicators, this is insufficient to conclude that 'this' is the variable that will be located as the common factor in a factor analysis." He is correct, and perhaps factor analysts have been slow to recognize this. However, factor analysts have distinguished between first-order and higher-order factors. The common cause of his example could be a second-order factor distinguished from any given first-order factor that confounds the second-order factor with some other cause by being a common cause of other first-order factors as well, which are indicated by different sets of manifest variables. So, the issue here is what sort of evidence would contradict the assumption that ξ_1 is the parent common cause (the common factor) of the variables $Y_2, \ldots, Y_5$?

Suppose we find what we believe is another indicator of ξ_1, say, Y_6, measured by some other method than methods used to measure $Y_2, \ldots, Y_5$ and include it with the variables $Y_2, \ldots, Y_5$ in a new confirmatory factor analysis of the hypothesis that $Y_2, \ldots, Y_6$ have a single common factor. If we do not reject the hypothesis, this is stronger evidence than the evidence we had with just the variables $Y_2, \ldots, Y_5$, especially if the loadings of the variables $Y_2, \ldots, Y_5$ on the current common factor of $Y_2, \ldots, Y_6$ are the same as they were on the common factor of the previous confirmatory factor analysis of just $Y_2, \ldots, Y_5$. But then again, there may still be the extraneous variable ε_1 confounded with ξ_1 in this case as in the previous case. If we reject the hypothesis of a single common factor for $Y_2, \ldots, Y_6$, we may be in a position to discover whether ξ_1 is distinct from the common factor of $Y_2, \ldots, Y_5$ while being a common cause of them and Y_6. What we can do is test a model as shown in Figure 8.4. The aim is to show that Y_6 is an effect of ξ_1 that is not confounded with ε_1 and thus may make it possible to distinguish ξ_1 from a common factor of $Y_2, \ldots, Y_5$.

If the model in Figure 8.4 fits to within sampling error, this strongly supports the hypothesis that ξ_1 is a common cause of $Y_2, \ldots, Y_6$ but not a common factor (parent common cause) of $Y_2, \ldots, Y_5$.

It is important to realize what rejecting the hypothesis that ξ_1 is a parent cause of Y_1 implies, when one embeds the latent variable and its indicators in a structural model. Consider the following pair of structural models.

In Model A of Figure 8.5 we see a model correctly representing reality. On the other hand, in Model B we see how a researcher represents that reality. He thinks that ξ_1, a common cause of variables $Y_1, \ldots, Y_4$, is also a cause of another latent variable η_2 that is a common cause of variables $Y_5, \ldots, Y_8$. He treats γ_{21} as a free parameter. His model will fit acceptably the data generated by Model A. But he will be wrong in asserting that ξ_1 is a common cause of η_2. There exists in the real world another method variable ξ_2 that is a common cause of both η_1 and η_2. This common method cause allows us to replace the fork between η_1 and η_2 in Model A with a directed arrow to obtain equivalent Model B. It is true that ξ_1 is also a cause of η_1. But it is not a common parent of $Y_1, \ldots, Y_4$. While performing the more stringent test of fixing the

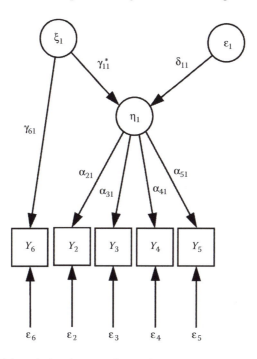

FIGURE 8.4 A model in which ξ_1 has an indicator that is not influenced by ε_1.

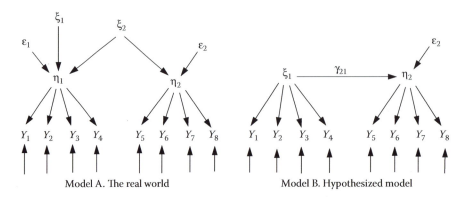

FIGURE 8.5 A hypothesized model may incorrectly support a latent as a cause of another latent.

variance of the latent exogenous variable ξ_1 to unity and the loading on the best indicator Y_1 to some prespecified value will test whether a common factor of $Y_1, \ldots, Y_4$ has that loading, rejecting this model may indicate that a latent variable defined by this effect may not be the true parent common factor, and further that there may be another factor present.

Still another way a researcher can tease out the distinction between ξ_1 and η_1 is to find another indicator of ξ_1 that is not an effect of the method factor ξ_2 and to include this with the other variables. Consider, for example, Model C in Figure 8.6.

Leaving a path between ξ_1 and η_2 with a free coefficient could allow the model to fit when there is no relation between ξ_1 and η_2.

In Model C of Figure 8.6, with reality still as in Figure 8.5 (Model A), variable Y_9 in the real world is a child of ξ_1 but has no influence from the method variable ξ_2. ξ_1 is a common cause of Y_9 and $Y_1, \ldots, Y_4$. But $Y_1, \ldots, Y_4$ also have η_1 as a common factor. Furthermore, Y_9 should be uncorrelated with

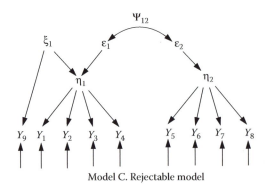

FIGURE 8.6 Indicator Y_9 permits distinguishing ξ_1 from η_1, allows for a possible common influence on η_1 and η_2, while testing that ξ_1 is *not* a cause of η_2.

variables $Y_5, \ldots, Y_8$, which would not be the case if it were an effect of η_1 and a descendent of ξ_2. This also establishes why contrary to Model B, ξ_1 is not a cause of η_2, since, if it were, this would make Y_9 correlated with $Y_5, \ldots, Y_8$, which it is not. On the other hand, variables $Y_1, \ldots, Y_4$ are correlated with variables $Y_5, \ldots, Y_8$. Thus the advantage of finding an indicator of ξ_1 that is independent of any other factors common at the first- or second-order level to the variables in $Y_1, \ldots, Y_4$, is that the hypothesis that it is *not* a cause of η_2 may be tested without a knowledge of the loading of ξ_1 on Y_1. The belief that ξ_1 is the latent variable one claims it is, is supported by Y_9 and $Y_1, \ldots, Y_4$, which the researcher has chosen because an examination of them or the manner by which they were constructed suggests that they all have ξ_1 as a common cause. But unless the value of the factor loading γ_{91} has been determined in a previous study by estimate, any fixed value given for it by the researcher in this situation is just an "educated" guess of the researcher. Furthermore, asserting that there is no path between ξ_1 and η_2 allows for a test. If we simply leave a path with a free coefficient between ξ_1 and η_2 in the model, and the estimated path coefficient turns out to be not significantly different from zero, the model would still be accepted. If we do not look closely at the estimated value and perform a further test of the path coefficient to see if it differs significantly from zero, we might be misled to think it is truly nonzero. Considerations like these lead to the ideas of the four-step procedure.

The Four-Step Procedure

If SEMs are formulated with multiple indicators for each latent variable, then a four-step procedure can be used in testing these models. A more detailed description of this procedure is given in Mulaik and Millsap (2000). Before we discuss specifically the four steps, we need to develop some background.

A nested sequence of models is a series of models for the same set of variables that have the same parameter structure or equivalent parameter structures in which successive models in the series introduce additional parameter constraints on top of those in the preceding model of the series. There are two kinds of model nesting: (1) *parameter nesting* and (2) *equivalence nesting*. In a *parameter-nested* series of models, the models all have the same parameter structure. Beginning with a least constrained model, the constraints on the parameters of this model are carried over to the corresponding parameters of the next model in the series, and additional constraints are assigned to some of the previously free parameters. Next, the constraints on the parameters of the second model are carried over to the corresponding parameters of the third model, and additional constraints are introduced to other free parameters from the previous model. This process can continue until all parameters are constrained, but it is not always necessary to go that far. In a series of equivalence-nested models, each model in the equivalence-nested series corresponds to a model in a corresponding nested sequence in having the same fit to the data as the parameter-nested model. The reason equivalence-nested models are considered is because they are often easier to obtain or more meaningful.

The nested sequence of models considered by the four-step procedure is (1) the *unrestricted model*, (2) the *measurement model*, (3) the *structural model*, and (4) models in which previously freed parameters are individually fixed either to zero or to nonzero values and tested. The assumption is that each latent variable of the structural model has at least four indicators, *by design*. Hence an unrestricted and a confirmatory factor analysis model, respectively, may be regarded as nested within the structural model. Furthermore, the restrictions introduced into the first three models of the series only involve fixing certain parameters to zero. In step (4) restrictions may be introduced that involve fixing parameters to zero or nonzero values.

The procedure begins with a breakdown of the full SEM into a series of four nested models. All nested models may be regarded as having the same or equivalent structure, but the models are nested by applying additional constraints to parameters in each successive model in the nested series. Beginning with the least constrained model, the next model in the series retains all of the constraints (fixed parameters, equality constraints among parameters) of the previous model and in addition adds additional constraints to other parameters. A property of nested sequences of models is that no more constrained model will fit better, and likely will fit worse, than any less constrained model in the nested sequence. Only if the constraints are "correct" will successive models continue to fit to within criteria of acceptable fit.

The unrestricted model of the *first step* was introduced in the previous chapter. To create such a model, all of the latent variables must be specified and their multiple indicators indicated. The following matrices in Figures 8.7a and b show how an unrestricted model is specified for the model in Figure 5.1. ?s are free parameters. 0s are zero fixed parameters. 1s are fixed 1.00 parameters.

The model in Figure 8.7a has only four latents (aside from the unique factors). The best indicator of each latent "factor" is given a free parameter on that factor while the indicator's loadings on the other factors are set to zero. The loadings of all other indicators on the factors are made free parameters. Each latent's variance is set to unity. Correlations between factors are free parameters. The unique variances are free parameters.

An alternative way of specifying the unrestricted model is the *lower-triangular-matrix method*. This is illustrated in Figure 8.7b.

The lower-triangular-matrix method begins by choosing h rows of the "factor pattern" matrix that one believes are a linearly independent set of row vectors. Then beginning with the first row, one frees the element in the first column and fixes all the remaining elements to zero. Then one takes the second chosen row and frees the elements in its first two columns and fixes the elements in the remaining columns to zero. In general, in the ith row, one frees

$$\Lambda = \begin{bmatrix} ? & 0 & 0 & 0 \\ ? & ? & ? & ? \\ ? & ? & ? & ? \\ ? & ? & ? & ? \\ 0 & ? & 0 & 0 \\ ? & ? & ? & ? \\ ? & ? & ? & ? \\ ? & ? & ? & ? \\ 0 & 0 & ? & 0 \\ ? & ? & ? & ? \\ ? & ? & ? & ? \\ 0 & 0 & 0 & ? \\ ? & ? & ? & ? \\ ? & ? & ? & ? \\ ? & ? & ? & ? \end{bmatrix} \qquad \Phi_{XX} = \begin{bmatrix} 1 & ? & ? & ? \\ ? & 1 & ? & ? \\ ? & ? & 1 & ? \\ ? & ? & ? & 1 \end{bmatrix} \qquad \Psi^2 = \begin{bmatrix} ? \\ ? \\ ? \\ ? \\ ? \\ ? \\ ? \\ ? \\ ? \\ ? \\ ? \\ ? \\ ? \\ ? \\ ? \\ ? \\ ? \\ ? \end{bmatrix}$$

FIGURE 8.7a Parameter matrices of an unrestricted model.

the first i elements and fixes any remaining to zero. Then one frees elements in all the other rows of the pattern matrix. For the $h \times h$ matrix of covariances among the exogenous latents, one fixes the diagonal elements to unities and the off-diagonal elements to zeros. In all, h^2 parameters are fixed and distributed across the pattern matrix and the matrix of covariances among the exogenous latents in a proper way. Jöreskog's method produces a solution

$$\Lambda = \begin{bmatrix} ? & 0 & 0 & 0 \\ ? & ? & ? & ? \\ ? & ? & ? & ? \\ ? & ? & ? & ? \\ ? & ? & 0 & 0 \\ ? & ? & ? & ? \\ ? & ? & ? & ? \\ ? & ? & ? & ? \\ ? & ? & ? & 0 \\ ? & ? & ? & ? \\ ? & ? & ? & ? \\ ? & ? & ? & ? \\ ? & ? & ? & ? \\ ? & ? & ? & ? \\ ? & ? & ? & ? \end{bmatrix} \qquad \Phi_{XX} = \begin{bmatrix} 1 & 0 & 0 & 0 \\ 0 & 1 & 0 & 0 \\ 0 & 0 & 1 & 0 \\ 0 & 0 & 0 & 1 \end{bmatrix} \qquad \Psi^2 = \begin{bmatrix} ? \\ ? \\ ? \\ ? \\ ? \\ ? \\ ? \\ ? \\ ? \\ ? \\ ? \\ ? \\ ? \\ ? \\ ? \\ ? \\ ? \\ ? \end{bmatrix}$$

FIGURE 8.7b Specifying an unrestricted model by the lower-triangular method.

$$
\Lambda =
\begin{bmatrix}
? & 0 & 0 & 0 \\
? & 0 & 0 & 0 \\
? & 0 & 0 & 0 \\
? & 0 & 0 & 0 \\
0 & ? & 0 & 0 \\
0 & ? & 0 & 0 \\
0 & ? & 0 & 0 \\
0 & ? & 0 & 0 \\
0 & 0 & ? & 0 \\
0 & 0 & ? & 0 \\
0 & 0 & ? & 0 \\
0 & 0 & ? & 0 \\
0 & 0 & 0 & ? \\
0 & 0 & 0 & ? \\
0 & 0 & 0 & ? \\
0 & 0 & 0 & ? \\
\end{bmatrix}
\qquad
\Phi_{XX} =
\begin{bmatrix}
1 & ? & ? & ? \\
? & 1 & ? & ? \\
? & ? & 1 & ? \\
? & ? & ? & 1 \\
\end{bmatrix}
\qquad
\Psi^2 =
\begin{bmatrix}
? \\ ? \\ ? \\ ? \\ ? \\ ? \\ ? \\ ? \\ ? \\ ? \\ ? \\ ? \\ ? \\ ? \\ ? \\ ? \\ ?
\end{bmatrix}
$$

FIGURE 8.7c Measurement model. 0s and 1s are fixed parameters, while ? denotes an estimated parameter.

that is a linear transformation of the lower-triangular method. Both produce the same reproduced covariance matrix and have identical fit.

Equivalently, one can perform a maximum-likelihood "exploratory" factor analysis of the manifest variables with the number of factors fixed at the number of latents in the structural model. This would be a linear transformation of the latents of the unrestricted model which does not change the fit. What is desired is the chi-square value, which should be the same (except for slight differences due to the different minimization criteria used by the programs) for each method of specifying an unrestricted model.

It is essential that the unrestricted model fits the data very closely, preferably to within sampling error (or if one uses indices of approximation, to a very close approximation—see Chapter 15). The chi-square test of fit tests the hypothesis that a common factor model with k common factors and uncorrelated unique factors fits the data. k is the number of latents in the previously specified structural model. [k is not determined by an exploratory factor analytic search for the "proper number of common factors" but by the number of latents specified by a substantive hypothesis in the structural model. This was a misunderstanding of the four-step procedure created by Hayduk and Glaser (2000) and further committed by Herting and Costner (2000) upon reading Hayduk and Glaser's critique of the four-step procedure and not having Mulaik and Millsap (2000) to see how they actually conceived of the four-step procedure. It is just that the "exploratory" common factor model with a fixed number of factors by hypothesis is a way of obtaining an estimate of the chi-square for an unrestricted model. No exploring for the "proper number of factors" is done.] The hypothesis does not specify any particular relations between the latents and the observed variables, which is why different rotations of the common factors would still yield the same degree of fit, as long as the latents have unit variances. The unrestricted model contains more constraints than the saturated model and fewer constraints than the measurement model that follows in the series.

If the unrestricted model does not fit the data acceptably, then none of the subsequent models in the nested sequence will fit either, since the constraints of this model are carried over into the more constrained models of the series. Lack of fit usually means that the hypothesized number of latents and/or the zero correlations among the unique factors are incorrect. So, a simultaneous test of the number of latents and the assumption of closure and self-containment is performed by this test. With a failing unrestricted model, the researcher should carefully reexamine his or her assumptions about the latent causes in the model. Does some other causal structure than a common factor structure that has more latents apply? Are there doublet factors (with loadings on only two variables) present but unrepresented in the model (say, by correlations among pairs of unique factors for the corresponding variables)? Unless one can resolve these considerations in an objective manner, one must not go forward to test the other models of the sequence.

If one does not resolve the lack of fit for the given number of factors hypothesized by introducing additional theory or knowledge about the variables and adjusting the model accordingly to a successful outcome, then one effectively enters an exploratory mode and no longer is following the four-step procedure of testing a hypothesis. It has been rejected. One can perform an exploratory factor analysis, reevaluate the number of factors, and rotate to oblique simple structure to see whether a better conception of what might be an appropriate model emerges.

Still, exploratory factor analysis is not the only technique one could or should use. Herting and Costner (2000) suggested one could use the TETRAD program of Glymour et al. (1987) to explore with some substantive constraints—if necessary—for a more likely causal structure underlying the data. This program has more flexibility than exploratory factor analysis in searching for causal structure. It will suggest not only possible causal connections, but also variables that might be dropped that confuse the causal picture.

One might suggest that the TETRAD program be used to develop the initial model against some other data set (so as not to confound an exploratory approach with a confirmatory one by using the same data to develop a hypothesis and then to test it).

On the other hand, if the unrestricted model fits acceptably, then one is able to go forward to the *second step* to test the next more constrained model in the nested series, *the measurement model*.

The measurement model gets its name from the idea that in this model one presumably is concerned with testing the relations between the latents and the indicators in one's structural hypothesis. The indicators measure the latents.

But contrary to common belief, it is not the relations between a particular latent and its indicators that are tested, but rather hypotheses about which latents are *not* related to certain indicators. These hypotheses are specified by introducing additional zero parameters into the factor loading matrix to indicate that certain indicators are not related to certain latent factors. Thus a measurement model specification looks like the following.

Note again that the measurement model leaves the correlations between the factors free. It is these correlations among the factors that will bear the relations between the latent causes in the structural model. The variances of the latent factors, however, are set to unity, simply to set the metric for each of them. However, an alternative method of setting the metric at this point is to leave the variances of the latent factors free and fix one loading on each respective factor to some fixed value, for example, unity. Although variances, covariances, and loadings will be different in this case, they will still reproduce the same covariance matrix for the observed variables and produce equivalent chi-squares. Also, it should be noted that there is some leeway here to free up more loadings in the factor loading matrix of the measurement model, if the researcher has good reason to believe that an indicator may not have a zero loading on some other factor. Perhaps a reason why many measurement models fail when based on prior factor analyses is that researchers fix to zero what corresponds to low, nonzero loadings in the prior analysis. So, if one does not believe that the loading should be zero, it should be free.

Failure of the measurement model usually means that at least one zero relation between an indicator and a latent variable is causing lack of fit. (Correlations between unique factors should not occur, if they did not occur in testing the unrestricted model.) One can apply Lagrange multiplier tests or modification index tests of the constrained zeros in the factor loading matrix to determine which contributes the most to the lack of fit by having the highest significant chi-square or modification index. That zero loading may then be freed and a new analysis performed. Zero loadings should be freed one at a time for each successive analysis. The researcher should also seek to provide a rationale for why an indicator does not have a zero relation to a given latent. The researcher hopefully will not free so many loadings as to end up with the unrestricted model. Each freed parameter represents a loss in a degree of freedom, and degrees of freedom indicate dimensions in which the data are free to differ from the model, providing dimensions in which the model can be tested. A model that is tested in only a few ways, relative to the potential number by which it could be tested, is an inferior model overall.

Furthermore, failure of the measurement model may force one into the exploratory mode, where again the TETRAD program or an exploratory factor analysis with rotation to oblique simple structure may be used to find a better measurement model. A conceptual reanalysis is also in order, because the researcher should not rely exclusively on the computer to do his or her thinking. As long as the measurement model obtained fits well (nonsignificant chi-square, very high index of approximation), even though it is derived in an exploratory manner, one may still believe the underlying latents are as before and continue to the next step to test the hypothesized causal structure between the latents. If one does not believe the modified model represents the same latent causes, then the new latent variables should be given an explicit meaning and one should respecify one's causal hypothesis for the next step—without looking at the correlations among the factors to suggest what the causal connections should be like.

If one arrives at an acceptable measurement model, then one moves on to the *third step* to test the *structural model*, which involves testing zero constraints on the parameters of causal paths between latent variables. Up to now relations between latents have been specified as estimated covariance bridges between the latents. In the measurement model, the matrix of covariances among the latent variables was saturated, meaning every covariance was free. They are now replaced with directed causal paths between the latents. Some of the paths are eliminated; the rest have free parameters.

One goes forward to the fourth test to test individual parameters left free in the previous model only if one obtains acceptable fit with the structural model (nonsignificant chi-square or a very high degree of approximation). There are several ways to proceed: (a) Test whether the coefficients of free parameters differ significantly from zero using zstatistics with the standard errors of the parameter estimates. These tests are not independent, so some procedure, like using more stringent significance levels as determined by the Bonferonni method, may be applied. (b) One can simultaneously test whether all parameters are equal to zero, which amounts to testing whether all of the variables are unrelated. This corresponds to a test of the *null model*, that the variance–covariance matrix of the observed variables is a diagonal matrix. But rejecting this test, which ordinarily one hopes to do, may not tell the researcher what specific path coefficients are nonzero, since this is a test about the covariances among the observed variables set equal to zero. Instead z statistics using the standard errors of the previously estimated values for the structural parameters could be used to test that each of these previously estimated path coefficients are zero. (c) Instead of testing whether the remaining path coefficients are zero, one may have hypotheses about specific, nonzero values for such coefficients. Such values may be suggested by previous studies with other data sets that provide estimates for these parameters. The estimates are then used as fixed parameters in formulating a model in this step. While the prior estimates may contain errors of estimate, they may yet provide the only nonzero values available to consider. If the resulting model fails to fit, the researcher can then use Lagrange multiplier tests or modification indices to determine which fixed nonzero parameters seem to be contributing to the lack of fit. In those cases the currently estimated values may be optimal. Or the researcher can conduct z statistic tests of the individual fixed values using the standard errors of the corresponding parameter estimates to see whether the estimated values differ significantly from the hypothesized values.

Some Problems That May Be Encountered with the Four-Step Procedure

As one moves from a less constrained to a more constrained model, as between a measurement model and a structural model, the full-information maximum-likelihood (FIML) estimation procedure may alter previously obtained values for free parameters obtained for the less constrained model. This is because, when using the FIML estimation procedure, parameter estimates are not independent of one another. Free parameter estimates are altered jointly in each iteration, and the direction of change in multidimensional space is in a direction that jointly improves to some degree the fit of the whole model and not just individual parameters.

Suppose you began with a measurement model that only freed those loadings of indicators on the factors of which they were supposed to be the principal indicators. Then suppose further that the loadings of these indicators on their designated factors are high, as expected, but still the measurement model does not fit acceptably. So, next you try freeing up cross-loadings of the indicators on other factors leaving covariances between factors free, guided perhaps by careful inspection of the indicator contents and the Lagrange multiplier tests or modification indices of individual fixed parameters. You obtain an acceptable measurement model as a result, and you next hope to explain the covariances between the latent factors by causal paths between them. But you do not want a causal model of the covariances between the latents to be saturated, because then nothing of the structural model is tested. So you place constraints, often zero path coefficients, between certain latents. This represents your original full SEM.

When constraints are placed on the path coefficients between latents in the structural model, in FIML estimation these constraints not only affect the free parameters relating latents to other latents of the structural model, but they can also affect the estimates of the factor loadings relating the latents to the manifest indicators, changing them from what they were in the measurement model. So, one might obtain an acceptable fit of the structural model that nevertheless involves a change in the measurement model that is inconsistent with one's measurement hypothesis. This can occur when the variables are highly correlated, and the factors underlying them are also strongly correlated, with numerous low-to-moderate free cross-loadings of indicators on factors other than just those they load highest on. Fixing certain parameters between the latents to zero (or some other values) can then affect the free loadings of the manifest variables on the latent variables by altering them considerably. It may even become impossible to achieve convergence of the estimation process.

We see in this case that beginning with just a structural model to test, we would not become aware of the inconsistency between our conception of the relationship of the indicators to the latent factors and our conception of the relations between the latent variables, because we would not have model results to compare. To be sure, it is possible still for the structural model to fit acceptably, and yet not in a way consistent with values of the fixed and free parameters of the less constrained measurement model. If one were to compute the reproduced covariances among the latents of the structural model, they might be different from the covariances obtained between the same number of latents in the measurement model. The additional constraints on the latents of the structural model affect (in FIML estimation) not only the covariances, but also the free loadings of the indicators of the latents. A possibility also is that the covariances among the latents cannot be accounted for simply by paths between the latents themselves, even after some of these path's coefficients are constrained to zero. Additional higher-order latents, not yet hypothesized, affecting the first-order latents, may be necessary to account for the correlations between the latents. So, ambiguity remains, and it is best to get it out in the open so as to deal with it.

One way of proceeding, if we have stronger faith in our measurement model than our structural model, is to fix the free loadings of indicators on the latents of the successfully fitting measurement model to the values obtained when estimating them for the measurement model. This effectively constrains not only the relations between indicators and latents, but also the covariances between the latents to the values obtained for the measurement model. So, any additional constraints on the path coefficients between the latents will only influence the fit of these to the covariances between the latents and not the relations of the latents to the manifest indicators. A nonsignificant chi-square difference test between the "fixed measurement model," which we will call the more constrained measurement model, and a corresponding "measurement-fixed" structural model obtained by introducing additional constraints on coefficients of directed paths between the latents will tell us whether the additional constraints on paths between the latents are inconsistent with our measurement model. The chi-square of the measurement model will have the same value it had in the original measurement model with free factor loadings of indicators on latents, but its degrees of freedom will be greater, reflecting the additional fixed parameters. The chi-square of the corresponding more constrained structural model will have to be computed anew.

But one might argue that the original structural model represents a stronger test than the measurement model, because it tests both the relations between the indicators and the latents and the relations of these indicators to other indicators as mediated by the latents. On the contrary, what is tested by chi-square tests are not the relations of paths with free coefficients between indicators and latents, and between indicators as mediated by free paths between latents, but the constraints on these paths, usually in the form of zero coefficients. So, hypothesizing that the path between two latents has a zero coefficient implies that the indicators of these latents should be unrelated, other things being equal. Such a hypothesis might not be incompatible with the original measurement model, involving hypotheses as to whether indicators have certain factors in common, since these are more fundamental. The relation between indicators and their common factors can stand whether or not the common factors themselves are causally connected or correlated in some way.

Testing Invariance across Groups of Subjects

Jöreskog (1971) developed ways of testing in factor analytic models whether parameters within a sample or across samples are equal (or invariant). Byrne (1994) has given a summary of this method. She notes that although originally Jöreskog recommended first testing the hypothesis that all group population covariance matrices are equal, Muthén (personal communication to Byrne, October 1988) indicates that this step is usually not necessary. It is always possible for there to be certain invariant parameters across groups, even though the population covariance matrices are not equal. For example, we have already determined that under restriction of range, factor pattern coefficients can be invariant across groups (Meredith, 1964), while variances and covariances among factors can vary across groups, with the groups subject to different forms of selection that influence the factor variables. So, one can begin with testing less general forms of invariance.

Central to Jöreskog's method was his discovery that he could estimate unconstrained free parameters individually within groups to minimize a maximum-likelihood fit function within each respective group. Or he could constrain a parameter to be equal across groups and need only estimate one value for that parameter that, when applied to the corresponding parameter in the respective group models, minimizes the weighted sum of all group fit functions across the groups. A total overall fit function value, weighting the individual group fit functions by sample sizes, is then obtained by summing across groups and used to obtain an overall chi-squared across all the groups. This procedure was first introduced in a simultaneous confirmatory factor analysis program, but it has subsequently been introduced in Jöreskog's *LISREL©* programs, and other commercial programs for structural equation modeling have followed suit.

The procedure to follow is analogous to the four-step procedure in using nested models. One will begin with the least restricted versions of a model and then introduce in a logical way additional constraints in subsequent models, until lack of fit is obtained. We will presume that the data come in several groups that differ in such things as sex, nationality, education level, occupation, prior experience, and cultural background. What we want to study is whether a causal structure is the same or different across the groups, and we would like to identify which parameters are invariant and which are not across the groups.

However, it is necessary to have a preliminary model to work with. Byrne (1994) suggests a procedure I use of first establishing a "baseline" or "calibration" model in a sample from one of the groups. I will presume that the researcher again has at least four indicators of each exogenous or endogenous latent. Then the four-step approach may be applied to the data in each one of the groups individually. The aim is to find a best-fitting model in at least one of the groups. This model will be used as the basis for comparisons across groups.

For the tests of equivalence, we will presume that such a baseline model has been found. We will now apply a four-step procedure to the multiple-group case. First, as suggested by Jöreskog (1971), we can test the unrestricted model, fixing the number of factors equal to the number of latents (excluding disturbances) in the baseline structural model. The test of the unrestricted model also simultaneously tests for uncorrelated uniquenesses, in other words, whether a common factor model of any kind is embedded in the structural model in connection with its specified number of latent variables and indicators of them. This model should be specified in the same manner in each group. No equality constraints are introduced, since these only concern estimated parameters, and the factor loadings of the unrestricted model have no special meaning. So, we are simply testing the same unrestricted model with the same fixed number of factors in each group. We are also testing the uncorrelated uniquenesses in each group and by implication the assumption of a self-contained model. If we obtain acceptable fit in each case to within sampling error or to a very high degree of approximation (in large samples), we can go forward. Failure to fit the unrestricted model in at least one group will require a pause to rethink whether using the same model for the manifest variables is still appropriate. It is possible to have different numbers of latents in different groups, even different causal structures, if some of the causal effects, measured by structural coefficients, are expected to be the same between certain variables across groups. There is considerable flexibility in modeling across groups. But we will set aside developing distinct models for the variables in each group and then testing invariance of certain parameters across groups, to consider a simpler case in which we assume we have the same or similar models across groups with the same number of latent variables.

Assuming one has acceptable unrestricted models in all the groups, one can then go to the measurement model tests. In the simultaneous test of measurement models across groups, variances of latent variables should remain free across groups. Suppose now that the measurement model in the calibration or baseline sample fixed the variances of latent factors to unity. How should we set the metric for each factor across the groups if we leave the variances of the latent factors free across the groups? We can do this for each factor by picking the largest estimated loading on each factor in the calibration sample and fixing all corresponding loadings on the respective factor across all groups to that value. This will insure that within the calibration group model the variances

will be estimated to equal unity, whereas the variances of corresponding factors in other groups can vary while in the same metrics. Differences in factor
variances across groups can have substantive import. Covariances between
factors should be left free across all groups.

Assuming that all measurement models with free factor loadings fit the
data acceptably, we may now consider constraining each corresponding factor loading across all groups to be equal. If these constraints do not yield
invariant loadings across the groups, then one should fall back to the previous unconstrained models. Technically with varying loadings in the same
metric across groups, the causes are not precisely the same, since a cause is
defined by its effect, and its effect is the degree to which a unit change in the
causal variable produces a change in the effect variable. In this case, one can
go on also to establish structural models within each group.

If one is only testing confirmatory factor analysis models, given equal
corresponding loadings across groups, a next step would be to constrain
corresponding unique variances to be equal across groups. Under restriction
of range with selection on variables that are causes of both the manifest and
the latent common factors, not only should the factor loadings be invariant across selected groups, but so should the unique variances because
of the invariance of errors of estimate under selection (Meredith, 1964,
1993). Covariances among factors need not be invariant under restriction
of range.

If the researcher has invariant measurement models across groups, he or
she can go on to test for the invariance of parameters corresponding to causal
paths between latents, by constraining these parameters to be equal, respectively, across groups. The social scientists rarely establish natural constants
and carry them over from one study to the next. To some extent this is because
their subject matter is much more complex and less well understood than the
subject matter of the physical sciences—although, at the cutting edge, in the
physical sciences there is just as much uncertainty and ambiguity as in the
behavioral and social sciences. The physical scientists have just been successful in gaining knowledge in some areas in their endeavors for a longer time
than in the social and behavioral sciences. But it may also be because the
behavioral and social sciences have not in fact searched for natural constants.
For example, is the factor loading of a judgment on a specific scale of friendliness invariant? It represents how much a unit change in the judgment of
friendliness will produce a change in the rating on the scale. Should that not
be invariant in any context in which the friendliness judgment is made on that
scale? If the numbers we obtain in our structural equation and factor analytic
studies are so ephemeral, what can we hope to generalize from the studies?
But we may hope for better.

In one study, Carlson and Mulaik (1993) estimated the loadings of certain scales on certain factors in a calibration condition, and then used these
loadings in later experimental conditions where other information was provided to influence judgments. For example, we estimated the loading on the

friendliness scale to be 0.902 (with the loading of the same judgment factor on the sympathy scale fixed to unity) in a calibration condition. This value was the value for this parameter in the best fitting model in this condition. Then we used the value of 0.902 as a fixed parameter value for the friendliness scale in two later experimental conditions using the same subjects. Similarly, other loadings on other scales were also estimated in a calibration condition and then carried over as fixed parameters in later experimental conditions applied to the same subjects.

We assumed that the semantics underlying verbal rating scales was stable and relatively invariant, so that the relationship between judgment and ratings on scales of that judgment would be invariant. With four indicators of each latent judgment, we were able to establish the objective validity of the latent judgment variable by confirmatory factor analysis, and we were able to use these fixed loadings successfully in different experimental conditions, as judged by the good fit for models using them as fixed values in these conditions.

So, suppose we used these same four scales of friendliness, with their fixed loadings on a friendliness judgment factor in another study that combined the judgments of friendliness with those of intelligence in a study of leadership. If we have natural constants for the loadings of the friendliness and intelligence scales on these dimensions, can we not use them as fixed values every time we use those scales? One major benefit is that we vastly increase our numbers of degrees of freedom, if we do. But another may be that we rule out equivalent models in those parts of our models involving these fixed parameters. The values will not fit just any data, but a specific kind of data for which they are applicable.

What we stipulate is that one cannot free those fixed parameters to get an equivalent model with different values for the constants using the current data. The fixed values are determined elsewhere in other studies with other data sets and experimental conditions. In those studies we have established these fixed values as conventions, as the way we are going to theorize and think about data of a certain kind from that point on, or until evidence arises that using them leads to major predictive failure. But as scientists, we also presume that they are invariants of nature and not just mere social conventions. So, if you want to offer an alternative model to the one I use with the specified fixed parameters obtained elsewhere, you have to carry out studies elsewhere also to find different fixed values and show that they also lead to successful predictions. But I will also insist, in addition to fit to within measurement error, that your model must have *much* better fit with nearly as many or more degrees of freedom than mine before I will concede to it.

So, my recommendation to researchers is that they seek to establish natural constants for structural coefficients, particularly in connection with the indicators of the measurement models. Use these as fixed values in other studies and see how far you can get with them.

Modeling Mean Structures

Up to now we have considered only covariance structures on variables where the means of variables are assumed to be zero. We will now consider the case where the means of the observed and common factor variables are not zero. Consider the model equation for a common factor analysis with nonzero means:

$$\mathbf{Y} = \mathbf{a}(1) + \mathbf{\Lambda X^*} + \mathbf{\Psi E}. \tag{8.2}$$

The $n \times 1$ vector $\mathbf{a}$ is a vector of constant intercepts for each of the n observed variables. (1) is the constant 1 which we will treat as if it is a variable whose only values are 1. For the present we will still assume that the means of the unique variables are zero, that is, $E(\mathbf{E}) = \mathbf{0}$. Furthermore, let us further assume that the random vector $\mathbf{X^*}$ is further decomposed into the following:

$$\mathbf{X^*} = [\mathbf{\mu}_X(1) + \mathbf{X}]. \tag{8.3}$$

We show a common factor model with nonzero means in Figure 8.8. The "variable" (1) is a constant, a degenerate variable, and may be regarded as the same "variable" as (1) in Equation 8.2. $\mathbf{\mu}_X$ may be interpreted as the factor pattern loadings of the variables on the variable (1). $\mathbf{X}$ is a random vector of deviations from the mean, that is, $E(\mathbf{X}) = 0$. In Equation 8.3 $\mathbf{X}$ may be

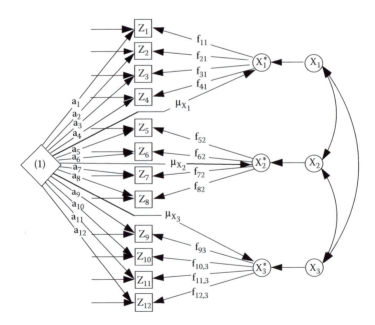

FIGURE 8.8 Common factor model with nonzero means.

treated as a vector of correlated disturbances on the common factors in $\mathbf{X}^*$. Equations 8.2 and 8.3 together define an SEM in which $\mathbf{X}^*$ is an endogenous, dependent variable whereas (1) is an exogenous "variable." To illustrate, in Figure 8.8 we show a common factor model for Equations 8.2 and 8.3.

Let us now take the expected value of both sides of Equation 8.2. This yields the equation

$$\mu_Z = \mathbf{a} + \mathbf{F}\mu_X. \tag{8.4}$$

The parameters of $\mathbf{a}$ and μ_X both contain unobserved values. They are $n + r$ in number. There are however only n observed values in μ_Y by which to determine the values of $\mathbf{a}$ and μ_X, and so, without additional constraints placed on the values of $\mathbf{a}$ and μ_X, these parameters are underidentified. Being the means of latent variables the values of μ_X are arbitrary. They could all be fixed to zero, in which case $\mathbf{a} = \mu_Y$. In any case the number of estimated parameters among the variable intercepts and latent factor means must be fewer than the number of observed variables (or observed sample means). On the other hand, all the elements of $\mathbf{a}$ could be given specified values, and then we would have the equation

$$\mathbf{\Lambda}\mu_X = \mu_Y - \mathbf{a}. \tag{8.5}$$

The solution for μ_X is then given as

$$\mu_X = (\mathbf{\Lambda}'\mathbf{\Lambda})^{-1}\mathbf{\Lambda}'(\mu_Y - \mathbf{a}).$$

Sample mean vectors may be substituted for the population mean vectors in obtaining estimates of the latent means. But if one distributes constraints across elements of $\mathbf{a}$ and μ_X, the estimates will have to be obtained by a structural equation modeling program. Generally, modeling intercepts and means of latent variables in single populations will not yield anything particularly meaningful unless there is well-developed theory to provide values for the means and intercepts.

Multigroup Comparisons of Mean Structures

Modeling mean structures is more meaningful in multigroup comparisons where relative differences between means can be considered. Ordinarily, with latent variables, absolute values for latent variable means are arbitrary and not very meaningful. One can, of course, test equivalence of intercepts and latent variable means along with tests of equivalence of other model parameters across groups.

Byrne (1994) described a multigroup model in which the model structure in each group is similar to the one described above. She recommended fixing the latent factor mean parameters [the loadings of the latent factors on the pseudovariable (1)] for one group to zero and leaving the corresponding mean parameters free in the other groups. The intercepts of each of the observed variables were constrained to be equal across groups. We show this in Figure 8.9, using a multigroup version of the model in Figure 8.8. The path diagram shown is applicable to each group. In the diagram, following a notation used by Byrne (1994), an * indicates a free parameter to be estimated separately within each group, *= indicates a parameter that is constrained to be equal across groups although its common value is free, and *0 indicates a parameter that is fixed to zero in one group and free in the others. 1.0 indicates a parameter fixed to 1.0 in each of the groups to set the metric of the latent factor. These constraints allow not only for the parameters to be identified and estimable, but also to overidentify them, allowing for tests. The latent mean values are fixed to 0's in one group and are free across the other groups.

It is important to note that constraining corresponding "factor loadings" and intercepts to be equal across groups is an essential assumption that must be met to be able to draw meaningful comparisons of the mean structures across the groups. The factors are essentially defined by their loadings on the observed variables (which are the same correspondingly across groups). If the factor loadings are not the same across groups, then the factor variables are not the same variables; hence the factor means are not comparable across

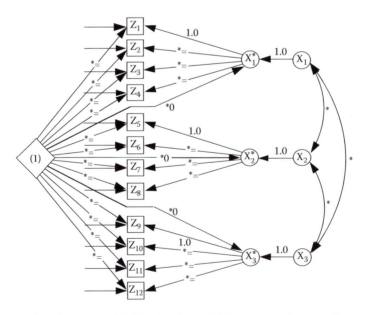

FIGURE 8.9 A multigroup model (showing the model in one group because all group model structures are the same). * indicates estimated parameter, *= indicates a parameter constrained to be equal, correspondingly, across groups, and *0 indicates a parameter correspondingly fixed to zero in one group and free in the other groups.

groups, because they correspond to different variables across the groups. Similarly, if the intercepts are not the same across groups, then the values of the factor means, which are determined, in part, by the intercept values and the observed means, will be arbitrary and, again, not comparable. The latent mean estimates, furthermore, are not in absolute terms, but relative to the group in which the latent means are fixed to zero.

9

Confirmatory Factor Analysis

Introduction

Confirmatory factor analysis is a special case of structural equation modeling. Although its conception has the same mathematical model as exploratory factor analysis, it differs considerably from exploratory factor analysis in how the model is used and the mathematics of its estimation. There is no searching of data for common factors via computing the eigenvalues and eigenvectors to get the estimates of factor loadings, factor correlations, and unique variances. Instead the researcher begins with a conception of a set of latent exogenous causal variables having specified effects on a set of endogenous manifest "indicator variables" and seeks to test his/her concept of the relation of the latent causes to the manifest indicators as a hypothesis. The hypothesis is formulated before seeing data on the manifest indicators. Usually the hypothesis is based in part on prior knowledge about variables such as the manifest indicators and their possible causes, but also possibly involves a new concept of how the causal variables combine together to be common causes of the indicators.

Relation of Common Factor Model to Structural Equation Model

The model equation for common factor analysis is usually given as

$$\mathbf{Y} = \mathbf{\Lambda}\boldsymbol{\xi} + \mathbf{\Psi}\boldsymbol{\varepsilon}$$

where $\mathbf{Y}$ is a $p \times 1$ vector of observed random variables, $\mathbf{\Lambda}$ is a $p \times n$ matrix of factor pattern loadings, $\mathbf{\xi}$ is an $n \times 1$ vector of common factor random variables, $\mathbf{\Psi}$ is a $p \times p$ diagonal matrix of unique factor pattern loadings, and $\mathbf{\epsilon}$ is a $p \times 1$ column vector of unique factor variables. The matrix of variances and covariances for $\mathbf{Y}$ is given by

$$\mathbf{\Sigma_{YY}} = \mathbf{\Lambda \Phi \Lambda'} + \mathbf{\Psi}^2$$

where $\mathbf{\Phi}$ is an $n \times n$ matrix of variances and covariances among the common factor variables while $\mathbf{\Psi}^2$ is a $p \times p$ diagonal matrix of unique factor variances. Because confirmatory factor analysis is a special case of a structural equation model, we rewrite the model equation in the notation of structural equation modeling as

$$\mathbf{Y} = \mathbf{\Gamma \xi} + \mathbf{\Psi \epsilon} \tag{9.1}$$

Compare this to the full structural equation model equation in Equation 5.7a:

$$\begin{bmatrix} \mathbf{\eta} \\ \mathbf{y} \end{bmatrix} = \mathbf{A} \begin{bmatrix} \mathbf{\eta} \\ \mathbf{y} \end{bmatrix} + \mathbf{\Gamma} \begin{bmatrix} \mathbf{\xi} \\ \mathbf{x} \end{bmatrix} + \begin{bmatrix} \mathbf{\Delta} & \mathbf{0} \\ \mathbf{0} & \mathbf{\Psi} \end{bmatrix} \begin{bmatrix} \mathbf{\zeta} \\ \mathbf{\epsilon} \end{bmatrix}$$

In the common factor model the dependent variables are observed only, hence no $\mathbf{\eta}$; no dependent variable is a function of other dependent variables, hence no $\mathbf{A} \begin{bmatrix} \mathbf{\eta} \\ \mathbf{y} \end{bmatrix}$; there are no manifest exogenous variables, hence no $\mathbf{x}$; and since there are no endogenous variables that are functions of other endogenous variables, there are no $\mathbf{\zeta}$ and thus no $\mathbf{\Delta}$. So, all that remains is what we see in Equation 9.1.

Early Attempts at Confirmatory Factor Analysis

L. L. Thurstone, one of the early developers and popularizers of factor analysis, did not believe in performing purely exploratory studies. He usually began a factor analytic study by formulating a hypothesis as to what common factors were likely to be found in the domain under study (Thurstone, 1951). He then either selected or constructed measures that he believed would reflect these common factors, making sure he had at least four measures of each factor anticipated so as to overdetermine the factor. He subjected the correlations among the measures to common factor analysis followed by rotation to oblique simple structure. Then he evaluated by eye the degree to which the resulting solution fulfilled his expectations.

Thurstone's use of exploratory methods to perform quasi-confirmatory studies had its limitations. The hypothesis was not explicitly specified in terms of model parameters. So it might not be clear to others reading

Thurstone's papers that a hypothesis was being tested as opposed to a simple exploratory study being done. There was also no explicit measure of the degree to which the hypothesis had been supported or disconfirmed by the data. That determination was totally subjective.

Multiple-Group Factor Analysis

A primitive form of confirmatory factor analysis is multiple-group factor analysis (Tryon, 1939; Holzinger, 1944; Thurstone, 1949), formulated initially as a multiple-factor extraction method to speed up the factor extraction process that ordinarily extracted one factor at a time. "Group" here refers to "group factor" and not to "group of subjects." Guttman (1952) explicitly described the multiple-group method as a confirmatory factor analysis method and worked out its algorithm in detail. It is surprising that despite its simplicity of implementation, the method was not used more than occasionally in the 1950s and 1960s. The idea was that the researcher could hypothesize what variables would be dependent on what factors and then proceed to find centroids among the groups of variables representing various factors, respectively. The centroids would represent the hypothesized factors. There has been, however, a revival of interest in this method as either a substitute for or a preliminary step of a confirmatory factor analysis procedure. Hunter and Gerbing (1982) advocated the method as an inexpensive computational alternative to full information maximum-likelihood estimation in both confirmatory factor analysis and linear structural equations modeling. Their method employed two stages: (1) a multiple-group factor analysis to obtain estimates of the factor pattern matrix and the factor correlation matrix for a hypothesized measurement model of a structural equation model and (2) two-stage least-squares estimation of the structural relations between the latent variables based on the correlations among the latent variables found in the first stage. Mulaik (1988) described an iterative variant of the multiple-group method. McDonald and Hartmann (1992) described a one-pass, noniterative variant of the multiple-group factor analysis method as a method for obtaining starting values for unknown parameters in structural equation modeling. The principles underlying the multiple-group method are also closely related to those of partial least squares (PLS) estimation as developed by Herman Wold (1975). An excellent comprehensive discussion of the PLS method with full references is given by Lohmöller (1989).

An Example of Confirmatory Factor Analysis

In Chapter 1, we were introduced to a structural equation modeling study by Carlson and Mulaik (1993). We are going to use data from that study to illustrate confirmatory factor analysis, although the original study involved a structural equation model. For details of the study refer back to Chapter 1. The correlations among the 15 five-point trait-rating scales of that study are shown in Table 9.1.

Now, what differs from an exploratory factor analysis at this point is that we specify a hypothesis, based on the information given above, involving how the factor pattern matrix will have certain loadings. Carlson and Mulaik (1993) were able to specify or *fix* zero loadings in certain positions of the factor pattern matrix. They, of course, implicitly constrained correlations between the common factors and unique factors to be zero and correlations among the unique factors to be zero. However, they did not have any knowledge of the magnitudes of certain other loadings, and so they left these loadings *free*. This meant the free parameters were to be estimated conditional on the fixed parameters. The values of the correlations among the common factors were also treated as free parameters, as were also the unique factor variances.

Table 9.2 shows the hypothesis specified for the factor pattern loadings, which in this case were only zero loadings. Free parameters are indicated by "?." Note that all unique factor variances are free, as are also correlations among the common factors. In this case the hypothesis only indicated what indicators were not related to what latent common factors by specifying zero loadings. Free parameters, on the other hand, are "filler" for the model and are not part of the hypothesis, since nothing is specified for them by freeing them. Free parameters are estimated by an iterative algorithm that seeks to find estimates for the free parameters that minimize an overall discrepancy or lack-of-fit function *conditional* on the fixed parameters of the model. The discrepancy function measures the degree of lack of fit between the observed correlation matrix for the observed variables and the model's reproduced correlation matrix for these variables. In short, one

TABLE 9.1

Correlations Among 15 Trait Scales (Carlson and Mulaik, 1993)

```
 1 1.000  Friendly
 2 0.777 1.000  Sympathetic
 3 0.809 0.869 1.000  Kind
 4 0.745 0.833 0.835 1.000  Affectionate
 5 0.176 0.123 0.123 0.112 1.000  Intelligent
 6 0.234 0.159 0.205 0.183 0.791 1.000  Capable
 7 0.243 0.155 0.187 0.186 0.815 0.865 1.000  Competent
 8 0.234 0.190 0.238 0.215 0.818 0.841 0.815 1.000  Smart
 9 0.433 0.319 0.321 0.435 0.174 0.209 0.239 0.258 1.000  Talkative
10 0.473 0.480 0.410 0.527 0.220 0.274 0.269 0.261 0.744 1.000  Outgoing
11 0.433 0.438 0.406 0.526 0.188 0.227 0.242 0.228 0.711 0.853 1.000  Gregarious
12 0.447 0.396 0.350 0.500 0.192 0.221 0.227 0.224 0.758 0.846 0.801 1.000  Extraverted
13 0.649 0.693 0.697 0.694 0.283 0.344 0.370 0.365 0.443 0.552 0.514 0.473 1.000  Helpful
14 0.662 0.692 0.676 0.679 0.311 0.345 0.375 0.351 0.431 0.557 0.514 0.493 0.740 1.000  Cooperative
15 0.558 0.543 0.510 0.632 0.213 0.289 0.287 0.287 0.745 0.886 0.820 0.830 0.631 0.626 1.000  Sociable
```

TABLE 9.2

Hypothesized Factor Pattern Loadings and Free Parameters for a Confirmatory Factor Analysis of the Correlations Among 15 Trait-rating Scales from the Carlson and Mulaik (1993) Study

Variable	Factor 1	Factor 2	Factor 3	Unique Variances
1 Friendly	?	**0**	**0**	?
2 Sympathetic	?	**0**	**0**	?
3 Kind	?	**0**	**0**	?
4 Affectionate	?	**0**	**0**	?
5 Intelligent	**0**	?	**0**	?
6 Capable	**0**	?	**0**	?
7 Competent	**0**	?	**0**	?
8 Smart	**0**	?	**0**	?
9 Talkative	**0**	**0**	?	?
10 Outgoing	**0**	**0**	?	?
11 Gregarious	**0**	**0**	?	?
12 Extraverted	**0**	**0**	?	?
13 Helpful	?	?	**0**	?
14 Cooperative	?	?	**0**	?
15 Sociable	?	**0**	?	?

Correlations among factors

Factor	1	2	3
1	**1.000**		
2	?	**1.000**	
3	?	?	**1.000**

Note: Hypothesized (fixed) parameters are shown in bold and free parameters as question marks.

seeks estimates of free parameters that minimize the lack of fit between the reproduced correlation matrix and the observed correlation matrix, consistent with the constraints on the fixed parameters. Within the constraints given by the fixed and constrained parameters the iterative algorithm is free to seek any values for the free parameters, which in some cases could be zero or of whatever sign. In this way, any lack of fit can then be attributed to the fixed or constrained parameters, because otherwise, without these constraints, and assuming that the model is still identified (a concept to be touched on shortly), the reproduced correlation matrix would then fit the sample correlation matrix perfectly, by mathematical necessity, but not empirical necessity.

Note also that Table 9.2 does not show fixed zero correlations between the common and unique factors or among the unique factors. These are fundamental assumptions of most applications of the common factor model. These fixed zero correlations among the unique factors and between the common and unique factors contribute to making the model identified as well as overidentified.

The model specified in Table 9.2 was then applied to the sample correlation matrix and analyzed using Bentler's EQS© program for structural equation modeling. Confirmatory factor analysis is but a special case of a structural equation model, which is a very general kind of model that allows considerable freedom in representing linear causal relationships. The result of the analysis is shown in Table 9.3.

The chi-square test for goodness of fit of the reproduced correlation matrix based on the above parameter values with the observed correlation matrix yielded a chi-square of 225.115 with 84 degrees of freedom, with a p value <0.001, which was significant. The model did not fit to within sampling error. (A nonsignificant chi-square is what is desired.) However, Bentler's CFI index of goodness of fit was 0.968, which is usually considered quite good as an approximation. However, the parsimony ratio of degrees of freedom

TABLE 9.3

Results of a Confirmatory Factor Analysis of the Model in Table 9.2 Applied to the Correlations Among the 15 Personality Rating Variables in Table 9.1

	Factor 1	Factor 2	Factor 3	Unique Variances
1 Friendly	0.849*	0	0	0.529*
2 Sympathetic	0.922*	0	0	0.386*
3 Kind	0.925*	0	0	0.381*
4 Affectionate	0.901*	0	0	0.434*
5 Intelligent	0	0.877*	0	0.480*
6 Capable	0	0.925*	0	0.381*
7 Competent	0	0.925*	0	0.381*
8 Smart	0	0.903*	0	0.430*
9 Talkative	0	0	0.803*	0.596*
10 Outgoing	0	0	0.948*	0.317*
11 Gregarious	0	0	0.891*	0.454*
12 Extraverted	0	0	0.898*	0.439*
13 Helpful	0.725*	0.222*	0	0.594*
14 Cooperative	0.715*	0.229*	0	0.603*
15 Sociable	0.172*	0	0.833*	0.342*

Correlations among factors

Factor	1	2	3
1	1.000		
2	0.223*	1.000	
3	0.556*	0.302*	1.000

Note: Estimates of free parameters are shown with asterisks following them. Coefficients are given for a solution where all variables, both observed and latent, are rescaled to have unit variances.

divided by $p(p + 1)/2$ was $84/120 = 0.700$, which is not indicative of a highly tested model. (A parsimony ratio of 0.95 or greater would be better, and combined with a high degree of fit, say, CFI > 0.95, this would then be a highly tested model with very good fit.) So, even if the goodness of fit is strong, as an approximation, the model awaits further development and specification of parameters to fully test it.

It is possible to evaluate the fixed parameters to determine if some are contributing significantly to the lack of fit. In Bentler's EQS program this is accomplished by the use of *Lagrange multiplier tests*. In LISREL©, Jöreskog's structural equation program, an analogous test uses what are called *modification indices*. Since the above data were analyzed using EQS, the Lagrange multiplier tests were performed on each fixed parameter, and these in turn were sorted in descending order of magnitude. The way to use the results of the Lagrange multiplier test is to free the fixed parameter with the largest single-degree-of-freedom chi-square among the Lagrange multiplier tests, and reanalyze the resulting model. This was done several times, each time freeing the fixed parameter remaining with the largest chi-square value. We should stop if the model achieves a nonsignificant chi-square. We should also use this procedure judiciously, in that we do not want to free too many fixed parameters, because freeing a fixed parameter results in the loss of one degree of freedom. Lagrange multiplier tests and modification indices are further discussed in Chapter 15.

A degree of freedom corresponds to a condition in which the model is free to differ from the observed correlation matrix. The degrees of freedom df for the model are computed as

$$df = p(p + 1)/2 - k$$

where p is the number of observed variables and k is the number of free parameters in the model. $p(p + 1)/2$ is the number of nonredundant elements on the diagonal and off one side of the diagonal of the correlation matrix that the model is fitted to. If no parameters are estimated, corresponding to a model with all fixed parameters, the degrees of freedom would be a maximum of $p(p + 1)/2$. This would correspond to a maximally testable model for the number of variables in the correlation matrix.

If as many parameters are estimated as there are distinct elements of the correlation matrix to be fitted, the degrees of freedom will be zero and the free model parameters will always be able to be adjusted to allow the model to fit perfectly to the observed correlation matrix. Such a model would be useless for a test, since no hypothesis has been asserted that could be disconfirmed by a possible lack of fit.

So, a degree of freedom corresponds to a condition by which the model is tested for goodness of fit. What we hope to attain is a model with numerous degrees of freedom, relative to the potential number of degrees of freedom (i.e., a high parsimony ratio) that we could have, and excellent fit. Freeing up fixed parameters results in a weaker model, since fewer hypothesized constraints remain by which the model potentially could be tested, which is indicated by fewer degrees of freedom.

Nevertheless, we were able, after freeing a number of fixed parameters as indicated by the use of Lagrange multiplier tests, to find parameters to free to obtain a model that fit with a nonsignificant chi-square. The results are shown in Table 9.4.

How would we interpret the final model? The fact that few zero coefficients in the factor pattern matrix contributed to lack of fit, as indicated by the

TABLE 9.4

Final Standardized Estimated Model Obtained by Freeing Eight Fixed Parameters According to Lagrange Multiplier Tests

Variables	Factor 1	Factor 2	Factor 3	Unique Variances
1 Friendly	0.834*	0	0	0.552*
2 Sympathetic	10.052*	0	−0.210*	0.360*
3 Kind	1.069*	0	−0.249*	0.364*
4 Affectionate	0.907*	0	0	0.421*
5 Intelligent	0	0.868*	0	0.497*
6 Capable	0	0.908*	0	0.418*
7 Competent	0	0.949*	0	0.316*
8 Smart	0	0.932*	0	0.363*
9 Talkative	0	0	0.807*	0.591*
10 Outgoing	0	0	0.948*	0.318*
11 Gregarious	0	0	0.889*	0.458*
12 Extraverted	0	0	0.898*	0.440*
13 Helpful	0.727*	0.205*	0	0.596*
14 Cooperative	0.717*	0.207*	0	0.607*
15 Sociable	0.193*	0	0.808*	0.336*

Correlations among factors

Factor	1	2	3
1	**1.000**		
2	0.251*	**1.000**	
3	0.633*	0.296*	**1.000**

Freed correlations among unique factors

$\rho(E3, E1)$	0.215*
$\rho(E9, E1)$	0.181*
$\rho(E10, E2)$	0.375*
$\rho(E8, E7)$	−0.619*
$\rho(E9, E8)$	0.236*
$\rho(E14, E13)$	0.281*

Note: Degrees of freedom have been reduced from 84 to 76. Parsimony ratio is now only 0.63. Chi-square is 91.73 for 76 df with $p = 0.113$. CFI = 0.997.

Lagrange multiplier tests, suggests that there is some support for the theory behind the three factors hypothesized. It seems that the low negative loadings of the variables *sympathetic*, and *kind* on outgoingness, suggest that a person with a higher degree of outgoingness will be modestly less likely to be seen as sympathetic and kind. Perhaps people who are sympathetic and kind are not seen to be as outgoing and sociable. However, we also note that there is a moderate correlation between friendliness and outgoingness. This leaves some freedom for an outgoing person to be a critic of others or garrulous but not as sympathetic and kind as it would seem. Carlson and Mulaik (1993) hypothesized in other parts of their study that *helpful* and *cooperative* would fall on a separate factor to some degree, since they believed ratings on these scales could be driven in part by information as to the actual helpfulness or cooperativeness of the person alone apart from information as to the person's friendliness and ability. So the correlation between $E13$ and $E14$ is not surprising.

Faceted Classification Designs

In psychology, factor analysis has made some of its most substantial contributions to the study of mental abilities as measured by tests of intellectual performance. Using exploratory factor analysis, psychologists have discovered that intellectual performance is highly complex and dependent on many factors. Guilford (1967), for example, claimed to have established 82 factors of the intellect but postulated the existence of at least 120 factors. Guttman (1965) went farther than Guilford in conjecturing that the potential number of factors of the intellect is almost unlimited. Paradoxically, then, the technique of factor analysis, which originally was developed by psychologists such as Thurstone to help simplify the conceptualization of intellectual processes, had by the late 1960s achieved quite the opposite effect of inducing psychologists to consider intellectual processes as even more complex than they originally believed. As a result, faced with the growing plethora of factors coming from psychological laboratories using factor analysis, many psychologists began to doubt the value of factor analysis as a technique, with the question: What kind of coherent theory of mental processes can we construct with so many factors?

In response to such criticism, both Guilford (1967), with his structure-of-the-intellect model, and Guttman (1965), with his faceted definition of intelligence, attempted to bring order out of chaos by organizing mental tests (and their associated factors) into classification schemes that have some explanatory, theoretical significance. In doing so, Guilford (implicitly) and Guttman (explicitly) took the position that a coherent theory of mental processes will not come automatically from the methods of exploratory factor analysis but will have to be constructed on the basis of other, theoretical grounds. For example, Guttman (1965) recommended that researchers seek to identify common external features of intellectual-performance tests, which may be used to classify these tests in the expectation that tests having the most classes in common will be intercorrelated the most, and those having the least in common will be intercorrelated the least [this is Guttman's (1959) contiguity metahypothesis]. Guilford (1967) appeared to use a kind of information-processing model of intellectual processes to guide him in specifying which tests elicit certain hypothetical intellectual processes. Guilford categorized tests according to their content (figural, symbolic, semantic, or behavioral), the mental operation involved (evaluation, convergent production, divergent production, memory, or cognition), and the kind of product elicited (units, classes, relations, systems, transformations, or implications).

In effect, drawing upon computer terminology, we might regard Guilford's system as classifying tests according to the kind of input involved, the kind of operations carried out, and the kind of output produced. In more elaborate forms, such an input–operation–output model may have some value in suggesting further research. In analogy with Guilford, who identified a factor of the intellect with each three-way combination of content, operation, and product, the input–operation–output model may identify a factor with each independent pathway through the system.

Interactions thus represent levels of functioning on pathways linking process centers, while main effects represent the levels of function of the centers themselves. The general level effect is a measure of the level of functioning of the whole system.

At this point the reader may see an analogy between a classification design and an analysis of variance design. In the present case variables are classified, while in analysis of variance, observations are classified. Nevertheless, the analogy suggests that we may anticipate common factors to correspond to the effects of the comparable analysis of variance design: there will be a general factor corresponding to the grand mean common to all observations; there will be main effect factors corresponding to main effects, and "interaction" factors corresponding to interaction effects in analysis of variance.

However, as in analysis of variance, not every effect anticipated by the design of the experiment may be present, and so not every common factor corresponding to an "effect" of the design may be present either. Thus the researcher will need to justify including a common factor suggested by the classification design, using substantive theory.

For further discussion of classification designs, see Mulaik (1975) and Mellenbergh et al. (1979).

Multirater–Multioccasion Studies

Researchers may be concerned in establishing objective evidence for the stability of personality characteristics over time. To add further to establishing objectivity, they will often use more than one rater on a given occasion to describe the subjects. And beyond this, each rater may make several ratings on scales designed to measure the same trait, which also contributes to the objectivity of the final result by having the rater make several ratings based on the same judgment.

The author collaborated with a psychologist working with a group of nursing educators in the western United States (Ingmire et al., 1967) who were seeking to evaluate a continuing education program they were conducting in several states. They had conceived a situational exercise in which their workshop participants were to interview a patient, played by a nurse serving as a patient-actor. After performing in the exercise, the patient-actor and an external observer made ratings of the participant on several five-point rating scales designed to rate the nurse's supportiveness of the patient. The situational exercise was given to the participants both at the beginning of the series of workshops (which were conducted several times during a year) and at the end of the year. The researchers then decided they also needed a control group, and obtained 352 otherwise eligible nurses who had not participated in the workshop to undergo evaluation in the same situational exercise with the same patient-actors and external observers. A possible source of uncontrolled variation was that in each state a different patient-actor and a different external observer were used. Like the workshop participants, the control group subjects were given the same exercise twice, a year apart. The research question we may ask is whether there is an enduring trait of "supportiveness" by which nurses may vary, yet be somewhat stable over a year's time within a given nurse. The control group would establish the stability. The table of correlations among the 22 variables used in this study is given in Table 9.5.

Several years later (Mulaik, 1988) I used data from the control group subjects to illustrate a confirmatory factor analysis involving a multirater–multioccasion model. I will now perform a new analysis of those data with an even more constrained model, with more degrees of freedom, that upholds essentially the same results as in Mulaik (1988).

A path diagram of the multirater–multioccasion model applied to the correlation matrix in Table 9.5 is shown in Figure 9.1. The diagram shows that on each of two occasions, year 1 and year 2, the rated subject's behavior causes the patient-actor and the external observer to form judgments of the subject's supportiveness, which in turn caused each of the raters to make their ratings on the respective rating scales. Three invariants were hypothesized: (1) *Stimulus to judgment invariance*: the subject's supportive behavior in year 1 would cause the same proportional degree of supportiveness judgment in the rater in year 2. Other aspects of the rater's judgment were idiosyncratic with the rater. Consequently the path coefficient from X_5 (subject's behavior in year 1) to X_1 (patient-actor's judgment in year 1) was constrained to equal the path coefficient from X_6 (subject's behavior in year 2) to X_2 (patient-actor's judgment in year 2). Likewise, the path coefficient from X_5 (subject's behavior in year 1) to X_3 (external observer judgment in year 1) was constrained to

TABLE 9.5

Correlations Among 22 Rating Variables of 352 Nurses' Supportive Behavior in a Situational Exercise Conducted on Two Occasions a Year Apart*

Patient-actor ratings year 1
1 100 Communicating understanding
2 64 100 Friendliness
3 73 65 100 Supportive
4 64 57 72 100 Personal involvement
5 71 61 71 62 100 Security as patient

Patient-actor ratings year 2
6 32 27 34 27 29 100 Communicating understanding
7 24 24 31 24 23 67 100 Friendliness
8 24 24 31 22 25 72 70 100 Supportive
9 19 24 26 23 21 64 62 71 100 Personal involvement
10 31 28 34 29 31 73 72 79 66 100 Security as patient

External observer ratings year 1
11 44 36 46 39 39 32 31 31 25 34 100 Respect for individual
12 48 40 47 39 44 36 30 31 23 31 68 100 Encourages patient to talk
13 45 39 51 40 42 31 25 26 20 28 71 70 100 Recognizes need for security
14 42 39 47 36 40 36 24 28 31 30 61 67 65 100 Information seeking
15 47 48 51 43 47 37 28 32 26 33 72 73 72 69 100 Supportive
16 49 43 51 43 49 33 27 29 23 33 73 76 75 69 88 100 Understanding

External observer ratings year 2
17 20 24 21 10 20 45 50 47 40 49 34 33 25 25 32 29 100 Respect for individual
18 21 19 24 17 21 50 47 50 42 52 31 39 32 31 34 34 68 100 Encourages patient to talk
19 24 27 26 16 25 45 41 46 38 52 29 33 33 27 31 31 66 71 100 Recognizes need for security
20 19 22 26 12 18 49 38 52 44 47 31 32 31 31 32 33 65 70 64 100 Information seeking
21 24 25 26 17 27 52 52 59 49 61 32 44 36 32 37 36 69 73 73 68 100 Supportive
22 20 25 27 17 27 54 51 57 49 61 32 39 33 32 36 34 71 75 74 72 87 100 Understanding

* Decimal points have been omitted.

---

equal the path coefficient from X_6 (subject's behavior in year 2) to X_4 (external observer's judgment in year 2). (2) *Semantic invariance principle*: the relation between judgment and rating for a given rating scale would be the same on each occasion a year apart. Hence, for example, the factor loading for variable Y_2 (friendliness) on factor X_1 was constrained to equal the factor loading for variable Y_7 (friendliness) on factor X_2. Comparable constraints of equality were made for all factor loadings for comparable pairs of corresponding rating variables. (3) *Invariance of variances of disturbances* of corresponding variables a year apart. Hence, for example, the error variance of ε_2 on variable Y_2 was constrained to equal the error variance of ε_7 on variable Y_7. Similarly, the

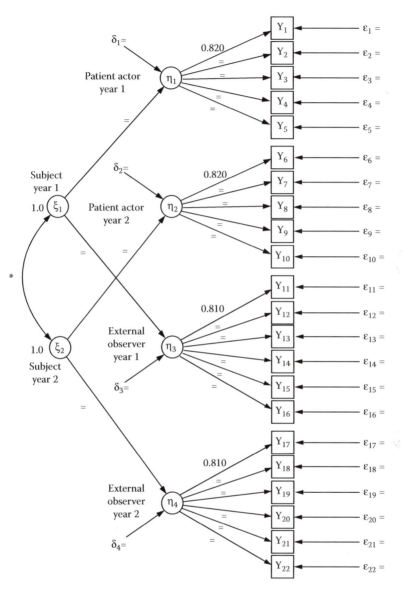

FIGURE 9.1 Path diagram for a model of nurses' ratings by a patient-actor and an external observer in a nurse-patient situational excise conducted twice, a year apart. Numerical values for coefficients indicate fixed parameter values; = denotes parameter constrained equal to its counterpart a year apart; * denotes a free parameter.

variance of the second-order disturbance δ_1 on X_1 was constrained to equal the comparable variance of the second-order disturbance δ_2 on X_2, and so on. The correlation between a subject's supportiveness behavior in year 1 with that in year 2 was a free parameter.

232

Linear Causal Modeling with Structural Equations

TABLE 9.6

First- and Second-order Factor Pattern Matrices, First- and Second-order Unique
Variances, and Correlation between Second-order Common Factors

	First-order factors					Second-order					Correlation	
	$\eta 1$	$\eta 2$	$\eta 3$	$\eta 4$	ψ^2		$\xi 1$	$\xi 2$		δ^2		$\xi 1$	$\xi 2$
1	**820**	**000**	**000**	**000**	306=	1	791=	**000**	1	407=	1	**100**	
2	762=	**000**	**000**	**000**	401=	2	**000**	791=	2	407=	2	58*	**100**
3	870=	**000**	**000**	**000**	219=	3	841=	**000**	3	246=			
4	763=	**000**	**000**	**000**	400=	4	**000**	841=	4	246=			
5	839=	**000**	**000**	**000**	274=								
6	**000**	**820**	**000**	**000**	306=								
7	**000**	762=	**000**	**000**	401=								
8	**000**	870=	**000**	**000**	219=	$\chi^2_{228} = 409.636$		$p < 0.001$					
9	**000**	763=	**000**	**000**	400=								
10	**000**	839=	**000**	**000**	274=	CFI $= 0.972$							
11	**000**	**000**	**810**	**000**	374=								
12	**000**	**000**	847=	**000**	316=	RMSEA $= 0.047$							
13	**000**	**000**	834=	**000**	336=								
14	**000**	**000**	792=	**000**	402=	PR $= 0.901$							
15	**000**	**000**	935=	**000**	167=								
16	**000**	**000**	952=	**000**	135=								
17	**000**	**000**	**000**	**810**	374=								
18	**000**	**000**	**000**	847=	316=								
19	**000**	**000**	**000**	834=	336=								
20	**000**	**000**	**000**	792=	402=								
21	**000**	**000**	**000**	935=	167=								
22	**000**	**000**	**000**	952=	135=								

Note: Decimal points have been omitted.
Chi-square, CFI index, and Parsimony Ratio (PR) are given.
Fixed parameters in bold. = denotes constrained equal; * denotes free parameter.

The metric of the factor loadings was set by fixing the loading of Y_1 on X_1
and the loading of Y_6 on X_2 to 0.820. Similarly, the loadings of Y_{10} on X_3 and
of Y_{17} on X_4 were both fixed to 0.810. These values were determined by an
exploratory factor analysis as being values consistent with an approximate
variance of unity for the common factor in question. But the values were also
set to be equal according to the semantic invariance principle. (Otherwise
setting the metric is arbitrary.)

A confirmatory factor analysis of the model with these constraints was
performed using the EQS program, Version 5.1, on an iMac computer. The
results are shown in Table 9.6.

Although the chi-square goodness-of-fit statistic is significant with $p <$
0.001, indicating significant lack of fit, the model is still a very good

approximation to the data, as indicated by the fit indices, CFI = 0.972 and Root mean square error of approximation (RMSEA) = 0.047. The model has plausibility in that it conforms well to a stimulus–response paradigm where the nurse-subject provides a common stimulus for two raters to rate, and their judgments as responses are then further expressed in ratings on several scales, which conform well to a single common factor among them. Lack of fit may be explained by small departures from causal homogeneity among the nurse-subjects and also by a few replacements of the patient-actor and external rater from one year to the next in a few state regions. Nevertheless, the model is highly tested with a parsimony ratio of 0.901. Thus the gem revealed by the study is an estimated correlation of 0.58 between the nurse-subject's objective supportiveness behavior measured twice over a year apart. This was also the finding in Mulaik (1988). The multirater–multioccasion paradigm seems to be an excellent research paradigm for studying trait stability. One may wonder why a multilevel analysis has not been done to deal with different regions and different actors and observers in the situational exercise. Unfortunately, the original data that would make that possible are no longer in existence.

Multitrait–Multimethod Covariance Matrices

In seeking to formulate methods for establishing the objective validity of certain trait-rating constructs, Campbell and Fiske (1959) recommended that researchers obtain ratings on the same set of trait scales by more than one method. They called the correlation matrix among the trait-rating scales, measured under the several methods, a multitrait–multimethod (MTMM) correlation matrix. They argued that the objective validity of the ratings would consist in the coherence of ratings on corresponding trait scales across the different methods. They described four conditions supporting the objectivity or *convergent validity* of the ratings: (1) Correlations between corresponding scales under different methods should be nonzero and statistically significant. (2) Correlations between corresponding scales under different methods should be higher than those between dissimilar scales compared across the different methods. (3) Correlations between corresponding scales measured under different methods should be higher than correlations measured within the same method between different scales. (4) The pattern of correlations between scales in different methods should be similar to the pattern of correlations between scales within the same method.

The limitations of the Campbell and Fiske (1959) approach lay principally in the subjective way of determining convergent validity by merely inspecting the correlation matrix. Also, correlations were used, which would often force corresponding scale ratings into different metrics, by dividing covariances between the respective scales by different standard deviations within and between different methods. Corresponding variables might also have different reliabilities under different methods, varying then the pattern of relative magnitudes of correlations across different methods, even when there is convergent validity. Nevertheless, various researchers sought to develop ways of analyzing MTMM matrices to provide more objective ways of assessing convergent validity. A good early review of this topic is given by Schmitt, Coyle, and Saari (1977).

Jöreskog (1974) demonstrated how an MTMM correlation matrix used by Campbell and Fiske (1959) to illustrate their method could be modeled by a confirmatory factor analysis model. Campbell and Fiske (1959) took data from a study conducted by Kelly and Fiske (1951) of the ratings of 124 clinical psychology students in a clinical setting on five different trait scales using three methods: staff ratings, peer ratings, and self-ratings. The correlation matrix among these scales is shown in Table 9.7.

Jöreskog (1974) hypothesized five trait factors, T1, ..., T5, one for each of the five traits rated. Corresponding scales were hypothesized to load on the same factor, so there were three indicators of each trait factor across the three methods of rating. Presumed nonzero loadings were treated as free parameters, no effort being made to prespecify a value to test for each of these loadings. Scales not corresponding to a factor in question were hypothesized to have zero loadings on the factor. Jöreskog also initially hypothesized three method factors, M1, M2, and M3, with scales rated by the same method presumed to load on the same method factor. Factor variances were fixed to unity to determine the metric of the solution. Correlations among trait factors were treated as free parameters, and correlations among method factors were also treated as free parameters. Correlations between trait factors and method factors were fixed to zero.

In his initial estimation of the free parameters of this model, Jöreskog obtained a solution in which method factors M1 and M3 were correlated 1.00 with each other. So, he reparameterized his model to postulate two method factors, with scales under method 1 (staff ratings) and method 3 (self-ratings) loading on the same method factor M1 (in the new model) and scales rated by teammates loading on method factor M2.

I have reestimated the parameters of Jöreskog's model for the correlation matrix given in Table 9.7 using Bentler and Wu's EQS 5 program, and I have obtained essentially the identically same solutions as Jöreskog (1974). Initially when I postulated three method factors, they yielded estimated correlations of unity between methods 1 and 3. So, I reparameterized, as did Jöreskog (1974), to have only two method factors and the results are shown in Table 9.8.

The chi-square test of goodness of fit with 64 degrees of freedom was 61.512, which was not significant. Several goodness of fit indices indicated excellent fit as well: Jöreskog's GFI index was 0.941, whereas Bentler's CFI was 1.00. Some of our enthusiasm for the good fit of this model should be tempered by the knowledge that the sample size was only 124, which, being less than 200, is currently regarded as inadequate for statistical inference purposes with chi-square statistics in confirmatory factor analysis. But parameter

TABLE 9.7

MTMM Correlation Matrix Based on Ratings of Clinical Psychology Students on Five Traits by Three Kinds of Raters (after Campbell and Fiske, 1959)

	1P	2P	3P	4P	5P	1T	2T	3T	4T	5T	1S	2S	3S	4S	5S
Staff ratings															
Assertive 1P	1.00														
Cheerful 2P	0.37	1.00													
Serious 3P	−0.24	−0.14	1.00												
Unshakeable poise 4P	0.25	0.46	0.08	1.00											
Broad interests 5P	0.35	0.19	0.09	0.31	1.00										
Teammate ratings															
Assertive 1T	0.71	0.35	−0.18	0.26	0.41	1.00									
Cheerful 2T	0.39	0.53	−0.15	0.38	0.29	0.37	1.00								
Serious 3T	−0.27	−0.31	0.43	−0.06	0.03	−0.15	−0.19	1.00							
Unshakeable poise 4T	0.03	−0.05	0.03	0.20	0.07	0.11	0.23	0.19	1.00						
Broad interests 5T	0.19	0.05	0.04	0.29	0.47	0.33	0.22	0.19	0.29	1.00					
Self ratings															
Assertive 1S	0.48	0.31	−0.22	0.19	0.12	0.46	0.36	−0.15	0.12	0.23	1.00				
Cheerful 2S	0.17	0.42	−0.10	0.10	−0.03	0.09	0.24	−0.25	−0.11	−0.03	0.23	1.00			
Serious 3S	−0.04	−0.13	0.22	−0.13	−0.05	−0.04	−0.11	0.31	0.06	0.06	−0.05	−0.12	1.00		
Unshakeable poise 4S	0.13	0.27	−0.03	0.22	−0.04	0.10	0.15	0.00	0.14	−0.03	0.16	0.26	0.11	1.00	
Broad interests 5S	0.37	0.15	−0.22	0.09	0.26	0.27	0.12	−0.07	0.05	0.35	0.21	0.15	0.17	0.31	1.00

TABLE 9.8

Factor Loadings and Correlations Among Factors in the MTMM Model (Jöreskog, 1974) for Correlations in Table 9.7

	Factor Pattern Loadings							
	T1	T2	T3	T4	T5	M1	M2	u_{ii}^2
Staff ratings								
Assertive	0.871	**0.000**	**0.000**	**0.000**	**0.000**	0.107	**0.000**	0.239
Cheerful	**0.000**	0.836	**0.000**	**0.000**	**0.000**	0.017	**0.000**	0.302
Serious	**0.000**	**0.000**	0.573	**0.000**	**0.000**	−0.296	**0.000**	0.583
Unshakeable poise	**0.000**	**0.000**	**0.000**	0.781	**0.000**	−0.253	**0.000**	0.318
Broad interests	**0.000**	**0.000**	**0.000**	**0.000**	0.689	−0.335	**0.000**	0.392
Teammate ratings								
Assertive	0.829	**0.000**	**0.000**	**0.000**	**0.000**	**0.000**	0.162	0.291
Cheerful	**0.000**	0.697	**0.000**	**0.000**	**0.000**	**0.000**	0.294	0.468
Serious	**0.000**	**0.000**	0.722	**0.000**	**0.000**	**0.000**	0.322	0.349
Unshakeable poise	**0.000**	**0.000**	**0.000**	0.213	**0.000**	**0.000**	0.533	0.674
Broad interests	**0.000**	**0.000**	**0.000**	**0.000**	0.599	**0.000**	0.439	0.429
Self-ratings								
Assertive	0.552	**0.000**	**0.000**	**0.000**	**0.000**	0.111	**0.000**	0.689
Cheerful	**0.000**	0.454	**0.000**	**0.000**	**0.000**	0.221	**0.000**	0.751
Serious	**0.000**	**0.000**	0.428	**0.000**	**0.000**	0.227	**0.000**	0.767
Unshakeable poise	**0.000**	**0.000**	**0.000**	0.429	**0.000**	0.380	**0.000**	0.678
Broad interests	**0.000**	**0.000**	**0.000**	**0.000**	0.697	0.622	**0.000**	0.168
Correlations among factors								
T1	**1.000**							
T2	0.559	**1.000**						
T3	−0.371	−0.438	**1.000**					
T4	0.381	0.662	−0.082	**1.000**				
T5	0.548	0.292	0.125	0.430	**1.000**			
M1	**0.000**	**0.000**	**0.000**	**0.000**	**0.000**	**1.000**		
M2	**0.000**	**0.000**	**0.000**	**0.000**	**0.000**	−0.208	**1.000**	

Note: Zeros and unities in bold are fixed parameters.

estimates should be relatively stable. Each of the trait factors had strong loadings on their respective factors. The method factors did not have strong influences on all ratings in all conditions presumed to be affected by them as evidenced by low loadings in many cases, and one near zero loading.

Jöreskog's demonstration of a confirmatory factor analysis model for the MTMM correlation matrix has engendered numerous attempts to fit the model to other MTMM matrices. These efforts have not always been successful. Frequently, Jöreskog's (1974) model was empirically underidentified when applied to certain correlation matrices. Sometimes the algorithms for estimating parameters of this model did not converge. On other occasions it was discovered that freeing up correlations between the trait factors and method factors led to the model's being underidentified (Wothke, 1984; Widaman, 1985). [This happens if one frees up the correlations between trait factors and method factors in Jöreskog's model (1974) applied to the Campbell and Fiske (1959) correlation matrix in Table 9.8.] The realization eventually emerged that Jöreskog's (1974) model was only one of several models that might be fit to this kind of data. For example, in some cases it was found that there were no trait factors, only method factors, and in other cases there were no method factors, only trait factors. On other occasions the number of trait factors was fewer than the number of trait scales. Wothke (1984) worked out mathematically necessary conditions for identification of the models and showed that in some cases the models could be empirically underidentified.

Widaman (1985) described sequences of "hierarchically nested" models that contain the MTMM model of Jöreskog (1974) as a special case, which might be applied to a given MTMM correlation matrix to find the best fitting model for that data. In developing these sequences he developed a way of classifying MTMM models according to the properties of their trait factors and their method factors. For the trait factors there were four cases:

1. no trait factors
2. t trait factors, fixed unit intercorrelations (or *one* trait factor)
2′. t trait factors, fixed zero intercorrelations among trait factors
3. t trait factors, free intercorrelations among trait factors.

Cases 2 and 2′ estimate the same number of parameters and are not strictly nested within one another. Widaman noted that it is not possible to perform a statistical test of the difference in fit between the two models, although one might prefer one to the other if its chi-square value were smaller. He also pointed out that cases designated by larger integers are more inclusive (or less restricted) than cases with smaller integers, so cases with smaller numbers are nested (special cases of) cases with larger case numbers.

Widaman (1985) also noted that a parallel set of cases could be formulated for the method factors of the MTMM models:

A. no method factors
B. m method factors, fixed unit intercorrelations (*one* method factor)
B′. m method factors, fixed zero intercorrelations among method factors
C. m method factors, free intercorrelations among method factors.

Again, cases B and B′ are not nested within one another as are cases 2 and 2′ above. Cases designated by letters that come earlier in the alphabet are nested within cases designated by letters that come later in the alphabet.

Widaman (1985) suggested that a taxonomy for MTMM models could be generated by cross-classifying the four trait cases with the four method cases. A model could then be designed by its method and trait factor structures, respectively, for example, 3B′ designates a model that has *t* freely correlated trait factors and *m* orthogonal method factors. Furthermore, one could determine that one model is nested within another if both the trait and method cases respectively were nested within the trait and method cases respectively of the other model. For example, model 2A is a more restricted model nested within 3A, 2C, and 3C. However, model 2A is not nested within 1C, since model 2A has a higher number than 1C. The implication of one model being nested within the other is that one can construct and test a nested sequence of models, beginning with the least restricted model. If the least restricted model does not fit the data, then one knows that the more restricted models will not either. On the other hand, if the least restricted model fits, then one can proceed to test more restricted models in the sequence until one reaches a model that does not fit.

In formulating his classification of models, Widaman (1985) excluded models in which correlations between method factors and trait factors were free parameters. Those cases led to underidentified models. He required trait factors to be uncorrelated with method factors. He also noted that some pairings of trait cases with method cases were problematic. A model 2B is a model with a single common factor in which method and trait factors are indistinguishable. A model 2′B′ would have a single trait factor and a single method factor, each orthogonal to the other, but loading on all variables. Such a model is not identified.

In many cases, researchers have had difficulties in finding and fitting method factors to their data. Marsh (1989) proposed not including common method factors in the model but instead freeing up the correlations between unique factors of trait scales in the same method while retaining fixed zero correlations between unique factors of variables under different methods. This is a less restricted way of introducing "method" effects that usually has better fit than models with a single common method factor for all scales in a given method. On the other hand, it abandons formulating and testing a hypothesis about the structure of the method effect within a given method. Still models with correlated uniquenesses within methods may be seen as less restricted models to compare with models that do impose method factor structure, using chi-square difference tests, to test the nature of the structure. But Byrne and Goffin (1993) note several possible limitations to Marsh's correlated unique factor model: the trait-factor loadings and correlations among traits in Marsh's correlated uniquenesses models are higher than they typically are in Jöreskog's common factor model when applied to the same data set. (This may not really be a limitation but a difference in representation.) But Byrne and Goffin believe that, as a result, Marsh's correlated unique factor model for handling method effects is biased toward providing stronger evidence for convergent validity and weaker evidence for discriminant validity. Perhaps a clearer limitation of the correlated unique factors models is Byrne and Goffin's (1993) observation that because unique factors of variables under different methods cannot be freed to correlate (otherwise the model may become underidentified), this introduces an untestable and unrealistic assumption that method effects are uncorrelated—which Jöreskog's (1974) MTMM factor analysis model can test, but only with well-defined common method factors. Testing the orthogonality assumption would require comparing the model in which these correlations are zero against an identified model in which they are free to be any permissible value. Jöreskog's (1974) model, on the other hand, has its own untestable and unrealistic assumption, that trait factors and method factors are uncorrelated, for, again, to test this orthogonality assumption would require estimating a model in which these correlations are free parameters, to yield a chi-square difference test when the chi-square of the less restricted model is subtracted from the chi-square of the more restricted model, and the difference chi-square has degrees of freedom equal to the difference in degrees of freedom of the original chi-squares.

Numerous excellent reviews have been written about MTMM models within the confirmatory factor analysis framework. Readers may wish to consult Wothke (1984, 1996), Browne (1984), Marsh (1988, 1989), Graham and Collins (1991), Marsh, Byrne, and Craven (1992), and Byrne and Goffin (1993). There have also been explorations of other kinds of models besides the additive model of factor analysis. Browne (1989) formulated a multiplicative model whereby traits and methods multiply rather than add together:

$$\Sigma = D(P_M \otimes P_T + E^2)D$$

where Σ is a $t \times m$ covariance matrix of t trait variables measured under m methods, D is a $t \times m$ diagonal matrix of true-score variances for the $t \times m$ variables, P_M is an $m \times m$ matrix of correlations among m method variables,

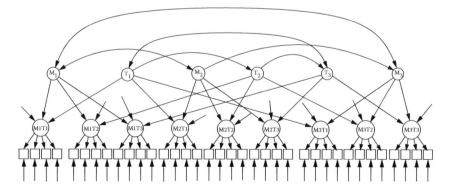

FIGURE 9.2 A facet design for a 3×3 MMMT model with four indicators per latent first-order factor representing a method/treatment combination. Method factors are mutually uncorrelated with trait factors.

and $\mathbf{P}_T$ is a $t \times t$ matrix of correlations among t trait variables, with $\mathbf{E}^2$ a $tm \times tm$ diagonal matrix of unique variances. We will not pursue this model further, but refer the reader to Browne's paper and to papers by Cudeck (1988) and Byrne and Goffin (1993).

To close this discussion of MTMM models, I would like to point out that the MTMM factor analysis models described so far can be seen as special cases of the faceted design model. If the faceted design models we described earlier are seen to be built on an analogy with factorial analysis of variance designs with repeated observations per cell, the MTMM model described by Jöreskog (1974) has a certain analogy with the randomized blocks design with one observation per cell. The weaknesses of the randomized blocks model rest, among other things, on there being only one observation per cell of the design. Many of the identification problems with the corresponding MTMM model arise because too few indicators are trying to do too much work. Having multiple indicators (at least four) per trait–method combination would eliminate many of the identification problems. An illustration of just such a model is shown in Figure 9.2.

10

Equivalent Models

Introduction

When a researcher formulates a model with estimated parameters, he or she must consider the possibility that there are equivalent but distinct models that could be formulated that would reproduce the same covariance matrix when estimated against a given sample covariance matrix. The existence of equivalent models when supported by empirical evidence other than simple fit to the same covariance matrix creates a problem for the researcher, who must then seek to eliminate these alternatives in some way in favor of his or her model. In this section we will discuss the meaning of equivalent models mathematically and will also consider techniques for generating them for consideration.

Stelzl (1986) was the first to consider ways of systematically generating equivalent models in structural equation modeling. Lee and Hershsberger (1990) simplified Stelzl's rules with a simple replacement rule. MacCallum et al. (1993) further refined these rules and illustrated their implications with models taken from the literature. We will draw heavily on these authors' papers for our discussion here.

In this discussion we presume that the models are recursive and represented by DAGS. We further presume that if a path linking two variables in the graph is shown, it corresponds to either a free parameter for that path or a parameter fixed to set the metric. The path may be either a directed arrow indicating a directed causal connection or a two-headed arrow indicating covariation. Latent variables will be presumed to have fixed variances of 1.00.

Definition of Equivalent Models

Given any arbitrary covariance matrix $\mathbf{S}$ for a given set of variables, two models, A and B, with constrained and free parameters are fit to $\mathbf{S}$ by estimating free parameters in such a way that the reproduced covariance matrix minimizes some specified discrepancy function. Let $\hat{\mathbf{\Sigma}}_A$ be the reproduced covariance matrix obtained for model A and $\hat{\mathbf{\Sigma}}_B$ the reproduced covariance for model B. We say that models A and B are equivalent if and only if $\hat{\mathbf{\Sigma}}_A = \hat{\mathbf{\Sigma}}_B =$ for all covariance matrices $\mathbf{S}$ to which models A and B may be fit. Naturally $\hat{\mathbf{\Sigma}}_A$ and $\hat{\mathbf{\Sigma}}_B$ will have the same degree of fit to $\mathbf{S}$. However, equal degree of fit does not imply that the models have equivalent reproduced covariance matrices for any covariance matrix to which they are fit. In some cases, for some $\mathbf{S}$ it may be possible to find two models that have an equal degree of fit, but $\hat{\mathbf{\Sigma}}_A \neq \hat{\mathbf{\Sigma}}_B$. Such models are not equivalent. It may even be possible to find some $\mathbf{S}$ for which by chance $\hat{\mathbf{\Sigma}}_A = \hat{\mathbf{\Sigma}}_B$, while this is not the case for all $\mathbf{S}$. These are not cases of equivalent models. We confine the concept of equivalent models to cases that are necessarily equivalent regardless of the $\mathbf{S}$ to which the models are applied.

Replacement Rule

Lee and Hershsberger (1990) proposed their "replacement rule" for generating equivalent models from recursive, acyclic models. Given a model with a graph G, one must first focus on breaking the model up into blocks of variables. The *focal block* will be the block of variables whose paths between them will be modified with the replacement rule. Variables within the focal block must be connected to others in the focal block either by directed arrows or by a covariance bridge between them, but not both. A *preceding block* will be a block of variables containing the parents of variables in the focal block. All variables in the focal block must have the same parents in the preceding block. The paths from the parents in the preceding block to their offspring variables in the focal block may not be changed in application of the replacement rule to the variables in the focal block. Finally, there may be a *succeeding block* of descendents of variables in the preceding and focal block (Figure 10.1).

Now, the *replacement rule* is this: In a given focal block of variables, a direct path $X \rightarrow Y$ may be replaced by (a) a correlation between the disturbances of X and Y or (b) a path arrow pointing in the opposite direction, that is, $X \leftarrow Y$, as long as the parents of Y are the same as or include the parents of X (MacCallum et al., 1993). By the same token, within a focal block a covariance between a pair of disturbances can be replaced by a directed path in either direction between the corresponding variables, as long as the effect variable

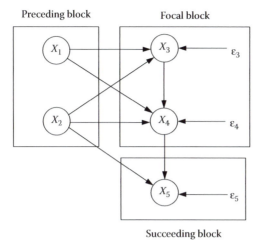

FIGURE 10.1 Illustration of the preceding block, focal block, and succeeding block in a model.

has the same or more parents in the preceding block than the source variable after the change (Figure 10.2).

In Figure 10.3, we show three equivalent models made with the replacement rule. Model B is made from model A by reversing the arrow between X_3 and X_4. Model C is made from either of the other two models by replacing the arrows between X_3 and X_4 with a correlated disturbance between X_3 and X_4.

If a preceding block and a focal block are saturated, meaning every variable in a block is linked to every other variable in the block, either with a directed arrow, or with a covariance bridge, or with correlated disturbances, then we can merge the preceding and focal blocks into a single, saturated block as in Figure 10.4. This is made possible because every variable in the focal block has a link to every variable in the preceding block. Consequently, every variable in the resulting saturated block is related to every other variable by some form of a link.

A saturated block provides more possibilities for forming equivalent models. We may replace the paths within the saturated block by any other saturated set of paths, replacing correlations with directed paths, paths

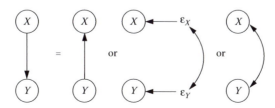

FIGURE 10.2 The replacement rule. Anyone of these may be replaced by anyone of the others.

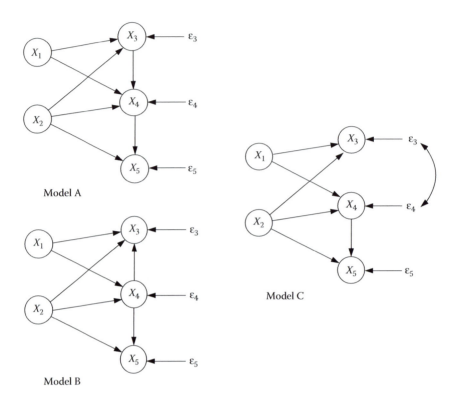

Model A

Model B

Model C

FIGURE 10.3   Three equivalent models. Models B and C are made from model A by applying the replacement rule.

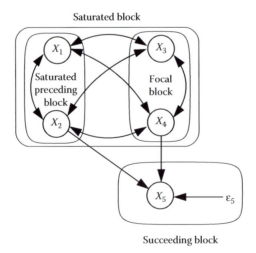

Saturated block

Saturated preceding block

Focal block

Succeeding block

FIGURE 10.4   A model with a saturated block made by merging a saturated preceding block with a saturated focal block, and a succeeding block.

with opposite directed paths, directed paths with correlations, and any endogenous variables of these with correlated disturbances. The resulting models will be equivalent. Examples of four equivalent models formed from the saturated block in Figure 10.4 are shown in Figure 10.5.

The case of saturated models with three variables should be committed to memory (Figure 10.6). With three links between three variables and four kinds of links (arrow in one direction, arrow in opposite direction, covariance bridge between disturbances, and covariance bridge between a pair of variables), one might think that there are $4 \times 4 \times 4 = 64$ permutations of the links. However, not all would apply if, of necessity, one or two variables are exogenous and the remaining endogenous. If only one is exogenous, then of necessity it is a cause of each of the other two (assuming recursive models in which exogenous variables are mutually correlated). There thus are three ways in which the remaining two endogenous variables can be related: The first of the two is a cause of the second, the second is a cause of the first, or neither of the two is a cause of the other, but their disturbances can be correlated. There would not be a covariance bridge between the two endogenous variables, since they are endogenous. Only exogenous variables have covariance bridges between them. Thus with three ways to pick the one exogenous variable and three ways in which the remaining two endogenous variables are related, there are $3 \times 3 = 9$ models of this kind. On the other hand, with two variables exogenous, then of necessity there will be a covariance bridge between them, and each will be a cause of the third and only endogenous variable. There are only

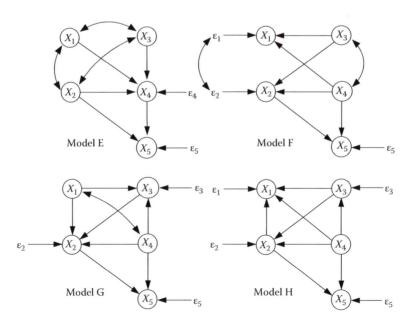

FIGURE 10.5 Four equivalent models formed from the model in Figure 10.4.

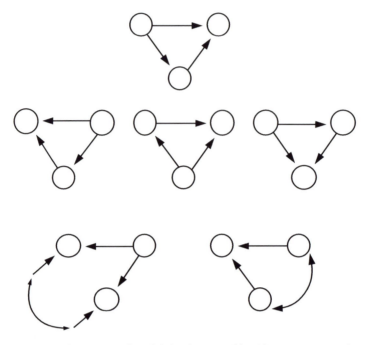

FIGURE 10.6 Equivalent saturated models for three variables. There are six more than shown.

three ways to pick the endogenous variable. So, there are only three equivalent saturated models of this kind. In all then, there are $9 + 3 = 12$ distinct recursive equivalent saturated models for three variables in which there are also exogenous and endogenous variables.

For four variables, when one is exogenous and three are endogenous, there are four ways to pick the exogenous variable, and $((3 \times 2)/2) = 3$ paths among the remaining three endogenous variables, with three kinds of paths for each of these paths. That yields $4 \times 3^{\binom{3}{2}} = 4 \times 3^3 = 108$ distinct saturated equivalent recursive models. But we also have to consider cases with two exogenous and three exogenous variables. For two exogenous variables there are $\binom{4}{2}$ combinations of two variables chosen from four to serve as exogenous variables. That leaves two variables as endogenous, and they would have only a $(2 \times 1)/2 = 1$ path between them. This one path can have three kinds of relations between them. Hence the number of saturated equivalent models among four variables with two exogenous variables is $\binom{4}{2} \times 3^{\binom{2}{2}} = 18$ models. For three exogenous variables, with only one endogenous variable having no paths to any other endogenous variable, there are just $\binom{4}{3} = (4 \times 3 \times 2 \times 1)/(1 \times 3 \times 2 \times 1) = 4$ ways to pick three variables as exogenous. Hence there are four saturated equivalent recursive models with but one exogenous variable. So in all there are $108 + 18 + 4 = 130$ saturated equivalent recursive models obtainable from four variables.

In general, for p variables and k, $1 \leq k \leq p - 2$, exogenous variables, there are $\binom{p}{k} \times 3^{\binom{p-k}{2}}$ saturated equivalent recursive models. Additionally there are p distinct models with one endogenous variable that has no other endogenous variables to have relations with, so all of the exogenous variables are the causes of the one endogenous variable and are connected between each pair by a covariance bridge. So, the total number of saturated equivalent recursive models for p variables is

$$\sum_{k=1}^{p-2} \binom{p}{k} \times 3^{\binom{p-k}{2}} + p.$$

As p increases, this number increases rapidly.

There are cases where the replacement rule of Lee and Hershsberger does not apply. Consider the following model with six variables. I have not been able to find an equivalent model for this model, because there are no focal blocks whose variables are linked and have the same set of parent causes in a preceding block.

In Figure 10.7 you cannot reverse the arrows $X_1 \rightarrow X_3$ and $X_2 \rightarrow X_3$ either singly or simultaneously, because X_1 and X_2 are uncorrelated, and would become correlated if either or both of these paths were reversed in direction. We cannot reverse the arrow $X_3 \rightarrow X_4$ because that would make X_3 a collider and make X_1 and X_2 uncorrelated with X_4, X_5, and X_6, whereas now they are correlated. Reversing $X_4 \rightarrow X_5$ would make X_4 a collider and make X_5

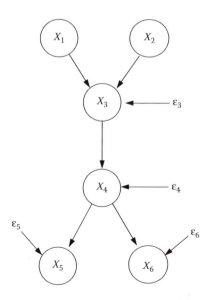

FIGURE 10.7 A model for which there may be no equivalent model obtained by reversing arrows or substituting correlated disturbances.

uncorrelated with X_1, X_2, and X_3, whereas now they are correlated. Similarly, reversing $X_4 \to X_6$ would also make X_4 a collider and make X_6 uncorrelated with X_1, X_2, and X_3, whereas now they are correlated. Reversing both $X_4 \to X_5$ and $X_4 \to X_6$ simultaneously makes X_5 and X_6 uncorrelated, whereas now they are correlated. In short, there are no pairs of variables with the same set of parent causes, and no focal blocks. If it is a question of direction of causality, and causal directions were hypothesized, the directions shown in the model are the only directions possible consistent with a pattern of correlations consistent with this model. This implies that any data fitting this model have no equivalent rival model postulating different causal directions consistent with it.

We also need to point out that reducing the number of equivalent models with respect to a given model results from fixing more path coefficients to zero. Equivalent models proliferate from freeing parameters.

MacCallum et al. (1993) searched the literature for covariance structure modeling and found 99 studies in three journals. Of these studies, they found few that considered the possibility of equivalent models. Perhaps more now do since the publication of their paper. They also took a model from educational psychology and a model from industrial–organizational psychology reported in the literature and worked out equivalent models for each of these models. For the educational psychology model with five variables, they found three other models that were equivalent to the original model reported in the literature. These equivalent models even reversed the direction of causal paths, drastically changing the theoretical implications from those of the authors.

The industrial–organizational models were sparser, with fewer parameters, but even then, for five variables MacCallum et al. (1993) were able to find three other models equivalent to the original model reported in the literature. In these equivalent models causal directions were reversed and estimated path coefficients changed sometimes dramatically, for example, from 0.53 in the original model to 0.30 in an equivalent model that had reversed another causal path. So, the implication is that the existence of equivalent models seemingly has serious implications for the theory, drawing support from the good fit of a particular model to an observed covariance matrix.

How then can we exclude equivalent models in favor of a given model? One thing we can do is consider temporal ordering among the variables. We might assume that the causes precede effects. A causal arrow from a variable measured later in time to a variable preceding it in time may not be plausible (although there may have been a delay in making the measurement of a quantity that has not changed since the supposed cause was applied, making this not an iron-clad rule). We can also consider embedding our models within a context where there are already known causal directions between other variables outside the variables of our models. How these causal directions affect relationships between variables in our model and outside the model can be an important clue as to whether or not a given causal direction within our models is plausible.

Equivalent Models That Do Not Fit Every Covariance Matrix

So far we have considered equivalent models that have the same number of degrees of freedom. Mulaik and Quartetti (1997) considered a different kind of equivalence, one that would not apply to every sample variance–covariance matrix. The issue concerns whether or not certain latent variables are first- or second-order factors. In 1993 Gustafsson and Balke (1993) published a paper in *Multivariate Behavioral Research* titled "General and specific abilities as predictors of school achievement." In this paper they described two models as alternative ways of representing the relationship of general intelligence to specific test scores. Both models contained a general factor, but in one model it was represented as a second-order general factor, with a causal influence on three first-order factors, and in the second it was represented as a first-order general factor uncorrelated with three other first-order factors. The model in which all factors are first-order factors is a variant of Holzinger and Swineford's (1937) bifactor model having a general factor and several uncorrelated group factors. Gustafsson and Balke (1993) called this a "nested factors model." They called the second model the "hierarchical model."

Gustafsson and Balke (1993) were motivated to display these two models because they were interested in showing that even when general intelligence appears as a second-order factor in a hierarchical model, it nevertheless has an impact on the observed variables, which the graph of the hierarchical model seems not to demonstrate. They reported that the nested model has some similarities to the Schmid–Leiman (1957) decomposition of a hierarchical model from an exploratory factor analysis with a second-order factor. But the difference here, Gustafsson and Balke (1993) asserted, is that the nested factors model and the hierarchical models are not mathematically equivalent (Figure 10.8). The hierarchical model places constraints on the relations between the general factor and the observed variables, but the nested factors model does not (Gustafsson and Balke, 1993, p. 416). Nevertheless the Schmid–Leiman procedure (1957) produces a table of correlations between the general factor and the second-order unique factors that looks superficially like a Holzinger and Swineford bifactor model of a general factor and group factors.

Mulaik and Quartetti (1997) created a hierarchical model with specified loadings and then, using path-tracing rules, produced the table of the Schmid–Leiman decomposition of the hierarchical model. The specific hierarchical model that they used is shown in Figure 10.9. The Schmid–Leiman decomposition of "loadings" of the observed variables on the second-order general

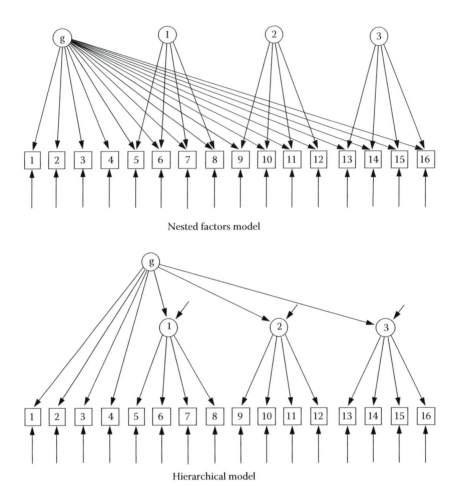

FIGURE 10.8 A nested factors model and a hierarchical model that in some circumstances generate the same covariance matrix.

and specific factors is given in Table 10.1. They then generated a random sample covariance matrix from the population correlation matrix produced by the model in Figure 10.9 and showed that one could fit, essentially perfectly, incompletely specified hierarchical models and nested factors models to this sample correlation matrix based on 1000 observations. For the incompletely specified hierarchical model chi-square was zero with 101 degrees of freedom. For the incompletely specified model chi-square was 0.03 with 92 degrees of freedom. In the incompletely specified models the nonzero structural path coefficients were represented by free parameters in both models. Fixed zero path coefficients corresponded to zero coefficients in the models. In the nested factors model the factors were specified as uncorrelated. What this showed was that both models would essentially reproduce the covariance matrix generated by a hierarchical model.

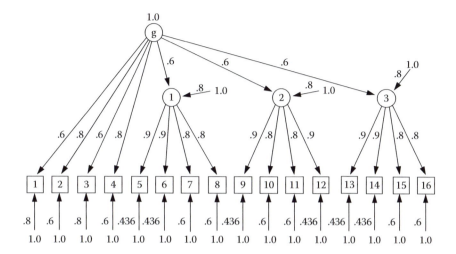

FIGURE 10.9 Hierarchical model with specified coefficients as population parameters.

We can see from comparing Figure 10.9 with Table 10.1 that the correlations between the factors and the observed variables are given by path-tracing rules. For example, the correlation between g and variable 2 is 0.8, between second-order unique factor 1 and variable 5 is $0.6 \times 0.9 = 0.54$, between second-order unique factor 2 and variable 9 is $0.8 \times 0.9 = 0.72$, and so on. Each correlation has to equal the product of the loading of the second-order factor with the loading of the observed variable on the corresponding first-order factor. This places a constraint on these correlations. Compare this to the case where each nonzero correlation between an observed variable and a first-order factor could be any nonzero value. That the incompletely specified nested factors model fit the hierarchical data perfectly is because it has no constraints on the parameters, which are free to have their values dictated by the sample covariance matrix from the hierarchical population.

Next, Mulaik and Quartetti (1997) created another population correlation matrix, this time using an unconstrained nested factors model in which the parameters were slightly altered in an unsystematic way from their corresponding values in Table 10.1. This time the same incompletely specified hierarchical and nested factors models were fit to the sample covariance matrix generated from this population. When the fit of the two models was compared, the hierarchical model's chi-square was nonzero (3.59 with 101 degrees of freedom) but nonsignificant. The nested factors model chi-square was zero with 92 degrees of freedom. Here we see that a hierarchical model may be an extremely close approximation to the nested model. In fact, in this case the power to reject the null hypothesis for the hierarchical model against the nested factors model generated data was only 0.18, which is not high at all. But if the nested factors model had been much different than a hierarchical model Schmid–Leiman decomposition, the chi-square could well have

TABLE 10.1

Schmid–Leiman Population Decomposition for The HO1 Model Showing Expected Loadings of Observed Variables on the Second-order Factors

	g	1	2	3
1	0.600	0.000	0.000	0.000
2	0.800	0.000	0.000	0.000
3	0.600	0.000	0.000	0.000
4	0.800	0.000	0.000	0.000
5	0.540	0.720	0.000	0.000
6	0.540	0.720	0.000	0.000
7	0.480	0.640	0.000	0.000
8	0.480	0.640	0.000	0.000
9	0.540	0.000	0.720	0.000
10	0.480	0.000	0.640	0.000
11	0.480	0.000	0.640	0.000
12	0.540	0.000	0.720	0.000
13	0.540	0.000	0.000	0.720
14	0.540	0.000	0.000	0.720
15	0.480	0.000	0.000	0.640
16	0.480	0.000	0.000	0.640

been significant. What this shows is that a hierarchical model may not fit data generated by a nested factors model, but it may still be a good approximation to these data and difficult to discriminate from the nested factors model in some cases.

We should note an important feature of these two models. The hierarchical model has more degrees of freedom—it estimates fewer parameters than does the incompletely specified nested factors model. So, when the two models are completely indistinguishable on fit, the recommendation is to prefer the model with more degrees of freedom, the hierarchical model, because it is more disconfirmable. The added degrees of freedom for the hierarchical model also give it the possibility of distinguishing itself from the nested factors model because the constraints on the loadings will lead the hierarchical model to differ considerably from data generated by a nested factors model, if the nested factors model deviates considerably from a hierarchical model.

A Conjecture about Avoiding Equivalent Models by Specifying Nonzero Parameters

In the physical sciences considerable effort is spent in determining values of natural constants such as the specific gravity of lead, the melting point of pure silver, and the atomic weight of sodium. The experiments designed to estimate these values are not testing a physical theory but assume its truth while leaving the parameter free to be estimated. To achieve high accuracy, a known theory of extraneous variables that affect measurements is used to control these extraneous variables. Testing comes in at a later stage, where these natural constants are used in various formulas combining several natural constants to predict certain outcomes in other experiments. If the predictions are not upheld, then something is wrong with the theory.

Behavioral and social scientists rarely establish natural constants and carry them over from one study to the next. To some extent this is because their subject matter is much more complex and less well understood than the subject matter of the physical sciences—although, at the cutting edge, in the physical sciences there is just as much uncertainty and ambiguity as in the behavioral and social sciences. The physical scientists have just been successful in gaining knowledge in some areas in their endeavors for a longer time than in the social and behavioral sciences. But it may also be because the behavioral and social sciences have not in fact searched for natural constants. For example, is the factor loading of a judgment on a specific scale of friendliness invariant? It represents how much a unit change in the judgment of friendliness will produce a change in the rating on the scale. Should that not be invariant in any context in which the friendliness judgment is made on that scale? If these numbers that we obtain in our structural equation and factor analytic studies are so ephemeral, what can we hope to generalize from the studies? But we may hope for better.

In one study, Carlson and Mulaik (1993) estimated the loadings of certain scales on certain factors in a calibration condition, and then used these loadings in later experimental conditions where other information was provided to influence judgments. For example, they estimated the loading of the friendliness judgment on the friendly scale to be 0.902 (with the loading of the same judgment factor on the sympathy scale fixed to unity) in a calibration condition. This value was the value for this parameter in the best fitting model in this condition. Then they used the value of 0.902 as a fixed parameter value for the friendliness scale in two later experimental conditions using the same subjects. Similarly, other loadings on other scales were also estimated in a calibration condition and then carried over as fixed parameters in later experimental conditions applied to the same subjects.

They assumed that the semantics underlying verbal rating scales was stable and relatively invariant, so that the relationship between judgment and ratings on scales of that judgment would be invariant. With four indicators of each latent judgment, they were able by confirmatory factor analysis to establish the objective validity of the latent judgment variable. And they were able to use these fixed loadings successfully in different experimental conditions, as judged by the good fit for models using them as fixed values in these conditions.

So, suppose we used these same four scales of friendliness, with their fixed loadings on a friendliness judgment factor in another study that combined the judgments of friendliness with those of intelligence in a study of leadership. If we have natural constants for the loadings of the friendliness and intelligence scales on these dimensions, can we not use them as fixed values every time they use those scales? One major benefit is that we vastly increase our numbers of degrees of freedom, if we do. But another may be that we rule out equivalent models in those parts of our models involving these fixed parameters. The values will not fit just any data, but a specific kind of data for which they are applicable.

What we stipulate is that we cannot free those fixed parameters to obtain an equivalent model with different values for the constants using the current data. The fixed values are determined elsewhere in other studies with other data sets and experimental conditions. In those studies we have established these fixed values as conventions, as the way we are going to theorize and think about data of a certain kind from that point on, or until evidence arises that using them leads to major predictive failure. But as scientists, we presume also that they are invariants of nature and not just mere social conventions. So, if you want to offer an alternative model to the one I use with the specified fixed parameters obtained elsewhere, you have to carry out studies elsewhere also to find different fixed values and show that they also lead to successful predictions. But I will also insist, in addition to fit to within measurement error, that your model must have *much* better fit with nearly as many or more degrees of freedom than mine before I will concede to it.

So, my recommendation to researchers is that they ultimately seek to establish natural constants for structural coefficients, particularly in connection with the indicators of the measurement models. Use these as fixed values in other studies to see how far you can get with them.

11

Instrumental Variables

Introduction

Subjects may be allowed to vary on extraneous variables if it is believed that their variation does not have any relationship to the causal variables of the study. Any influence on the dependent variables of such variation by extraneous variables will then enter as components of the disturbances, but the causal relations will be unaffected. In such cases the causal relations will be estimable without bias. But even in the cases where some of the variation in individuals is correlated with causes of these variables, it may be possible to still estimate the causal relations as long as they are endogenous and there are exogenous variables, this time known as *instrumental variables*, that are causes of the endogenous causes, but not directly of their effects. These instrumental variables must be independent of the disturbances on these endogenous effect variables.

For example, consider Figure 11.1, which shows three manifest variables, X_1, Y_1, and Y_2, where X_1 is a cause of Y_1, and Y_1 in turn is a cause of Y_2. However, the disturbances of Y_2 are correlated with endogenous cause Y_1 via the covariance between the disturbances, but not with exogenous X_1. We will assume that the disturbances have unit variances.

According to Heise (1975), X_1 in Figure 11.1 is an instrumental variable for the $Y_1 \rightarrow Y_2$ relationship if X_1 has no direct effect on Y_2 while at the same time affecting Y_1 either directly or indirectly through other intervening variables that themselves have no direct effect on Y_2. Furthermore, it must be the case

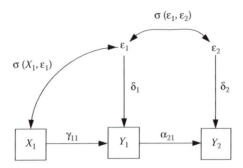

FIGURE 11.1   A system of variables with an instrumental variable X_1 and correlated errors.

that neither Y_1 nor Y_2 has a direct or indirect effect on X_1. And there can be no other variable that affects both X_1 and Y_2 both directly or indirectly. It is possible for X_1 to be correlated with the error ε_1 of Y_1, but not with the error ε_2 of Y_2. Finally, any other variable W that is only correlated with Y_1 but not with Y_2 is an instrumental variable for the $Y_1 \rightarrow Y_2$ relationship if it has the other properties of an instrumental variable.

To show that the coefficient of interest α_{21} is estimable without bias, consider that this system is given by the equations

$$Y_1 = \gamma_{11}X_1 + \delta_1\varepsilon_1, \quad E(X_1) = E(\varepsilon_1) = 0, \quad \sigma(X_1, \varepsilon_1) \neq 0, \quad \sigma(X_1, \varepsilon_2) = 0$$

$$Y_2 = \alpha_{21}Y_1 + \delta_2\varepsilon_2, \quad E(\varepsilon_2) = 0, \quad \sigma^2(\varepsilon_1) = \sigma^2(\varepsilon_2) = 1$$

from which we may derive the following expressions for the covariances of these variables:

$$\sigma(X_1, Y_1) = E[X_1(\gamma_{11}X_1 + \delta_1\varepsilon_1)] = \gamma_{11}\sigma^2(X_1) + \delta_1\sigma(X_1, \varepsilon_1)$$

$$\sigma(X_1, Y_2) = E[X_1(\alpha_{21}Y_1 + \delta_2\varepsilon_2)] = \alpha_{21}\sigma(X_1, Y_1) + \delta_2\sigma(X_1, \varepsilon_2) = \alpha_{21}\sigma(X_1, Y_1)$$

with the consequence that we are able to get

$$\alpha_{21} = \frac{\sigma(X_1, Y_2)}{\sigma(X_1, Y_1)}.$$

Unless $\sigma(X_1, \varepsilon_1) = 0$, γ_{11} is unidentified, or biased if estimated as $\gamma_{11} = \sigma(X_1, Y_1)/\sigma^2(X_1)$. But that does not matter, because our goal was to get α_{21}. In contrast, if only the variables Y_1 and Y_2 are given and the erroneous assumption made that the disturbance on Y_2 is uncorrelated with Y_1, then estimating α_{21} as $\alpha_{21} = \text{cov}(Y_1, Y_2)$ would yield a biased estimate, because $\text{cov}(Y_1, Y_2)$ is a function of both the causal effect of Y_1 on Y_2 and the correlation between the disturbances $\sigma(\varepsilon_1, \varepsilon_2)$. On the other hand, formulating a model in which Y_1 is correlated with the disturbance of Y_2 would require estimating the covariance between Y_1 and the disturbance of Y_2. But this would require estimating two parameters from only one observed covariance, and so the parameters of this model would be underidentified.

In summary, the instrumental variable X_1 has the advantage of introducing an additional variable, with additional observed covariances to work with, which allows us to obtain an unbiased estimate of the causal effect of Y_1 on Y_2. It also yields a test of whether there is a causal effect of Y_1 on Y_2. If the correlation between X_1 and Y_2 is significantly different from zero, while the partial correlation between X_1 and Y_2 given Y_1 is zero, this supports the existence of a causal relation between Y_1 and Y_2 by the d-separation criterion.

The problem, of course, with the instrumental variable technique is how to find instrumental variables. This involves a careful analysis of the variable Y_1 to establish possible causes of it to include in the model. But measuring and guaranteeing that these potential causes would be uncorrelated with any other causes of Y_2 may be problematic. On the other hand, the instrumental variable X_1 may be introduced as an experimentally manipulated variable that affects the values of Y_1. For example, Y_1 may be a judgment variable that a subject makes of a stimulus or situation, which in turn influences a variable measuring a response or action taken by the person as a consequence of that judgment. By initially performing an analysis of the stimuli to be presented, one may be able to assign quantitative values to the attributes of each stimulus. In fact, one may then be able to construct various stimuli with known values for the stimulus attributes, and generate stimuli by varying the values of the stimulus attributes at random. In effect, one is creating values of instrumental variables that are possible causes of the judgments made by the experimental subjects. This randomization procedure will then lend support to the idea that the stimuli presented to the subjects to judge and respond to vary in ways independently of any external causes of the response variable(s) Y_2. Although they did not recognize at the time that they were using an instrumental variable technique, Carlson and Mulaik (1993) effectively conducted such a study of personality ratings by randomly generating descriptions of persons to rate, with known values for their personality, and showing that the person-stimulus variables describing the persons rated drove the latent judgments raters made on several dimensions, and these in turn drove their ratings on four indicators, respectively, of each of these judgment variables.

Instrumental Variables and Mediated Causation

Since the mid-1980s, an issue debated back and forth in the literature concerns how to test (a) whether a variable X is a direct, unmediated cause of another variable Y, (b) whether the effect of X on Y is only mediated through an intermediate variable M, known as *complete mediation*, or (c) whether X has both a direct effect and a mediated indirect effect on Y, known as *partial mediation*. The three cases are illustrated in Figure 11.2. Controversy has centered on how one should test each of these cases.

Baron and Kenny (1986) advocated a four-step procedure based on the path model shown in Figure 11.2c: (1) test to see if the correlation $r(X, Y)$ is equal to zero; (2) test to see if $r(X, M)$ equals zero. If one rejects both null hypotheses in (1) and (2), this sets the stage for the next test by showing that X is related to both Y and M. (3) Regress Y on both X and M. Test the resulting regression coefficient d' of Y on M (with X held constant) to see if it differs significantly from zero. (Equivalently test whether the partial correlation between M and Y with X held constant is zero.) If one deems there is a relation between M and Y with X held constant, then (4) use the estimate of b' in the regression of Y on X and M to test the hypothesis that it is equal to zero. If so, one has complete mediation. If not zero, one has partial mediation.

The Baron and Kenny (1986) approach is equivalent in structural equation modeling to estimating the parameters of the just-identified model in Figure 11.2c and then testing them to see if they each differ from zero. Such tests, of course, are not statistically independent and they mix both confirmatory and exploratory procedures. James and Brett (1984) recommended instead that one begin with a complete mediation model as a structural equation model to test as shown in Figure 11.2b. In this case the complete mediation model is overidentified. Accepting the complete mediation model against the alternative that it is false is equivalent to showing that $\mathrm{cov}(X, Y) = cd = \mathrm{cov}(X, M)\, \mathrm{cov}(M, Y)$ after showing that $\mathrm{cov}(X, Y)$ is not equal to zero. It is also equivalent to showing that the partial correlation $\rho(X, Y \mid M)$ between X and Y with M held constant is zero. If one rejects the complete mediation model, contrary to the view of the partial mediation advocates, this does not logically require you to accept the partial mediation model nested within the original complete mediation model.

There are other equivalent models implying different causal directions or unmeasured common causes that would also be consistent with the data fit

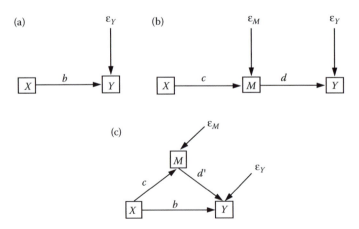

FIGURE 11.2   Three cases of unmediated and mediated causation: (a) unmediated causation, (b) complete mediation, and (c) partial mediation.

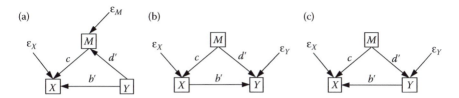

FIGURE 11.3 Three models equivalent to the partial mediation model of Figure 11.2c.

by the partial mediation model in question (Stone-Romero and Rosopa, 2004). Examples of some of these models are shown in Figure 11.3.

One way to reduce the indeterminacy of the three-variable partial mediation models is to introduce a fourth, instrumental variable.

In Figure 11.4a, we have a partial mediation model like that in Figure 11.3c to which has been added the instrumental variable Z. Z is selected as a cause of X that is independent of the disturbances on X, M, or Y. We must have reason to assume that $\rho(Z, X) \neq 0$. Better still, if Z is a manipulated, randomized cause of X, this usually (but not always) breaks any correlation that Z might have with other inputs to X. To establish that the causal connections in the model are correctly specified, there are a number of tests that we can perform: (1) X is a cause of Y, either direct or indirect or both, if $\rho(Z, Y) \neq 0$ and $\rho(Z, Y \mid X) = 0$. Conditioning on X blocks access of Z's variation to that of Y. (2) X is a cause of M if $\rho(Z, M) = b_{XZ}b_{MX} \neq 0$. (3) If $\rho(M, Y) \neq 0$ and $\rho(Z, Y) = \rho(Z, X) \cdot \rho(X, M) \cdot \rho(M, Y)$, then X is not a direct cause of Y but an indirect cause mediated by M,

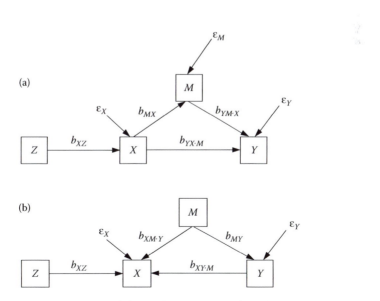

FIGURE 11.4 A partial mediation model (a) with an instrumental variable Z, and an equivalent three-variable model for X, M, and Y, but with addition of Z as a cause of X.

which is also a cause of Y. (4) If $\rho(Z, Y) \neq 0$ and $\rho(Z, Y \mid M) = b_{XZ}b_{YX \cdot M} \neq 0$, then X is a direct cause of Y.

The instrumental variable Z allows us to determine the direction of causation with respect to variable X. Consider the model in Figure 11.4b. Z is uncorrelated with M and the disturbances on X and Y. Z cannot be correlated with M or Y because there are no causal paths from Z to these variables. X is a collider on any causal path between Z and Y, and Z and M. Thus, if Z is correlated with these variables, then X is also a cause of them, either directly or indirectly, because X mediates the causal effect of Z to these variables (Scheines et al., 2001).

Conclusion

The topic of instrumental variables now has an extensive literature, mostly in the econometric literature. But instrumental variables are gaining recognition in the behavioural, medical and social sciences as well. Instrumental variables are used in natural settings when controlled experiments are not possible to estimate causal effects. They are used to establish causal direction, consistent estimates of causal effects, causal effects in the presence of omitted relevant causes and random error. The major problem in their use is finding instrumental variables for a given application. Sometimes the endogenous cause Y_1 of effect variable Y_2 is also an effect of some natural shock, X_1 such as variation in the weather, which may be reasonably regarded as having no relation to the disturbances on the endogenous effect variable. These natural, random shocks can be treated as an exogenous instrumental variable. For example, one may be interested in the effect of the number of hours spent shopping at the mall in a week on the amount in dollars of goods purchased in that week at the mall. As an instrument one may use a variable that measures the severity of local daytime weather during that week. Variation in weather may cause some variation in the number of hours spent shopping and in turn the amount of purchases.

12

Multilevel Models

Introduction

A common form of data found in the behavioral and social sciences has subjects nested within one or more hierarchical levels of categories. In an educational setting, subjects may be nested within classes, each with a different teacher, and the classes in turn may be nested within schools, and these in turn within school districts. In an organizational behavior study, subjects may be employees studied within stores, stores in turn being classified within cities, cities in turn within regions. A hospital researcher may study patients within wards, the wards in turn are in hospitals, and hospitals may be within hospital systems, and these in turn in regions of the country. If at each level of categorization there is variation in the units within that category, this may represent potential sources of causal variation that has an influence on the observed variables of the study. For example, variations in store managers may bring different leadership styles to bear on the employees working under them and influence their performance on various variables. In turn, different cities within which the stores are located may introduce variations in the culture, transportation systems, health systems, and housing availability that may influence differently the people working in each city and influence rates of tardiness, turn-over, absenteeism, wage, and health insurance demands. The cities in turn may be placed under the supervision of different higher level managers, who establish different policies in the regions within which the cities are located. These variations in policy may also influence the variation in behavior of the employees within the regions.

To study data hierarchically organized like this has led to the development of models of hierarchical multivariate analysis of variance and regression. More recently, hierarchical models with latent variables have been developed within the context of structural equation modeling. The topic is now well developed and several new textbooks on the subject have been published (Hox, 2002; Little, Schnabel, and Baumert, 2000; Reise and Duan, 2003; Bock, 1989; Heck and Thomas, 2000). We will only be able to provide a brief survey of this topic here.

Multilevel Factor Analysis on Two Levels

For this discussion, I am indebted to Hox (2002), who in turn was indebted to Muthén (1989). Suppose we have data on individual students in 150 classrooms of a given grade in a state school system. Each class, say, consists of approximately 35 students. In all, just for illustrative purposes, suppose there are 5350 students. By g we will denote a particular classroom containing N_g students. There are $G = 150$ classrooms in all. The total number of students is $N = 5350$. By $\mathbf{Z}_{jg}$ we will designate the jth student's scores on the n variables in classroom g. Now, let $\bar{\mathbf{Z}}_g = (1/N_g) \sum_{j=1}^{N_g} \mathbf{Z}_{jg}$ be the sample mean vector of the scores in classroom g. Let us denote by $\bar{\mathbf{Z}} = (1/N) \sum_{g=1}^{G} \sum_{j=1}^{N_g} \mathbf{Z}_{jg}$ the grand mean vector of all score vectors across all subjects in all classrooms. In a manner analogous to the analysis of variance, let us now designate a score for an observation vector as consisting of two components:

$$\mathbf{Z}_T = \mathbf{Z}_B + \mathbf{Z}_W.$$

Here $\mathbf{Z}_T$ represents the combining of two deviation scores, $\mathbf{Z}_B = \bar{\mathbf{Z}}_g - \bar{\mathbf{Z}}$ and $\mathbf{Z}_W = \mathbf{Z}_{jg} - \bar{\mathbf{Z}}_g$. These two components can be shown to be mutually orthogonal (but we will not do so here). Keep in mind, however, that the orthogonality property here is not empirical but mathematical.

Hox (2002), following Muthén (1989), suggests that there exist "orthogonal" population covariance matrices that combine as

$$\Sigma_T = \Sigma_B + \Sigma_W.$$

These in turn may be estimated by sample variance–covariance matrices

$$\mathbf{S}_T = \mathbf{S}_B + \mathbf{S}_W.$$

Now, for Σ_B and Σ_W we may seek to fit to each, respectively, a distinct structural model, for example, a common factor model. The first model fit to Σ_B represents a modeling of the covariation due to factors at the classroom level across classes. Variation in teacher effectiveness in different subjects and classroom resources may be the basis for some of the variation in the students' scores across classes. The second, fit to Σ_W, represents modeling of the covariation due to factors at the individual student level, perhaps "individual difference" factors such as verbal, spatial, and quantitative abilities.

Now, if N_g were the same in every class, it would seem natural to estimate Σ_W by

$$\mathbf{S}^*_{PW} = \frac{\sum_{g=1}^{G} \sum_{j=1}^{N_g} (\mathbf{Z}_{jg} - \bar{\mathbf{Z}}_g)(\mathbf{Z}_{jg} - \bar{\mathbf{Z}}_g)'}{N - G}, \tag{12.1}$$

the pooled within-classes variance–covariance matrix, and

$$\mathbf{S}^*_{B} = \frac{\sum_{g=1}^{G} N_g (\bar{\mathbf{Z}}_g - \bar{\mathbf{Z}})(\bar{\mathbf{Z}}_g - \bar{\mathbf{Z}})'}{G - 1}, \tag{12.2}$$

the between-groups sample variance–covariance matrix.

Muthén (1989, 1990), however, showed that although Equation 12.1 is an unbiased estimate of Σ_W, Equation 12.2, on the other hand, is an unbiased estimate of $\Sigma_W + c\Sigma_B$, where c is equal to the common size of all groups. In other words,

$$\mathbf{S}^*_W = \hat{\Sigma}_W \tag{12.3}$$

and

$$\mathbf{S}^*_B = \hat{\Sigma}_W + c\hat{\Sigma}_B, \tag{12.4}$$

where the hat over the Σ's indicates maximum-likelihood estimator of the matrix under the hat.

These results are correct as long as the sample size within groups is the same for each group. Of course, this is often unrealistic to assume. Classes in schools do not all have exactly the same sizes. They may differ by having slightly fewer or slightly more students than some mean size. But if the differences in sample size are relatively small across groups (relative to class size), then Muthén (1989, 1990) suggested using the following approximate value for c:

$$c^* = \frac{N^2 - \sum_{g=1}^{G} N_g^2}{N(G - 1)}. \tag{12.5}$$

The approximation only works well when both the number of students *and* the number of classes is large (Hox, 2002). Hox suggests only attempting the between-groups modeling when $G > 150$, otherwise parameter estimates and significance tests may not be stable or accurate for the between-groups model.

Hox (2002) suggests a multistep series of tests. First, one constructs a model for the within-groups variance–covariance matrix Σ_W estimated using the full set of students, with $N - G$ degrees of freedom. Because $\mathbf{S}_W^*$ in Equation 12.3 is estimated independently from $\hat{\Sigma}_W$ in Equation 12.4, Hox suggests performing an exploratory factor analysis to find the best factor model for Σ_W from $\mathbf{S}_W^*$, and then using this model in the next step in combination with a model for Σ_B, fitting the combined model to $\mathbf{S}_B^*$. Of course, one could always begin with an a priori model for Σ_W and test it against $\mathbf{S}_W^*$, but if this model is

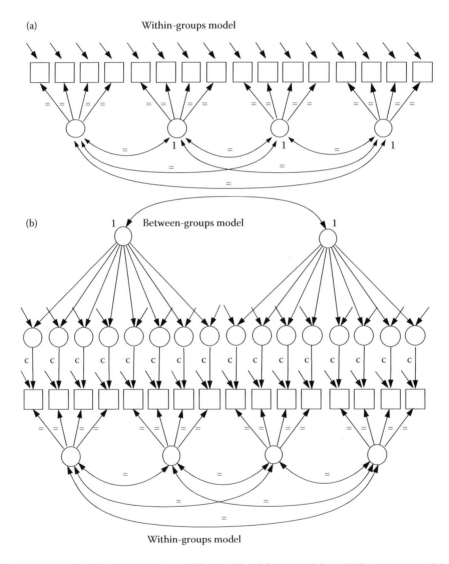

FIGURE 12.1 Simultaneous analysis of the multilevel factor model. (a) Within-groups model fit to $\mathbf{S}_W^*$. (b) Combined within-groups and between-groups models fit to $\mathbf{S}_B^*$. Corresponding parameters constrained equal across $\mathbf{S}_W^*$ and $\mathbf{S}_B^*$ indicated by =. Variances of latents fixed to unity indicated by 1's next to them.

rejected, there is no point in going on with this rejected model to combine it with a hypothesized Σ_B, fitted to $\mathbf{S}_B^*$, because the combined model would be naturally rejected also. In any case, one can test this model with a structural equation modeling program, using the options for simultaneous analyses in two groups. See Figure 12.1.

The first "group" here has the covariance matrix $\mathbf{S}_W^*$ and the second "group" the covariance matrix $\mathbf{S}_B^*$. The model for Σ_W fitted to $\mathbf{S}_W^*$ is also a part of the model fitted to $\mathbf{S}_B^*$. The free parameters in the within-groups model for both models can be constrained to be equal, correspondingly, across groups. In the between-groups model part of the model fit to $\mathbf{S}_B^*$, the scaling factors on select indicators should not be fixed to unity but to the scaling factor computed in Equation 12.5. All of this is best seen in Figure 12.1.

It should be pointed out that usually the within-groups covariance matrix $\mathbf{S}_W^*$ will be estimated based on numerous individual cases, whereas the between-groups model is based on the much smaller number of groups. The large number of individual observations will make the power of the chi-square test for the test of the within-groups model exceedingly large, leading frequently to rejection of the model due to small discrepancies not anticipated. Researchers thus often rely on indices of approximation to determine whether to accept (provisionally) a model or not. (More about this is given in Chapter 15.)

In many respects, the multilevel factor analysis model has similarities to the MTMM model, and should occasionally display similar problems of empirical underidentification and failures to converge. On the level of theory, the multilevel factor analysis model is often heavily exploratory in its formulation. The meaning of group factors may be largely artifactual since they do not have distinct indicators to give them independent grounding separately from the within-group factors, and theory may not be sufficient to give them clear meaning.

Multilevel Path Analysis

An alternative to multilevel factor analysis is to conduct multilevel path analysis, using distinct indicators for group-level latent causes. Every subject within a given group will receive the same score on a between-groups variable. For example, teachers may be measured for their college GPA, for their knowledge of the subjects taught as given by scores on examinations, and be rated by principals (based on observations) for their ability to stimulate interest in their classroom, for their course preparation and organization. One may furthermore have a causal model as to how prior education, as indicated by a teacher's grades in certain courses, causes the levels of knowledge, as indicated by subject matter examinations, to influence teacher class preparation, as judged by several expert observers of classroom behavior. Every individual within the class of such a teacher will be assigned the same teacher's scores on these teacher variables, in addition to scores on individual achievement, aptitude, and/or interest tests. The bottom-line issue will be to determine the extent to which variation in average student performance across classrooms on specified student variables is caused by variation in the teacher variables, as manifested in the data across classrooms. Additionally, there will be interest in the extent to which individual aptitude and interest test scores determine individual achievement on course outcome variables. The way to model data of this kind is with a bilevel SEM as shown in Figure 12.2.

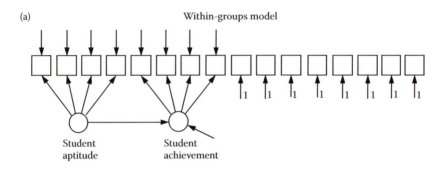

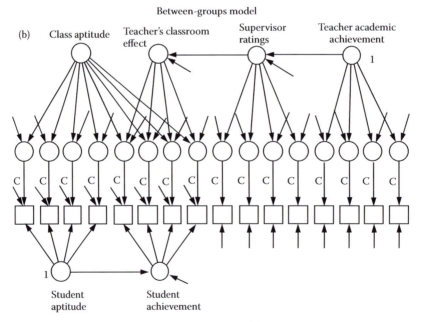

FIGURE 12.2 An illustrative bilevel SEM.

The reasoning behind how to formulate and test a bilevel SEM is essentially the same as that given for the bilevel factor analysis model. However, an important difference is that variables that vary only at the second level, for example, teacher variables, do not have within-group variance. To see this, we will follow the development of the total, between-group and within-group score vectors, given for the multilevel factor model. By $\mathbf{Z}_{jg} = \begin{bmatrix} \mathbf{Z}_{jg1} \\ \mathbf{Z}_{jg2} \end{bmatrix}$ we will designate the jth student's scores on n variables in classroom g. These n variables consist of individual student scores on aptitude tests and achievement tests $\mathbf{Z}_{jg1}$ as well as group scores $\mathbf{Z}_{jg2}$ corresponding to scores on the teacher's and other group-level variables. ("1" indicates level 1 or within-group scores on which individuals vary within the group and "2" indicates level 2 or between-group scores, which are the same for every individual within the group.) Every student in a given class obtains the same scores for the teacher/group-level variables $\mathbf{Z}_{jg2}$. Now, let $\bar{\mathbf{Z}}_g = (1/N_g) \sum_{j=1}^{N_g} \mathbf{Z}_{jg}$ be the sample mean vector of the scores in classroom g. On the other hand, by $\bar{\mathbf{Z}} = (1/N) \sum_{g=1}^{G} \sum_{j=1}^{N_g} \mathbf{Z}_{jg}$ let us denote the grand mean vector of all score vectors across all subjects in all classrooms. In a manner analogous to the analysis of variance, let us now designate a score for an observation vector as consisting of two components:

$$\mathbf{Z}_T = \mathbf{Z}_B + \mathbf{Z}_W.$$

Here $\mathbf{Z}_T$ represents the combining of two deviation score vectors, $\mathbf{Z}_B = \bar{\mathbf{Z}}_g - \bar{\mathbf{Z}}$ and $\mathbf{Z}_W = \mathbf{Z}_{jg} - \bar{\mathbf{Z}}_g = \begin{bmatrix} \mathbf{Z}_{jg1} \\ \mathbf{Z}_{jg2} \end{bmatrix} - \begin{bmatrix} \bar{\mathbf{Z}}_{jg1} \\ \mathbf{Z}_{jg2} \end{bmatrix} = \begin{bmatrix} \mathbf{Z}_{W_1} \\ \mathbf{0} \end{bmatrix}$. $\mathbf{Z}_B$ will be essentially the same as in the multilevel factor analysis model. But $\mathbf{Z}_W$ differs in that the within-group deviation scores on the group-level variables are all zero, since their raw scores are all the same within a given group and hence equal to their mean.

As before, this leads to a composition of the total variance and covariance matrix in the population as

$$\Sigma_T = \Sigma_B + \Sigma_W.$$

But in this case, with the 0's of within-group deviation scores on the group-level variables, this turns out to be

$$\Sigma_T = \begin{bmatrix} \Sigma_{11} & \Sigma_{12} \\ \Sigma_{21} & \Sigma_{22} \end{bmatrix} = \begin{bmatrix} \Sigma_{B_{11}} & \Sigma_{B_{12}} \\ \Sigma_{B_{21}} & \Sigma_{B_{22}} \end{bmatrix} + \begin{bmatrix} \Sigma_{W_{11}} & \mathbf{0} \\ \mathbf{0} & \mathbf{0} \end{bmatrix}.$$

These in turn may be estimated by sample variance–covariance matrices

$$\mathbf{S}_T = \mathbf{S}_B + \mathbf{S}_W$$

with

$$\mathbf{S}_{PW}^* = \frac{\sum_{g=1}^{G} \sum_{j=1}^{N_g} (\mathbf{Z}_{jg} - \bar{\mathbf{Z}}_g)(\mathbf{Z}_{jg} - \bar{\mathbf{Z}}_g)'}{N - G} \tag{12.1}$$

and

$$\mathbf{S}_B^* = \frac{\sum_{g=1}^{G} N_g (\bar{\mathbf{Z}}_g - \bar{\mathbf{Z}})(\bar{\mathbf{Z}}_g - \bar{\mathbf{Z}})'}{G - 1}. \tag{12.2}$$

As indicated earlier, when group sizes N_g are all the same, Muthén (1989, 1990) showed that Equation 12.1 is an unbiased estimate of Σ_W, whereas Equation 12.2 is an unbiased estimate of $\Sigma_W + c\Sigma_B$, where c is equal to the common size of all groups. In other words,

$$\mathbf{S}_W^* = \hat{\Sigma}_W \tag{12.3}$$

and

$$\mathbf{S}_B^* = \hat{\Sigma}_W + c\hat{\Sigma}_B. \tag{12.4}$$

Again, if the groups do not vary much in size, we may still obtain a good approximate solution by using the following value for c:

$$c^* = \frac{N^2 - \sum_{g=1}^{G} N_g^2}{N(G - 1)}. \tag{12.5}$$

We may then begin by proceeding to fit a within-groups structural model to $\mathbf{S}_W^*$.

However, if the software does not make the proper adjustments, the variables on which there is only between-groups variation should be modeled in the within-groups analysis as variables with uncorrelated disturbances having unit variances (Hox, 2002). This is necessary to keep the covariance matrix the same dimension as the between-groups covariance matrix and to keep it also from being singular. This will augment the degrees of freedom for the within-groups chi-square, and its degrees of freedom will have to be adjusted by hand. Next, we fit the resulting within-groups model along with a between-groups structural model to $\mathbf{S}_B^*$ as we did with the multilevel factor analysis model.

13

Longitudinal Models

Introduction

Models considered up to now assume that measurements are obtained simultaneously at a given point in space/time. They take no account of ordering among the variables, such as ordering in time or ordering across space. For example, a common factor model ordinarily has the assumption that subjects are measured on all variables simultaneously, so there is no ordering to be given among them representative of anything in the world. (They may be given numbers but these are not representative of anything other than that one variable is different from another.) But when measurement variables are ordered in time or space, this changes how we model their relationships, because then we must take into account this ordering. We will call models that take into account ordering of variables in space or time "longitudinal models." The field of longitudinal modeling is now quite well developed and encompasses many approaches to modeling. SEMs are only one approach to longitudinal modeling. We can thus only refer briefly to some representative longitudinal models.

Simplex Models

Although others preceded him in the mathematical development of simplex models, Marshal B. Jones (1959, 1960) was one of the first in the behavioral sciences to call attention to them. In fact, he can be considered a precursor to the development of structural equation modeling in the behavioral sciences. Jones was an aviation psychologist studying the effects of training on naval pilots at the U.S. Naval School of Aviation Medicine at Pensacola, Florida. In the 1950s, factor analysis was the principal tool used in studying correlations among variables. But Jones came to question whether factor analysis was the proper tool for dealing with training data. Nevertheless there were those who were using factor analysis at that time to study relations between training and performance. In the learning laboratories there were also those who were using factor analysis to study the relationships between performance at different stages in sets of learning trials—which is still training. What bothered Jones was that the technique of factor analysis was principally exploratory, concerned with discovery, while he believed one had to go beyond that to test specific hypotheses about how variables were related. Implicit in his thinking is causality. Training men to become naval aviators is a complex undertaking. There are numerous skills that must be learned, and some must be learned before one can undertake to learn others. And if you do poorly in learning basic skills, this will show up in poorer performance in the learning of more complex skills dependent on having mastered the basic skills. You will not succeed and likely will kill yourself and destroy a million dollar airplane if you try to learn how to land on an aircraft carrier in a heavy sea before learning the more basic skills of flying the plane. So the learning of basic piloting skills is causally related to the complex skills involved in landing a plane on an aircraft carrier in the ocean. In fact we can break down the skills needed to pilot a plane itself into more fundamental skills, and show that there is a hierarchy of skills that must be learned in a certain order in the process of learning a complex skill. And this is true not just in the case of naval aviators, but in the case of learning any complex set of skills, such as in becoming a concert pianist, a great golfer or baseball player, or a professional mathematician. Measures of performance along the way are ordered in time and this ordering must be taken into account in formulating and testing any theory of how the complex skill is learned.

In the process of trying to formulate models for this kind of data, Jones was exposed to some of the work of Louis Guttman (1954) on simplex and circumplex models. A simplex model accounts for correlations among trials in learning experiments to a high degree. Performance at any trial in a study of acquiring a complex skill depends only on abilities acquired up to the previous trial as well as new abilities acquired in the current trial. At each successive trial, old abilities are retained and new abilities acquired. Jones thus sought to find examples of learning data in the literature that would exhibit the kinds of dependencies he expected from a simplex. He found it in a study of correlations among stages of practice of subjects in the Discrimination Reaction Time Test (Fleishman and Hempel, 1955), which we reproduce in Table 13.1.

TABLE 13.1

Correlations among Eight Stages of Practice in the Discrimination Reaction Time Test (Fleishman and Hempel, 1955)

	1	2	3	4	5	6	7	8
1	1.00							
2	0.74	1.00						
3	0.71	0.82	1.00					
4	0.69	0.78	0.83	1.00				
5	0.62	0.74	0.79	0.80	1.00			
6	0.59	0.68	0.72	0.74	0.77	1.00		
7	0.57	0.66	0.72	0.72	0.73	0.74	1.00	
8	0.56	0.64	0.71	0.74	0.77	0.80	0.79	1.00

Before we go on to provide a simplex model for these correlations, using structural equation modeling, note one important feature of this correlation matrix: It has "superdiagonal form." With exceptions that may be due to sampling error, in any row the highest correlation is between the variable in question and the variable immediately preceding it, found in the diagonal adjacent to the principal diagonal. Furthermore, the correlations drop off in magnitude as you go away from the principal diagonal toward the left or down. The lowest correlation is found in the lower left corner. Also, the farther apart in the order of the variables any two variables are, the lower will be the correlation between them. When you encounter a correlation matrix with these properties, you need to think that a simplex model applies.

Now the path diagram for a simplex model for these data is shown in Figure 13.1. In this model the V's represent scores in successive trials of the experiment. The V's thus have an order in time. The F's are latent variables, as are also the D's and E's. The E's are error variances of the observed

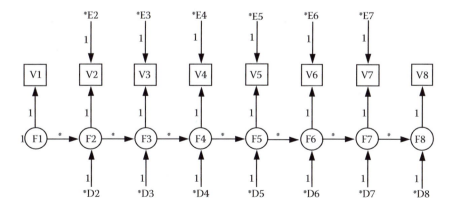

FIGURE 13.1 A simplex model for the correlation matrix in Table 13.1 for EQS.

variables, the V's. An asterisk by an arrow indicates a free structural parameter. An asterisk by the label for a latent variable indicates that its variance is a free parameter. The 1's are fixed parameters. $F1$ has a variance of 1 because it is equivalent to $V1$, which has a variance of 1 in the correlation matrix. $F8$ also has a variance of 1 determined by the variance of $V8$. Both $V1$ and $V8$ have no error variances, which we must assume to achieve identification of the model.

In this model $F1$ is presumed to be what is used or acquired in the first trial. $F1$ then has an effect on the latent variable $F2$, as does also $D2$. So, $F2$ represents the accumulated effect of $F1$ and $D2$ on $V2$. $F2$ in turn has an effect on $F3$ as does also $D3$. From here on the D's represent additional components learned. An immediately preceding F represents the accumulated learning of earlier components to be combined with the D learned in the current trial. Prior learning only enters in via an immediate preceding F and this only influences directly the next F in the series. This constraint on what variables the F's have an effect on is what gives this model a fair number of degrees of freedom.

Jones did not have a structural equation modeling program to fit a simplex model to these data. He was, nevertheless, able to construct a close approximation to what the correlations would be among the variables if a simplex model applied. However, I have the program EQS of Bentler and Chu (1995), and I fitted the above simplex model to the correlation matrix in Table 13.1. The results are shown in Table 13.2.

If we use these parameter estimates and path-tracing rules, we can see why variables that are farther apart have lower correlations in a simplex model. For example, variable $V1$ and variable $V8$ have correlations due to the product of all the coefficients on the arrows between the F's, all of which are less than unity. In fact, the correlation between any pair of variables is just the product of the coefficients on the arrows between the corresponding F's. Any pair of variables farther apart will have more arrows between the F's to pass through, and since the associated coefficients are less than unity, their correlation, equal to the product of these coefficients, will be less than the correlation of a pair of variables between them.

The fit of the simplex model in Figure 13.1 to the correlation matrix in Table 13.1 is indicated by the EQS program in Table 13.3.

The chi-square of 9.407 with 16 degrees of freedom is not significant, indicating fit to within sampling error. The goodness-of-fit indices (GFIs) such as the CFI are 1.00 or greater. The RMSEA is estimated to be 0, and the confidence interval on the RMSEA contains 0. So, this fit is about as good as it gets.

The next example of simplicial data does not concern ordering in time but rather in space. The correlation matrix also does not have the superdiagonal form, but rather what Jones called a "doubly concave" form. The model will be said to have a "bisimplex" form involving a simplex on two sets of variables. The data were collected by a dentist, Dr. John H. Manhold, on a sample of 600 naval cadets. Manhold determined the Bodecker index, a measure of cavity formation, for each tooth, excluding the four wisdom teeth. The index for each tooth was correlated with the index for every other tooth. A high correlation

TABLE 13.2

Model Equations with Parameter Estimates for
the Simplex Model in Figure 13.1 Applied to
the Correlation Matrix in Table 13.1

Standardized Solution

$V1 = 1.000F1 + 0.000E1$
$V2 = 0.923F2 + 0.384E2$
$V3 = 0.936F3 + 0.353E3$
$V4 = 0.923F4 + 0.385E4$
$V5 = 0.913F5 + 0.407E5$
$V6 = 0.889F6 + 0.457E6$
$V7 = 0.866F7 + 0.500E7$
$V8 = 1.000F8 + 0.000E8$
$F2 = 0.801F1 + 0.598D2$
$F3 = 0.949F2 + 0.317D3$
$F4 = 0.964F3 + 0.265D4$
$F5 = 0.953F4 + 0.304D5$
$F6 = 0.947F5 + 0.322D6$
$F7 = 0.977F6 + 0.215D7$
$F8 = 0.912F7 + 0.410D8$

Note: Obtained using the EQS Program of Bentler and
Chu (1995).

between two teeth indices meant that if you had cavities in one tooth, you
were likely to have them in the opposite one also. Jones then recognized that
the correlations had a simplicial basis, and he sought to fit a model to the
variables. These data appear in Jones (1960). Teeth and their labels are shown
in Figure 13.2.

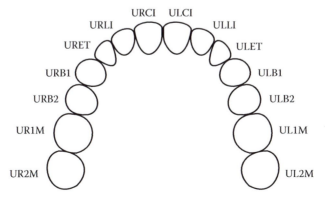

FIGURE 13.2 Diagram of the upper arch of teeth with standard dental labels for each tooth.

TABLE 13.3

Goodness-of-Fit for the Simplex Model Fit to the Correlation Matrix in Table 13.1

Goodness-of-fit summary

Independence model chi-square = 1579.169 on 28 degrees of freedom

Chi-square = 9.407 based on 16 degrees of freedom

Probability value for the chi-square statistic is 0.89570

The normal theory recursive least squares (RLS) chi-square for this ml solution is 9.280

Bentler–Bonett normed fit index = 0.994

Bentler–Bonett non-normed fit index = 1.007

CFI = 1.000

Bollen (IFI) fit index = 1.004

McDonald (MFI) fit index = 0.017

Lisrel GFI fit index = 0.988

Lisrel AGFI fit index = 0.974

Root mean square residual (RMR) = 0.012

Standardized RMR = 0.012

Root mean square error of approximation (RMSEA) = 0.000

90% confidence interval of RMSEA (0.000, 0.029)

Given the correlation matrix in Table 13.4, I formulated a SEM for it, which I show in Figure 13.3.

The variables in Figure 13.3 are ordered from the rear to the front of the mouth, from second molar to first molar, to second bicuspid to first bicuspid, to eye tooth to lateral incisor and then central incisor. UR means "upper right"

TABLE 13.4

Correlations between Upper Teeth in Terms of Degree of Dental Caries

1.00	UR2M												
0.54	1.00	UR1M											
0.43	0.42	1.00	UR2B										
0.32	0.34	0.62	1.00	UR1B									
0.22	0.23	0.31	0.38	1.00	URET								
0.25	0.22	0.24	0.29	0.42	1.00	URLI							
0.18	0.24	0.23	0.24	0.36	0.46	1.00	URCI						
0.14	0.19	0.20	0.20	0.34	0.41	0.70	1.00	ULCI					
0.21	0.25	0.20	0.27	0.37	0.59	0.40	0.43	1.00	ULLI				
0.21	0.26	0.20	0.31	0.46	0.37	0.33	0.31	0.43	1.00	ULET			
0.37	0.40	0.47	0.52	0.31	0.27	0.22	0.27	0.35	1.00	ULET	UL1B		
0.40	0.43	0.59	0.48	0.28	0.23	0.23	0.15	0.23	0.28	0.62	1.00	UL2B	
0.50	0.58	0.42	0.33	0.20	0.21	0.24	0.17	0.24	0.25	0.47	0.47	1.00	UL1M
0.58	0.51	0.42	0.42	0.26	0.31	0.21	0.20	0.27	0.28	0.48	0.44	0.55	1.00	UL2M

Source: Jones, M. B. (1960). Molar Correlational Analysis. Monograph series no. 4. Pensacola, FL: U.S. Naval School of Aviation Medicine.

Note: N = 600.

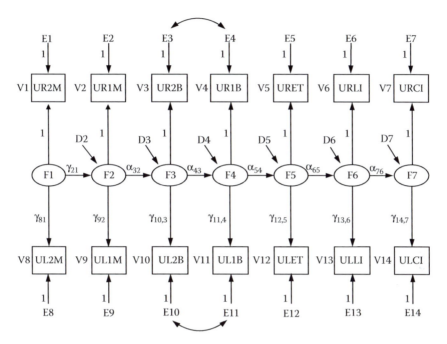

FIGURE 13.3 A bisimplex model with two manifest variables for each latent *F* variable for the dental caries data.

and UL means "upper left." The corresponding teeth on each side of the mouth are paired to serve as indicators of the same latent *F* variable. This suggests that physical contiguity between teeth is not the only basis for similarity in the degree of dental caries, but corresponding teeth that perform similar functions also are most similar in their degree of dental caries. In fitting the model of a true bisimplex, I initially specified that all error variables on the manifest variables were to be uncorrelated. But the model did not fit exactly, and Lagrange multiplier tests revealed that the errors on the two bicuspids on a given side were correlated. Hence we show in the model the bidirectional links between *E*3 and *E*4 and between *E*10 and *E*11 to indicate the correlation between these variables. Jones (1960), using his own methods, also found that the bicuspids on any one side were more correlated than suggested by a simplex model. The initial bisimplex model did not fit sufficiently well, but with the freeing of the error correlations between the bicuspid teeth, the model became acceptable as an approximation, although there was still significant residual covariance that remains unexplained. The large sample of 600 subjects allowed for the detection of small deviations from the bisimplex model. In any case, the chi-square statistic for goodness-of-fit with 69 degrees of freedom was 135.252, which was significant beyond the 0.001 level. The average absolute standardized residual was 0.023. The CFI index was 0.981, which indicates an excellent approximation. The GFI of LISREL was 0.969, while

TABLE 13.5

Correlations among Five Objective Tests given
1064 Sophomores at Bucknell College in 1931

1. Spelling	1.00				
2. Punctuation	0.62	1.00			
3. Grammar	0.56	0.74	1.00		
4. Vocabulary	0.48	0.50	0.58	1.00	
5. Literature	0.39	0.46	0.47	0.69	1.00

the RMSEA index was 0.04, which is an acceptable degree of approximation, with the 90% confidence interval on the RMSEA index given as (0.03, 0.05). So, while a bisimplex model based on order of teeth is an excellent approximation, it does not account for all of the correlation between teeth, which may be due additionally to the special shape and function performed by the respective teeth.

Jones (1960) also showed correlation matrices displaying simplicial form between variables that were not ordered in time or space, but in terms of nesting of abilities and knowledge. Consider the correlation matrix in Table 13.5 between five test scores obtained from 1046 sophomores at Bucknell College in 1931 (Bigham, 1932).

According to Jones (1960), the first three tests were based on one long passage that the students were to correct for spelling, punctuation, and grammar. The vocabulary test involved items that required the subject to show which of four words was synonymous with a given word. On the literature test the subject gave the names of English authors and books most connected with a certain theme or topic. Jones (1960) notes that "... for the most part, that is, 80% of the variance, the abilities required in spelling are included among those which are used in punctuation and grammar tests, that the same things which make for success in punctuation also make for success in grammar, and so on" (p. 25). He also notes that there seems to be more than just contiguity involved in this ordering of variables, but complexification of functions as well. The ordering could also be accounted for by the recency in which the skills were acquired. Spelling was learned before punctuation, and in turn grammar in grammar school and high school. Vocabulary at the college level involves learning many new technical terms. Vocabulary acquisition would take place in the first and second year. Literature may be explored in the second year, and its understanding and the learning of names associated with themes and topics would depend on the vocabulary one had acquired up to that point.

I now show another correlation matrix from Jones' (1960) monograph to illustrate that a simplex need not apply just to data ordered in time and space. The variables in this example involve the various stages of Ravens Progressive Matrices Test, which is a nonverbal test of general intelligence or analytic,

TABLE 13.6

Correlations among the Six Stages of the Ravens Progressive
Matrices Test

	1	2	3	4	5
1. Continuous patterns	1.00				
2. Analogies of figures	0.57	1.00			
3. Development of figures	0.53	0.64	1.00		
4. Combination of figures	0.47	0.54	0.66	1.00	
5. Resolving figures into parts	0.32	0.35	0.39	0.52	1.00

rule-inferring ability based on geometrical figures and patterns. There are five stages, and the test at each stage is designed to be more complex than the test in a preceding stage and measures whatever abilities were required in preceding stages, as well as requiring additional abilities. I leave it to the reader to provide the fitting of a simplex model to this and the preceding correlation matrix (Table 13.6).

A central theme in Jones' (1959, 1960) monographs concerns how exploratory common factor analysis often yields a good fit to simplicial correlation matrices with a smaller number of factors, but with factors that are difficult to interpret. Jöreskog in Jöreskog and Sörbom (1979) shows that for three and four variables, one can always fit a common factor model exactly to data generated by a simplex. Bast and Reitsma (1997) show that the common factor model will still be a close approximation for cases with more than four variables—as long as there is freedom to take enough factors. Mulaik and Millsap (2000) make a similar observation. For example, they show that an exploratory common factor model with three common factors will fit 10 variables generated by a simplex very well. The factors represent early, intermediate, and late effects, respectively, in the ordering of the variables, but seem to be artifacts. Jöreskog and Sörbom (1979) also show other forms of bisimplex models and should be consulted for further details about these models.

Latent Curve Models

Factor analysis was applied early on (Tucker, 1958; Rao, 1958; Scher, Young, and Meredith, 1960) to learning and growth curves. Meredith and Tisak (1990) in a seminal paper synthesized preceding developments in this area and suggested ways of generalizing these approaches to structural equation modeling. We will now survey some of these developments here.

We are going to consider cases where individuals i are measured successively in time or serially in space on certain variables. We will keep things simple by considering that the same single variable is measured on each of several repeated occasions, but the method can be generalized to several variables measured serially or longitudinally. We will initially draw heavily on Meredith and Tisak (1990).

Let $z_i(t)$ be a realization of a random variable Z in individual i's measured behavior on some characteristic, at t, where t is a nonrandom variable that may be discrete or continuous. Usually, in applications, t is discrete and Z is always continuous. The variable t can represent "time, age, grade, trial number, degree of arousal, experimental condition, test form or stimulus intensity, and might even be unordered and multivariate, as in the dummy coding of unordered experimental conditions" (Meredith and Tisak, 1990, p. 107).

Next, we will assume that for each $z_i(t)$ there is a function that relates Z to t. The function relating $z_i(t)$ to t itself will be thought of as a composite of a set of several common "basis functions" $g_k(t)$ which are themselves functions of t. These basis functions may be presumed to be observed or latent and even unknown. In other words, the relation between $z_i(t)$ and t is to be given by

$$z_i(t) = \sum_{k=1}^{r} w_{ik} g_k(t) + e_i(t). \tag{13.1}$$

The expression $e_i(t)$ denotes a realization of an error random variable $E(t)$ which may encompass error of measurement and even errors of approximation. w_{ik} is also a realization of a random variable W_k on which individuals i vary, representing the degree to which or the salience to which individual i "uses" the value of the kth basis function, $g_k^{(t)}$, at t in determining his/her response. The parameter w_{ik} is thus an individual difference parameter.

We will further consider that by varying t, implying that we observe the variable Z in each individual i at each of several discrete points t, we relate each $z_i(t)$ (t now also varying) to the same basis function $g_k(t)$, but the weights w_{ik} for individual i remain the same for each t, while the value of the kth basis function, $g_k(t)$, varies with t.

Meredith and Tisak (1990) then consider that the random variable Z is measured in each individual i at each of the points $t_1, t_2, \ldots, t_p$ of t. They then introduce a change in notation that brings out the similarities of the model to the common factor model. Let $z_{ij} = x_i(t_j)$ be individual i's score on occasion t_j, $\gamma_{jk} = g_k(t_j)$ be the value of function $g_k(t)$ at t_j, and $e_{ij} = e_i(t_j)$ be the error in Z at t_j. Over all the p occasions of t, let us gather these expressions into vectors representing the realizations of the random variables of a given individual. Let $\mathbf{z}_i' = [z_{i1}, z_{i2}, \ldots, z_{ip}]$, $\mathbf{w}_i' = [w_{i1}, w_{i2}, \ldots, w_{ir}]$, and $\mathbf{e}_i' = [e_{i1}, e_{i2}, \ldots, e_{ip}]$. We now write in matrix form an equation for a particular realization of individual i's responses as

$$\mathbf{z}_i = \Gamma \mathbf{w}_i + \mathbf{e}_i.$$

In terms of random variables varying over i, this becomes

$$\mathbf{Z} = \mathbf{\Gamma W} + \mathbf{E}. \tag{13.2}$$

We now consider the means and variance–covariance matrices for the random variables in question and then will introduce some assumptions that provide constraints on the model generally.

The expected value of $\mathbf{W}$ is given by $E(\mathbf{W}) = v$. In general, we assume this is *not* equal to zero. Similarly, $E(\mathbf{WW'}) = \mathbf{Y}$ and $E(\mathbf{EE'}) = \mathbf{Y}$. We will assume that $E(\mathbf{E}) = \mathbf{0}$ and $E(\mathbf{EW'}) = \mathbf{0}$. From these assumptions we may deduce that

$$E(\mathbf{Z}) \equiv \mu = \mathbf{\Gamma}v. \tag{13.3}$$

We introduce the additional assumptions that $\mathbf{W}$ and $\mathbf{E}$ are independent and do not merely have zero covariances, and further assume that the errors on different occasions are independent, implying that $\mathbf{\Psi}$ is a diagonal matrix of uncorrelated errors.

We may also further deduce the fundamental theorem of this model:

$$E(\mathbf{ZZ'}) \equiv \Omega = \mathbf{\Gamma Y \Gamma'} + \mathbf{\Psi}. \tag{13.4}$$

This obviously has the form of a common factor model. However, $E(\mathbf{Z}) \neq \mathbf{0}$ and $E(\mathbf{W}) \neq \mathbf{0}$.

If $\mathbf{Z}$ is a $p \times N$ matrix of p observations of Z on N subjects, then

$$\hat{\mu} = \frac{1}{N}\mathbf{Z1}$$

is an estimate of the mean vector for $\mathbf{Z}$, where $\mathbf{1}$ is an $N \times 1$ column sum vector.

This represents the average growth curve over the p occasions. On the other hand,

$$\hat{\Sigma} = \frac{1}{N}\mathbf{ZZ'}$$

is an estimate of the second moment matrix. Meredith and Tisak (1990) indicated that Tucker (1958) would perform an exploratory factor analysis of $\hat{\Sigma}$ and rotate the factor solution to obtain a solution (not unique) for $\hat{\mathbf{\Gamma}}$ and $\hat{\mathbf{Y}}$. However, the rotated solution was not a simple structure solution. Meredith and Tisak (1990) however suggested modeling the partitioned matrix in a confirmatory context as

$$\hat{\mathbf{M}}\begin{bmatrix} \hat{\Omega} & \hat{\mu} \\ \hat{\mu}' & 1 \end{bmatrix} = \begin{bmatrix} \hat{\mathbf{\Gamma}} & \mathbf{0} \\ \mathbf{0}' & 1 \end{bmatrix}\begin{bmatrix} \hat{\mathbf{Y}} & \hat{v} \\ \hat{v}' & 1 \end{bmatrix}\begin{bmatrix} \hat{\mathbf{\Gamma}}' & \mathbf{0} \\ \mathbf{0}' & 1 \end{bmatrix} + \begin{bmatrix} \hat{\mathbf{\Psi}} & \mathbf{0} \\ \mathbf{0}' & 1 \end{bmatrix}$$

$$= \begin{bmatrix} \hat{\mathbf{\Gamma}}\hat{\mathbf{Y}}\hat{\mathbf{\Gamma}}' + \hat{\mathbf{\Psi}} & \hat{\mathbf{\Gamma}}\hat{v} \\ \hat{v}'\hat{\mathbf{\Gamma}}' & 1 \end{bmatrix}. \tag{13.5}$$

They suggest that one could specify the matrix Γ by requiring its first column to be all 1's, the second to be $t_1, t_2, \ldots, t_p$, the third to be $t_1^2, t_2^2, \ldots, t_p^2$, and so on. Thus the basis functions would be simple polynomials. These, however, are not mutually orthogonal. We could, instead, use sets of orthogonal polynomials. Other functional forms, such as the negative exponential, are possible. To be identified, at least r^2 parameters across Γ, $\mathbf{Y}$, and $\mathbf{v}$ must be properly specified.

The matrix Ψ could be simplified to $\Psi = \psi\mathbf{I}$, testing whether all of the error variances are equal. One could allow some of the error variables to be correlated, for example, between adjacent variables only, so that Ψ is modeled as a tridiagonal matrix with free parameters on the principal diagonal and the adjacent diagonals.

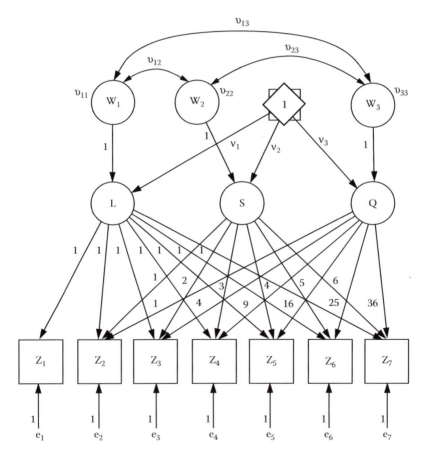

FIGURE 13.4 A latent curve model with L, S, and Q latent basis functions for seven variables representing a variable measured at seven points in time. Each of the v_{ij} is a free parameter as are v_1, v_2, and v_3.

In Figure 13.4 we show a latent curve model of seven variables, each representing an observation on some variable at some successive point t in time. The values of t are 0, 1, 2, 3, 4, 5, and 6. This is a model of a curve using variable level, slope, and quadratic effect. W_1, W_2, and W_3 are individual weight random variables, each having a zero mean to which is added a constant mean given respectively by v_1, v_2, and v_3. L represents the constructed level function latent variable, S the constructed slope or linear function latent variable, and Q the quadratic function latent variable. Note that L, S, and Q have no disturbances and are completely dependent on W_1, W_2, and W_3 and the coefficients v_1, v_2, and v_3.

The first column of Γ is a column of 1's, indicating that the value of the function at each point t in time is the same for a given individual. The second column of Γ has the coefficients 0, 1, 2, 3, 4, 5, 6. These establish a linear trend among the observed variables. The third column has coefficients 0, 1, 4, 9, 16, 25, 36, which establish a quadratic trend among the observed variables. These are indicated as fixed values on the paths connecting the respective latent function factors to the observed variables. Another thing to notice is that the model has the form of a common factor model. We will now show a model that is more in keeping with a SEM.

The previous example showed a model with latent exogenous individual weight variables that conformed to a common factor model. Suppose that the weight variables were themselves determined in part by manifest variables. We show this case in Figure 13.5, where the variables X_1 and X_2 have been introduced into the previous model to be additional sources for the variation in the individual level and slope parameters, respectively. X_1 affects only W_1 whereas X_2 affects both W_1 and W_2.

Reality or Just Saving Appearances?

To my mind, there is something problematic about many applications of longitudinal models (and other models as well). It is not that they attempt to model longitudinal data, but rather than that they tend to do so in ways where the mathematics seems to outrun the science of what is represented by the models. To what extent do these models correspond to reality?

There is a tradition in the history of science that goes back to the conflict between Galileo and the Roman Catholic Church and beyond (Losee, 1980). It concerns the role of mathematics in science. Does mathematics simply provide models that allow one to "save the appearances," that is, reproduce the relations in the observed data, without any pretense of representing reality? Or do mathematical models seek to represent reality? The Church in Galileo's time was committed to interpreting as true certain passages in the Bible that stated that the earth did not move. The Church was also committed to tolerating

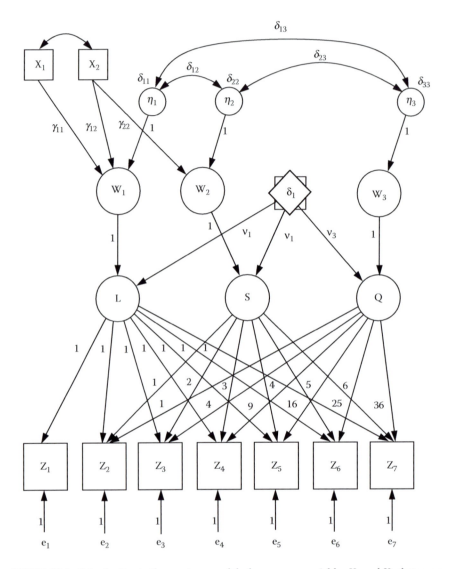

FIGURE 13.5 Introduction to the previous model of exogenous variables X_1 and X_2 that serve to moderate the level and slope parameters of the respective observed Z variables.

scientists' development of hypotheses that "saved the appearances," as long as they did not assert them as reality when they contradicted sacred text. The second-century Greek astronomer Ptolemy had developed a complex geocentric system of off-centered "deferent" orbits around the earth and planetary circular orbits called "epicycles" that were centered on the deferent orbits while following the path of the deferents around the earth. The system allowed for predictions of planetary positions, and accounted for retrograde motion observed in the planets. But because the system contradicted Aristotle's theory that the stars and planets were on concentric crystalline spheres centered on the earth, Ptolemy only claimed that his model merely "saved the appearances," providing a useful tool for calculating and predicting positions of planets, and was not a representation of reality. Even when Copernicus' much simpler heliocentric planetary system was put forth, with the earth one of the planets moving around the sun, it was tolerated by the Church as merely a "hypothesis" that saved the appearances. But Galileo argued that Copernicus' system was more than a hypothesis, it was reality. The earth really moved. That was the source of trouble Galileo had with the Church. But science has tended to side with Galileo in regarding scientific models as about reality and not merely "saving of the appearances."

Modern scientists, such as Galileo before them, regard as more objective (real) models that are simpler in requiring estimation of fewer parameters relative to the number of observations to be accounted for. Additionally, the model is invariantly supported across different laboratories by different researchers. External independent evidence for the entities of a model is also obtained. Objective models also make successful predictions of new phenomena.

Although the simplex models and the latent curve models that we have considered have positive degrees of freedom, their number relative to the number of observed parameters that they span is still relatively low. The ratio of the 16 degrees of freedom of the simplex model in Figure 13.1 to the 36 observed variances and covariances it is designed to account for is only 0.444. The ratio for the bisimplex model with two indicators per latent is higher, 0.66. The latent curve models have ratios in the 0.60's. But common factor models with many indicators per latent common factor and equality constraints on parameters often have ratios up in the high 0.80's and even 0.90's. These ratios decrease with each estimated parameter. We have argued in this chapter that equivalent models are possible because of estimated parameters in a model. And the existence of equivalent models raises the question of whether a given model is not just "saving the appearances" as opposed to asserting something objectively "real" about the world. More than "goodness-of-fit" is required.

It may be that quantitative psychologists tend to be more focused on the mathematics than the science of their models' applications. But ultimately for science's sake, they need to attend to the real-world meaning of entities and relations in their models, so that they are not just "saving the appearances." For example, what is the psychological meaning of each latent variable at each stage of the learning simplex? In the latent curve models, what is the psychological status of the L, S, and Q latents? Are these not merely mathematical devices for approximating a function with the first three terms of a power series? Granted, the curve is an interesting generalization of how general individuals' scores change across time. But do we know anything more about the physical or psychological mechanisms generating the function through time if we are able to fit a latent curve model to it? Is the latent curve just an artifact of the particular selection of variables and times at which they are measured? Or would it be invariant in impacting other similar variables measured at the same times? That will mean trying to find corroborating and external validation and support for the curve and for the latent variables of the models. Having four or more indicators per latent variable may also improve the situation. Corroboration of the existence of a latent among four or more variables gives it an independent existence. Showing that some estimated parameter values are natural constants for certain applications, will also lead to some currently estimated parameters becoming fixed parameters in these models, which will also increase degrees of freedom.

To a considerable extent, at this writing, longitudinal modeling is new in the behavioral sciences. This perhaps explains the emphasis on developing all-purpose models for describing change, which often do not consider multiple indicators of their latent variables or natural constants for certain applications. Time will tell whether these models will have more than mathematical interest in scientific applications in leading to real insights into the nature of change.

14

Nonrecursive Models

Introduction

So far, we have considered only recursive models, models with flow graphs that have no loops, no feedback, and no reciprocal effects between pairs of variables. We will now consider nonrecursive models with loops. I base much of this section on material in Heise's (1975) *Causal Analysis*. But an excellent source is also given in Chen (1983) and in the introduction to Wyatt (2004). A connected series of causal arrows that begins and ultimately ends at the same variable is a *loop*. In Figure 14.1, we show some basic kinds of loops in nonrecursive models.

Observe that in the case of each of the loops in Figure 14.1 there is an external variable impinging on one of the variables of the loop. This is essential to provide a value for the loop to begin with.

Nonrecursive models can have any number of loops and any number of variables in them. It will be important to identify each of the loops in such a model. I will use the convention in describing a loop by listing in order each variable in the loop in parentheses. I will deviate from Heise's convention by listing the start variable twice, at the beginning and the end. This is to avoid confusion with open paths that do not form loops. So in Figure 14.1a the one-variable loop is $(Y_1 Y_1)$. The reciprocal effects loop in Figure 14.1b is $(Y_1 Y_2 Y_1)$. The three-variable loop in Figure 14.1c is $(Y_1 Y_2 Y_3 Y_1)$. It is possible to begin a loop with any one of the variables in the loop, as long as that variable has an external variable impinging on it.

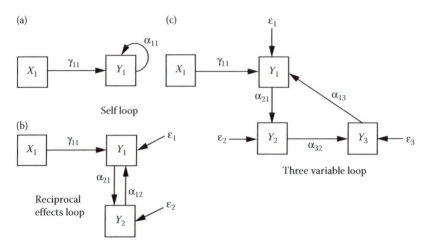

FIGURE 14.1   Three basic kinds of loops: (a) self loop; (b) reciprocal effects loop; and (c) three variable loop. (Adapted from Heise, D. R. (1975). *Causal Analysis.* New York: Wiley, p. 56.)

Flow Graph Analysis

The following discussion is based on principles of signal-flow graph analysis developed by the engineer Samuel Mason (1953, 1956). The theory of nonrecursive models is well understood by electrical engineers, who must analyze current flow in complex electrical circuits and estimate feedback effects (of loops). But the principles of flow graph analysis have been extended beyond electrical engineering to other fields of engineering where it is sometimes called "analysis of feedback control systems." It even has counterparts in hydrology. After World War II, mathematicians recognized that the varying graph theories developing in diverse fields had common properties and worked out and extended the theory of graphs as an abstract subject. Heise (1975) is a sociologist who adapted the theory of flow graph analysis to sociology.

We have already covered some principles of flow graph analysis in our earlier discussion of path-tracing rules for finding variances and covariances among variables in linear structural models. What is new is the use of the rules of flow graph analysis to condense or reduce path diagrams. In Figure 14.2, there are some common ways of reducing or condensing graph models either by absorbing variables or consolidating parallel paths. In Figure 14.2a, X has an indirect effect on the variable Z through an intervening variable Y. The intervening variable Y can be absorbed into X. The path coefficient then for X's effect on Z is the product of coefficients along the path from X to Z. In Figure 14.2b, if there are two parallel paths from X to Y through intervening variables U and V, we can consolidate these two paths into one direct path from X to Y. The path coefficient for the direct effect of X on Y is the sum of

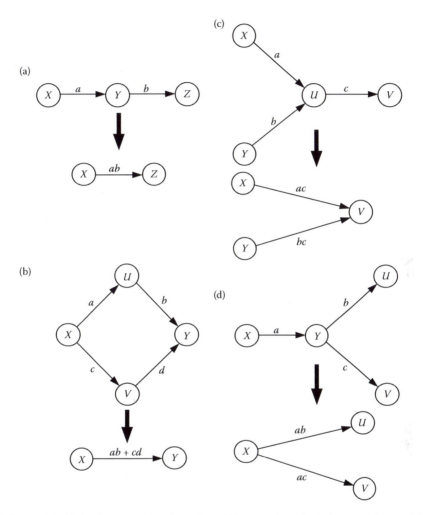

FIGURE 14.2 Reduction or condensing rules of flow graph analysis for recursive models. (a) Absorbing a variable; (b) consolidating paths in parallel; (c) absorbing a variable; (d) absorbing a variable. (Figure (a)–(c): From Wyatt, G. (2004). *Macroeconomic Models in a Causal Framework.* Edinburgh: Harmony House. With permission.)

the products of the path coefficients in the two parallel paths. In Figure 14.2c, X and Y are two immediate causes of a variable U, which in turn is an imme-diate cause of a variable V. We can absorb U into X and Y, respectively, by making X and Y immediate causes of V, where each has a structural coeffi-cient equal to the product of the coefficients along the path from X through U to V. Finally, in Figure 14.2d, a variable X may have an immediate effect on Y, which has two paths branching from it to the variables U and V, respectively. But we may eliminate Y by taking X as a direct common cause of both U and V. In that case, the path coefficients on the direct paths from X to U and V

are the products of the path coefficients on the paths from X to Y to U and V, respectively. There will be other absorbing and consolidating rules that involve path diagrams with loops, and we will display them shortly.

Open and Closed Paths

We now need to make a distinction between open and closed paths. We are already familiar with open paths, because they are the kind found in graphs of recursive models. An open path is a connected sequence of arrows with no reversal of arrows along the path that passes through no variable more than once and begins and ends with different variables (Heise, 1975). A closed path is a loop, that is a connected series of arrows that have no reversals along the path of the loop that begins and ends with the same variable. The beginning and ending variable can be chosen arbitrarily.

In Figure 14.3, I show a model with three loops: $(Y_1Y_2Y_1)$, $(Y_1Y_2Y_3Y_1)$, and $(Y_1Y_2Y_3Y_4Y_1)$.

Open paths are designated in a manner similar to loops, by indicating the ordered series of variables in the path, including both the starting variable and the ending variable. For example, in the nonrecursive model of Figure 14.3, there is an open path: $Y_1Y_2Y_3Y_4$. Another is Y_4Y_1. Another is X_1Y_1. $Y_2Y_3Y_4$ is even another. The effect of an open path, which, following Heise (1975) will be designated as E, is simply the product of the structural coefficients along the open path. $E_{Y_1Y_2Y_3Y_4} = \alpha_{21}\alpha_{32}\alpha_{43}$, $E_{Y_2Y_3Y_4} = \alpha_{32}\alpha_{43}$, and $E_{Y_1Y_4} = \alpha_{14}$.

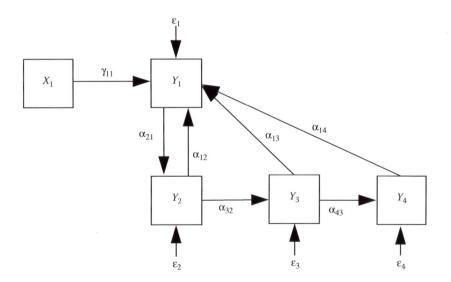

FIGURE 14.3   A nonrecursive model with three loops. (Adapted from Heise, D. R. (1975). *Causal Analysis*. New York: Wiley, p. 57.)

On the other hand, according to Heise (1975) the "return effect" or loop gain of a loop is represented as L and is equal to the product of the coefficients along the route of the loop. So, the return effect of the loop $(Y_1Y_2Y_1)$ is $L_{Y_1Y_2Y_1} = \alpha_{21} \cdot \alpha_{12}$. The return effect of the loop $(Y_1Y_2Y_3Y_1)$ is $L_{Y_1Y_2Y_3Y_1} = \alpha_{21} \cdot \alpha_{32} \cdot \alpha_{13}$. For the loop $(Y_1Y_2Y_3Y_4Y_1)$ the return loop is $L_{Y_1Y_2Y_3Y_4Y_1} = \alpha_{21} \cdot \alpha_{32} \cdot \alpha_{43} \cdot \alpha_{14}$. Heise (1975) indicates that the *return effect* of a loop "... indicates how much a variable in a loop will change after just one cycle around the loop" (p. 59).

Touching paths

In analyzing a nonrecursive flow graph, one must identify the loops and also whether they "touch" one another or an open path. According to Heise (1975) two loops are said to "touch" if the loop identifiers contain one or more variables in common. A loop is said to touch an open path if any variable in the open path except the first also appears in the identifier of the loop. In Figure 14.3, the open path $Y_2Y_3Y_4$ touches the loops $Y_1Y_2Y_3Y_1$ and $Y_1Y_2Y_3Y_4Y_1$. $Y_2Y_3Y_4$ does not touch the loop $Y_1Y_2Y_1$ because Y_2 is at the beginning of the open path. Also the loops $Y_1Y_2Y_3Y_1$ and $Y_1Y_2Y_3Y_4Y_1$ touch each other by sharing $Y_1Y_2Y_3$ in common.

The concept of touching is useful because we can use it to define *relevant feedback*, which will be important in Mason's (1953, 1956) formula for computing the total effect of an input variable on any dependent variable (which we will consider shortly). Consider that in Figure 14.4 X_1 is a (indirect) cause of

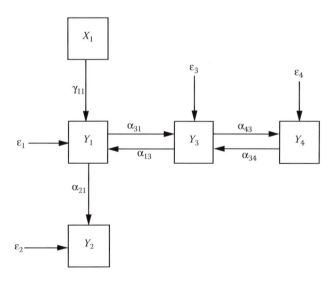

FIGURE 14.4 At Y_1 the open path $X_1Y_1Y_2$ touches the loop $Y_1Y_3Y_1$, which in turn touches the loop $Y_3Y_4Y_3$ at Y_3. The open path Y_1Y_2 does not touch the loop $Y_1Y_3Y_1$. (Adapted from Heise, D. R. (1975). *Causal Analysis.* New York: Wiley, p. 61.)

Y_2 via the open path $X_1Y_1Y_2$. But the loop $Y_1Y_3Y_1$ touches the path $X_1Y_1Y_2$ at Y_1 and by feedback can amplify the effect of X_1 on Y_2 by amplifying Y_1. So, the loop $Y_1Y_3Y_1$ is relevant feedback to the open path. But the loop $Y_3Y_4Y_3$ touches the loop $Y_1Y_3Y_1$ at Y_3, so it indirectly provides relevant feedback to the open path $X_1Y_1Y_2$ also. So relevant feedback is provided by any loops that directly or indirectly touch on the open path in question, often by touching other loops. The effect of loops is to amplify or dampen the effects in open paths.

Condensing Loops

Any multivariable loop may be condensed to a self-loop, with the return effect of the self-loop equal to the return effect of the multivariable loop. This is just a generalization of the condensing rule for open paths to closed paths. The importance of this is that we can analyze how loops function in terms of their operation as self-loops.

Consider the graph in Figure 14.5 of a three loop. (A loop with three variables is a "three loop.") The loop $YUVY$ with return effect bcd may be replaced by a self-loop on Y with return effect bcd (Wyatt, 2004). Furthermore, the self-loop can be absorbed into the XY open path when considering the total effect of X on Y. We will provide more details about that shortly.

What we now need to develop is a better understanding of what happens when a loop is in a model. We are going to assume that the flows within loops cycle around their loops, producing changes in variables within the loop very rapidly. At the same time, changes in variables having causal inputs to variables within the loops do not change (*stationarity*) while the loop seeks an *equilibrium*. This implies that the variable Y in Figure 14.4 should not be measured until the cycling of effects around the loop has reached a point of equilibrium; otherwise, an incorrect value for Y will be obtained. Fortunately

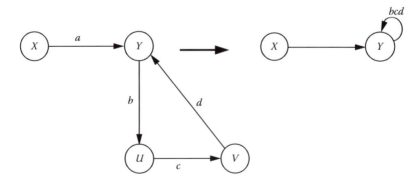

FIGURE 14.5 Any multiple variable loop may be condensed to a self-loop. (From Wyatt, G. (2004). *Macroeconomic Models in a Causal Framework*. Edinburgh: Harmony House, p. 6. With permission.)

in almost all applications this occurs after a very short period of time. And we will have to assume that a potentially infinite number of cycles are possible within the loop to reach equilibrium, even though practical equilibrium is obtained quickly.

Suppose now that the variable X in Figure 14.5 has gone from a value of 0 to a value of 1 and that the structural coefficient a equals 1. Suppose Y prior to this change in X was also 0 in value. Then the value of Y after X becomes 1 also becomes 1 (because $Y = 1X$) and the effect directly impacting on Y from X will remain at 1 as long as X does not change. However the change of 1 unit in Y goes from Y into the self-loop and begins a cycle of changes in the loop. Let $L_Y = bcd$ be the return effect of the self-loop. Then when the original change of 1 unit in Y returns to Y after its first cycle around the loop, it is multiplied by the return effect L_Y of the loop and becomes an increment added to the original value of Y, so that $Y = 1 + L_Y$. It will be maintained at that value because, as Heise puts it, the increment L_Y is also indirectly dependent on X. The input into the loop of 1 from the effect of X has already entered the loop. The input to Y of L_Y, being new, then becomes a new input to the loop. So, the increment of L_Y passes again around the loop, and comes out again at Y, having been multiplied at that point by the return effect L_Y of the loop to yield an increment $L_Y \cdot L_Y$, which is then added to $1 + L_Y$. So now, the value of Y is $1 + L_Y + L_Y^2$. It is important to understand that the cycling process takes place through time and that the input to Y from the self-loop is coming effectively from a different variable distinguished by its time of input. That is why it is being added to the other effects on Y. Again because the gain of L_Y^2 is dependent also on X, it will be maintained at that value as long as X does not change. We indeed assume X will not change while these cycles are taking place.

Now, the cycles through the loop do not stop at this point. At this point the gain of L_Y^2 now enters the loop as a new input and cycles around and comes out multiplied again by L_Y, meaning it produces a new increment of L_Y^3 to be added to Y. So, this process continues without end. But as long as $-1 < L_Y < +1$, the value of Y will converge to some finite value. (If L_Y is outside these limits, the process will diverge and the value of Y will blow up to infinity.) This value is the sum of the terms in the infinite series

$$1 + L_Y + L_Y^2 + L_Y^3 + L_Y^4 + L_Y^5 + \cdots .$$

Fortunately, mathematicians have worked out the value of this sum when L_Y satisfies the constraint that it is <1 in absolute magnitude. It is the finite sum of a well-known infinite power series and it equals $1/(1 - L_Y)$. Because this is a value >1, Y ends up with a gain >1. We are also fortunate that because L_Y is raised to increasingly higher powers, the terms of the series rapidly become very small. Consider the case where $L_Y = 0.30$. Then we get the power series

$$1 + 0.30 + 0.09 + 0.027 + 0.0081 + 0.00243 + 0.000729 + \cdots .$$

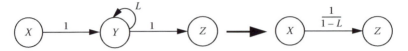

FIGURE 14.6 Y and its loop may be absorbed so that the effect of X on Z is $1/(1 - L)$. (Adapted from Heise, D. R. (1975). *Causal Analysis.* New York: Wiley, p. 62.)

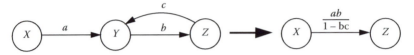

FIGURE 14.7   A general feedback loop and its collapse to yield the total effect of X on Z.

So, we see that after only a small number of cycles, the additional terms become negligible. Then the value of Y in the present case becomes $1/(1 - 0.30) = 1.428571\ldots$.

In Figure 14.6 we show how a mediating variable Y with a self-loop may be absorbed to yield the resulting effect of X on Z. Figure 14.7 shows how a general feedback loop may also be absorbed to yield the effect of X on Z.

Using the previous equivalences, we may find the total effect of any causal variable on any effect variable downstream from it by successively absorbing interior variables until only the cause and the effect variable remain.

In Figure 14.8a, we show a graph with two loops. We seek the total effect of X on Y. At V we can treat V as a branching node and apply the absorption rule of Figure 14.8d. One branch loops back to U and the other continues to Y. So, U can be absorbed by creating a self-loop on V with the return effect bd and the effect of U on Y becomes bc. We show the result of these changes in Figure 14.8b. Next we can eliminate the branch from X to U and then the self-loop on U by generalizing the rule in Figure 14.6. That leaves the effect of X on U equal to $a/(1 - bd)$. The result is shown in Figure 14.8c, which now has the form of the case shown in Figure 14.7, which can absorb U and the feedback loop from Y to U. In Figure 14.7, a corresponds to $a/(1 - bd)$ in Figure 14.8c. b in Figure 14.7 corresponds to bc in Figure 14.8c. c in Figure 14.7 corresponds to e in Figure 14.8c. The final result is shown in Figure 14.8d.

We illustrate another case with a self-loop in Figure 14.9. To find the total effect of X on Z, we first absorb the self-loop on Y in Figure 14.9a while replacing the effect of X on Y as $a/(1 - e)$. That leaves a fully recursive graph in Figure 14.9b. On the open path XYZ we multiply coefficients and add the product to the coefficient on the parallel path XZ. The result is $ac/(1 - e) + b$.

Mason's Direct Rule

Samuel Mason (1953, 1956) provided a general rule for finding in a causal network of variables the total effect of any causal variable on any other effect

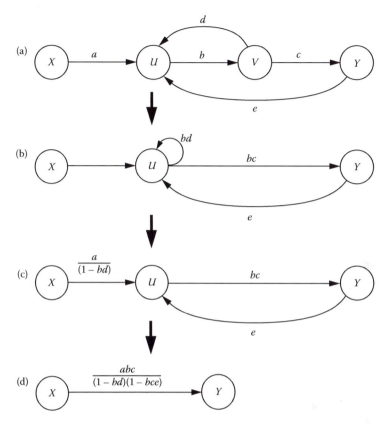

FIGURE 14.8 Showing how to obtain the total effect of *X* on *Y* by successive absorption of interior variables until only *X* and *Y* remain. (From Wyatt, G. (2004). *Macroeconomic Models in a Causal Framework*. Edinburgh: Harmony House, p. 13. With permission.)

variable directly or indirectly dependent on it. Heise (1975) presented this rule in his work, and I will present it here in a slightly modified form:

Mason's Rule: In a causal network of variables the total effect *T* of a causal variable *X* on an effect variable *Y* in the network dependent on *X* is determined as follows: Let $E, E', E'', E''', \ldots$ be the distinct open parallel paths from *X* to *Y* in the network. Let $L, L', L'', L''', \ldots$ be the return effects for all distinct loops providing relevant feedback to the total effect. Then

$$T = \left[\frac{(E + E' + E'' + E''' + \cdots) \cdot (1 - L) \cdot (1 - L') \cdot (1 - L'') \cdot (1 - L''') \cdot \cdots)}{(1 - L) \cdot (1 - L') \cdot (1 - L'') \cdot (1 - L''') \cdot \cdots)} \right]^*,$$

where * denotes a special operation in which the multiplications in the numerator and in the denominator are carried out before division; terms are deleted if they multiply the effects of touching paths, and division is performed only after such terms have been deleted.

To illustrate, let us take the causal network in Figure 14.10 and apply Mason's rule to it. We seek the total effect of *X* on *Z*. First, there is only

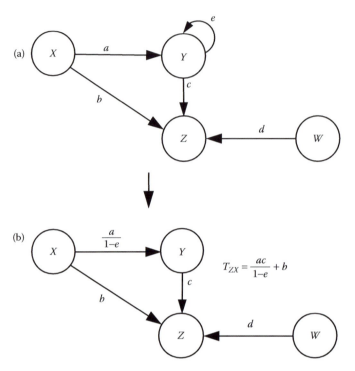

FIGURE 14.9   Finding the total effect T_{XZ} of X on Z when an intervening variable Y has a self-loop. (Adapted from Heise, D. R. (1975). *Causal Analysis*. New York: Wiley, p. 65.)

one open path from X to Z: $XUVZ$. The effect of X over this open path is $E_{XUVZ} = abe$. There is only one loop touching this open path, the loop UVU with return effect bc. Hence Mason's rule dictates that we write

$$T_{ZX} = \left[\frac{abe \cdot (1 - bc)}{(1 - bc)}\right]^* = \left[\frac{abe - abe \cdot bc}{(1 - bc)}\right]^*.$$

Because the open path from X to Z and the loop at V are touching, the operation * dictates that we delete the term $(abe \cdot bc)$ from the expression, leaving the result

$$T_{ZX} = \frac{abe}{(1 - bc)}.$$

We can obtain the same result from Figure 14.10a using simple graph reduction rules. We will first obtain the total effect T_{ZX}. Consider the Figure 14.10b at the lower left. We can use the collapsing rule in Figure 14.7 because the two loop UVU with X as an input to it is a case of the general feedback loop, which we can replace with a simple open path with effect $ab/(1 - bc)$. This makes the figure in (b) a straightforward recursive graph in which the total effect T_{XZ} is

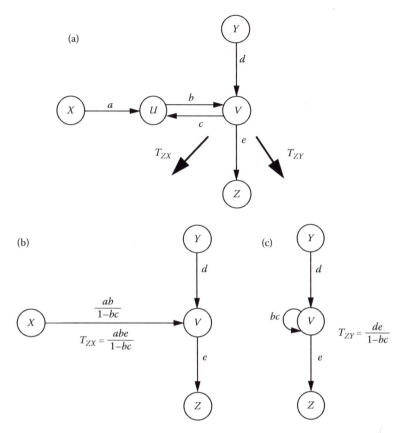

FIGURE 14.10 A graph with a two loop and the aim is to find the total effects T_{ZX} and T_{ZY}. (Adapted from Heise, D. R. (1975). *Causal Analysis*. New York: Wiley, p. 65.)

simply the product of $ab/(1 - bc)$ times e. Hence $T_{ZX} = abe/(1 - bc)$. On the other hand to find T_{YZ} the open path is YVZ, but a two loop touches this path at V. The two loop has an amplifying effect on the open path YVZ, but X is irrelevant to this amplifying effect. We can replace the two loop with a self loop as in Figure 14.10c. The graph now corresponds to part of another graph we have already seen for which we have the solution. Consider Figure 14.9a. The open path XYZ has a self loop on Y, which we said could be absorbed by eliminating the loop and replacing the effect of X on Y with $a/(1 - e)$. We can do the analogous thing here. We eliminate the loop on V and replace the effect d of Y on V as $d/(1 - bc)$, then we carry out the multiplication of $d/(1 - bc)$ times e to obtain $T_{ZY} = de/(1 - bc)$.

Let us now consider a more complex example shown in Figure 14.11. The aim is to find the total effect of X on Y. There are two open paths from X to Y: $XUVY$ with effect abd and $XWZY$ with effect *efh*. The path $XUVY$ touches the

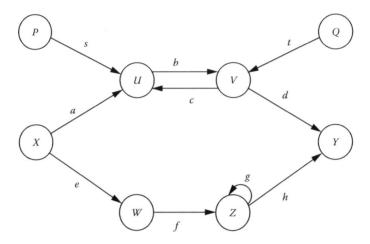

FIGURE 14.11 The aim is to find the total effect of X on Y by Mason's rule. Note that there are two parallel open paths to Y from X. There are also two loops, one touching each of these paths.

two loop between U and V, while the open path $XWZY$ touches the self-loop at Z. Mason's rule says that we should write

$$T_{YX} = \left[\frac{(abd + efh) \cdot (1 - bc) \cdot (1 - g)}{(1 - bc) \cdot (1 - g)} \right]^{*}.$$

The first thing we must do is expand the expressions in the numerator by removing parentheses through multiplication. So, we get

$$T_{YX} = \left[\frac{(abd + efh) \cdot (1 - bc - g - bcg)}{(1 - bc)(1 - g)} \right]^{*},$$

which is further expanded by further multiplying:

$$T_{YX} = \left[\frac{abd + efh - abdbc - efhbc - abdg - efhg + abdbcg + efhbcg}{(1 - bc)(1 - g)} \right]^{*}.$$

Now in the numerator let us eliminate any products of effects of a path with any return effect of a loop touching that path. Note that $abdbc$ multiplies the open path effect abd of $XUVY$ with the return effect bc of the loop UVU. So, we eliminate $abdbc$. The product $abdg$ is not a product of a path effect with a touching loop return effect. So, we retain that term. But $efhg$ should be eliminated because the self-loop on Z with return effect g touches the open path $XWZY$ with effect efh. The remaining terms $abdbcg$ and $efhbcg$ should also be eliminated because they contain products of open path effects with return effects of touching loops: $(abdbc)g$ and $(efhg)bc$. This now leaves in the numerator

$$T_{YX} = \left[\frac{abd - abdg + efh - efhbc}{(1 - bc)(1 - g)}\right] = \left[\frac{abd \cdot (1 - g) + efh \cdot (1 - bc)}{(1 - bc)(1 - g)}\right].$$

This result can now be simplified further to yield

$$T_{YX} = \frac{abd}{(1 - bc)} + \frac{efh}{(1 - g)}.$$

Note that we would find the same result if we used our previous reducing rules for flow graphs. Take each of the parallel paths separately. Over the open path $XUVY$, X would be shown to have the effect on Y of $abd/(1 - bc)$. Over the open path of $XWZY$, X would have the effect on Y of $efh/(1 - g)$. Since we sum all of the effects of X on Y, we obtain $T_{YX} = abd/(1 - bc) + efh/(1 - g)$, which is the same as the result obtained using Mason's rule. In some ways, the simple flow graph reduction rules are easier to apply and less prone to algebraic errors than Mason's rule.

Covariances and Correlations with Nonrecursive Related Variables

As a general rule if $X \xrightarrow{a} Y$ then $\text{cov}(X, Y) = \text{var}(X) \cdot a$. When one or both of the variables are in a nonrecursive loop, then we need to reduce the model to eliminate the loops and replace the structural coefficients with total effect coefficients.

Consider the model in Figure 14.12. We wish to find the covariance between Y and Z. The model at the top has a two loop between Y and Z. To properly treat the reciprocal relation between these two variables, the model is reduced to recursive paths with total effect coefficients based on the reciprocal loop's return effect. Y and Z are now functions of the exogenous variables X and W and the disturbances ε_Y and ε_Z. In obtaining the correlation, we need to obtain both expressions for the covariance between Y and Z and the variances of Y and Z. These are readily obtained with path tracing rules.

Because all of the coefficients in both the numerator and the denominator of the correlation coefficient have a common divisor of $(1 - cd)$, this factor may be factored out and (in this case) only the numerator terms are

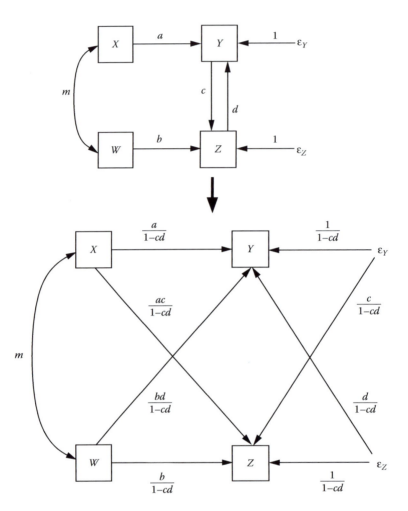

FIGURE 14.12   A model with a reciprocal causation loop is reduced with total effect coefficients so that the correlation between Y and Z may be obtained by path-tracing rules. (Adapted from Heise, D. R. (1975). *Causal Analysis*. New York: Wiley, p. 140.)

dealt with:

$$\rho(Y,Z) = \frac{\sigma(Y,Z)}{\sigma(Y)\sigma(Z)},$$

$$\sigma(Y,Z) = \frac{1}{(1-cd)^2}\left(a^2 c\sigma_X^2 + b^2 d\sigma_W^2 + amb + bdmac + c\sigma_{\varepsilon_Y}^2 + d\sigma_{\varepsilon_Z}^2\right),$$

$$\sigma_Y^2 = \frac{1}{(1-cd)^2}\left(a^2\sigma_X^2 + b^2 d^2\sigma_W^2 + \sigma_{\varepsilon_Y}^2 + d^2\sigma_{\varepsilon_Z}^2 + 2ambd\right),$$

$$\sigma_Z^2 = \frac{1}{(1-cd)^2}\left(b^2\sigma_W^2 + a^2 c^2\sigma_X^2 + \sigma_{\varepsilon_Z}^2 + c^2\sigma_{\varepsilon_Y}^2 + 2bmac\right).$$

Hence

$$\rho(Y,Z) = \cfrac{a^2 c \sigma_X^2 + b^2 d \sigma_W^2 + amb + bdmac + c \sigma_{\varepsilon_Y}^2 + d \sigma_{\varepsilon_Z}^2}{\sqrt{a^2 \sigma_X^2 + b^2 d^2 \sigma_W^2 + \sigma_{\varepsilon_Y}^2 + d^2 \sigma_{\varepsilon_Z}^2 + 2ambd} \\ \times \sqrt{b^2 \sigma_W^2 + a^2 c^2 \sigma_X^2 + \sigma_{\varepsilon_Z}^2 + c^2 \sigma_{\varepsilon_Y}^2 + 2bmac}}.$$

Heise (1975) shows that the correlation between variables in a loop is strongly affected by the variances of the instruments, for the correlation can range from positive to negative simply by varying the variances of the instruments. So, the correlation between variables in a loop, he says, "... implies nothing about the nature of their causal relations" (p. 141).

Heise (1975) goes on to argue that much confusion results in science when scientists attempt to interpret correlations between variables in loops. From the above formula we can see that if the loop effects c and d are nearly equal and opposite in sign, and further a and b and the variances of X and W are nearly the same, then the numerator of the correlation above will be diminished greatly relative to the standard deviations in the denominator. The reciprocal effects tend to cancel one another, reducing the correlation. On the other hand, if both are positive, they will tend to reinforce one another and increase the correlation. The positive effects will be even further increased with higher variances on the instrumental variables.

Identification

We have already considered how instrumental variables can assist in determining causal direction in recursive models. They are essential in many situations with nonrecursive models in providing identification of parameters in the models. Since most work with SEMs use commercial programs like EQS© and LISREL©, which are full-information estimation programs, we need here only state minimal conditions for achieving identification with models having nonrecursive relations in them. So, we need to show some graphs in which the models are identified.

With only two variables Y and Z, the parameters c and d on their reciprocal paths are underidentified. There is only one covariance between them and two parameters to be identified. If, as in Figure 14.13a, we introduce a single instrumental variable X as input to Y, we are now able to identify c, but with the free covariance h between the disturbances, d remains unidentified, so the model is underidentified. In Figure 14.13b, fixing the covariance between the disturbances to zero, we now have all parameters of the model identified. However, the assumption that the disturbances are uncorrelated is untestable, because we would have to be able to free that parameter to test the hypothesis that it is zero, and in doing so, the covariance and the model

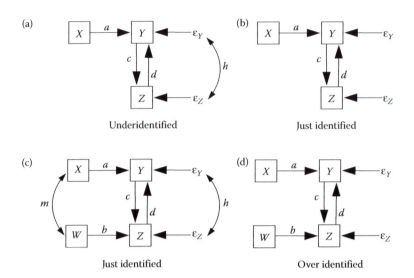

FIGURE 14.13 Use of instrumental variables to identify models with reciprocal causation. ((a) and (b) adapted from Heise, D. R. (1975). *Causal Analysis*. New York: Wiley, pp. 178 and 181.)

become underidentified. So, unless there is some external, independent justification for assuming that the covariance is zero, fixing it to zero is simply a way of making the model identified and useable.

In Figure 14.13c, we introduce a second instrumental variable W, this time as an input to the variable Z. If we leave the covariances between X and W and between the disturbances free, we have a just-identified model. But the second instrument makes possible estimation of d, even though the model itself is not testable. But if we fix the covariances between X and W and between the disturbances to zero, as in Figure 14.13d, we now have an overidentified model with two degrees of freedom. We now can test the assumptions about the covariances being zero.

Heise (1975, p. 175) states several rules for the identifiability of parameters in nonrecursive models. These are as follows.

A sufficient condition for indentifiability with nonrecursive causes is that if in a nonrecursive system of variables each nonrecursive causal variable is an entry point for an instrumental variable that has no other entry points in the nonrecursive system, then all of the structural coefficients in the system are identified.

Suppose a variable is in a loop and its disturbances are correlated with the values of its causes. All the structural coefficients in its structural equation will be identified only if the following two conditions hold:

(a) There is an instrumental variable for each nonrecursive relation in the equation.

(b) Suppose now that the equation contains several nonrecursive causes of the variable in the loop. Suppose their number is K. Then there must be at least K different instruments for the relations between the variable in question and its causes. (It is possible for each of the instrumental variables to be instruments for more than one of the K relations.)

If each nonrecursive cause in the equation is associated with an instrument for relations that pertain to that source only, then the structural coefficients are always theoretically identified.

We have only touched on the basics of identification with nonrecursive models. For treatment in more depth consult Heise (1975).

Estimation

Because of the presence of loops in nonrecursive models, ordinary least squares estimation was not able to obtain unbiased estimates of parameters. In the 1970s econometricians and sociologists used instead two-stage least squares estimation with nonrecursive models. With the introduction of latent variables into SEMs, the estimation methods became full information estimation, where all free coefficients in the system were estimated simultaneously. In contrast, ordinary least squares and two-stage least squares are what Heise (1975) calls "single-equation methods" because they estimated coefficients in one equation at a time. But with variables in complex causal networks constrained by theory, full information estimation is more efficient and accurate (when the theory is correct) in obtaining estimates. However, when the theory is wrong, errors in specification in one part of a model can be disseminated to erroneous estimation of other parameters of a model. So, there are drawbacks to full information estimation. Nevertheless, the commercial computer programs principally use full information estimation. The point is that we can use full information estimation programs like EQS, LISREL, and AMOS© to perform analyses of nonrecursive models. But we must remain mindful that if our models are seriously misspecified, the parameter estimates will also be seriously misspecified.

Heise (1975) notes a number of problems specifically concerning estimates of parameters of nonrecursive models. High collinearity between variables can produce erroneous estimates, especially in small samples. So, large samples are recommended. (Generally $N > 200$ is recommended for SEM studies, and the number should be higher when numerous variables are involved.)

If an instrumental variable has weak effects on its effect variable in a nonrecursive net, this will produce inaccurate estimates of its effect and that of others. Though difficult to find, researchers should seek instrumental variables that have strong effects. Or multiple instruments of a given effect variable in a nonrecursive loop may increase accuracy. Multiple indicators of a latent effect in a nonrecursive loop will also increase the accuracy of estimates. Again large samples aid in reducing sampling error.

Applications

The philosopher Immanuel Kant (1787/1996) held that our conceptions of objects concerned three levels of relations, each relation is a concept built by a synthesis of the previous levels' relational concepts: (1) *Inherence*: the relation of an object to its attribute. (2) *Causation*: the relation by which attributes of an object determine other attributes of an (often different) object. (3) *Community*: the mutual and reciprocal determination of the attributes of objects in a collection or community of objects. In the behavioral sciences, we begin by being preoccupied with identifying behavioral attributes of persons (traits, desires attitudes, and abilities) and then seek measurements of these. Here we function at the level of Kant's *inherence*. Next we become concerned with what determines these behavioral attributes, and the issue of causality becomes central. Behaviorism sought to find the causes in attributes of things in the external environment. Trait psychology sought to show how certain person traits determined other person traits, such as intelligence as a determiner of school achievement. Here psychologists functioned at the second conceptual level of *causation*. With multivariate structural equation modeling, psychologists, sociologists, and educational researchers had a method that allowed them to move to the third level, *community*, to the study of complex systems of individuals and variables measuring traits, behaviors, attitudes, abilities, achievements, motives, and their mutual causal interrelationships. Nonrecursive models represent the completion of the development of the third conceptual level of *community* in the behavioral sciences in providing one of the first of many techniques for studying the reciprocal mutual effects upon attributes of entities in these systems. A classic example is Bandura's (1989) triadic reciprocal relations between the person's behavior, perceptions, and the environment the person functions in. But mutual causal relations between individuals' perceptions and their attitudes and behavioral tendencies is another example of an application of *community*. Relations between parents and children, between employers and employees, between variables describing different sectors of a community, and between regions and nations are other potential applications. Causation no longer needs to be one way but reciprocal.

Most psychological variables used in rating scales involve judgments. We do not have direct measurement of a judgment formulated in the brain, but we do have indicators of these judgments. We presume they represent single latent variables, even though they may have a number of causes that accumulate in forming the judgment. If we can have multiple indicators of them, in that they have a common factor among them, we have a better reason to believe in their objective existence. Judgments in turn can be influences on one another in reciprocal relationships. After all, they are formed in the same brain. Behavioral scientists tend to assume all individuals are alike—perhaps unrealistically—but all will still go well if most individuals are quite similar, so that the causal mechanisms involved in their judgments are for all practical purposes the same. Because judgments take place in the same brain but in different centers, they can reciprocally influence one another very rapidly, implying that the neural pathways allow these to move around in loops very rapidly relative to the essentially static inputs of their causes. All of these considerations are compatible with studying relations among judgments with nonrecursive models with causal loops.

To illustrate, in Figure 14.14 I have laid out a hypothetical model of a model of the reciprocal causal relations between degree of job satisfaction and degree of intent to quit (the job). This model is inspired by James and Jones (1980) with further developments by Lance (1991). These variables are treated as latent variables, but we suppose that we can find at least four manifest indicators of each of them. We illustrate that with four indicators each on the right of the path diagram.

Most everyone would agree that as one's satisfaction with one's job goes up, the intent to quit the job goes down. So, we run a causal arrow from degree of job satisfaction to degree of intent to quit the job and assign a negative sign to its coefficient. We also presume that if one intends to quit one's job, that will produce a decrease in one's satisfaction with the job. So, we draw an arrow from degree of intent to quit to degree of job satisfaction and put a negative sign for its coefficient. We now presume that if a person has made various judgments about the job, these, being mostly anchored in the external environment, will not change, and so will present constant input to the degree of satisfaction judgment. Similarly, we presume that the degree to quit one's job is influenced by a number of external perceptions, which will also not change as degree of job satisfaction and degree of intent to quit mutually reach some equilibrium between them.

The task we have in setting a model with reciprocal causation is to discover variables that serve as instrumental variables for the variables in the nonrecursive loops. We should have at least one for each causal relation in the loop. And finding good instrumental variables is perhaps the most important issue for any researcher planning to study reciprocal causation. They are not always easy to find. And without good ones, your conclusions are subject to severe criticism. So, let us review what constitutes a good instrumental variable (Heise, 1975, pp. 160, 161): (1) It has no direct effect on the effect of the causal variable it influences. If the instrumental variable has an arrow to a causal variable in a loop, it must not have any arrow to any other variable that is an effect of this causal variable. (2) An instrument may affect its target variable in the loop indirectly and if it does, the mediating intervening variable must have no direct effect on the effect variable of the target

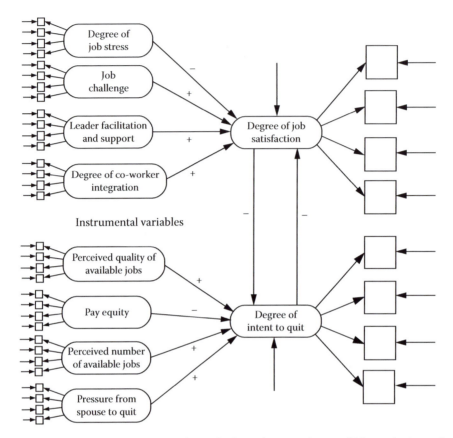

FIGURE 14.14 A model of reciprocal causal relation between degree of job satisfaction and degree of intent to quit. (Adapted from Lance, C. E. (1991). *Multivariate Behavioral Research*, 26, 137–162.)

variable in the loop. (3) In the loop, neither the target cause nor the effect of the target in the loop may have an effect, directly or indirectly, on the instrumental variable for the target variable. (4) No unspecified, unmeasured variable may jointly affect the instrumental variable and the effect of the target variable in the loop. The instrumental variable may not be correlated with the disturbance on the effect of the target variable in the loop. Finally, any other variable that is merely correlated with the target variable in the loop may also be an instrument if it satisfies the other conditions prescribed above.

Now, this is not a question of just drawing out one's path diagram with instruments satisfying the above conditions graphically. The graph must be your best judgment as to what is the case in the real world. So, you must be concerned with finding good instruments, for only with them will you be able to establish causal direction, and strength of the causal effect of a target variable on another variable in the loop.

To accomplish this you must already have achieved a good understanding of the situation you are modeling. You must have attained some understanding of the possible causes interacting in the situation. To some extent that is why you believe two judgment variables may be reciprocally related causally. But it should also be why you believe that there are certain causes of these variables, among which are variables that will function as instruments.

Now, most people have some experience with being in a job, of those things that made the job satisfying and those things that led them or others they observed to consider quitting the job. They know that a job was satisfying if they liked the co-workers, and they found the job challenging, interesting, and fulfilling. They also found it satisfying if their boss was a good person to work for, provided good, sound leadership, and gave them support both materially in good equipment to get the job done and emotionally in giving encouragement and praise for good work. But they may not have found the job satisfying, despite these other things, if the amount of work to be done was too much to handle, too difficult, and too stressful.

They know that if the job is not satisfying in some way, they will consider quitting. But they also know that the decision to quit will be influenced in part by the perceived alternative jobs available, their quality, and their quantity. If you quit, you want there to be a high probability of finding another job soon and it should be a better job than what you now have. Now, you may also be influenced to quit if you feel that the pay you are getting for your job is not fair compared with similar jobs in other places. And you may be influenced by your spouse to quit your job, independently of the nice things you say at home about your job, if the spouse has a job offer for an even better job in another city, or the spouse wants to move to a better school district for the kids. Or the spouse wants you to earn more money than you are getting in your present job.

Now, if we have this kind of knowledge, we should be able to formulate a model of the causal relations between degree of job satisfaction and degree of intent to quit. From among the causes of these two variables, we should be able to find some causes that would serve as instrumental variables for them. If a variable is an instrumental cause of degree of job satisfaction, it must not also be a direct cause of degree of intent to quit. It must not be correlated with the disturbances on degree of intent to quit, although it can be correlated with the disturbances on the degree of job satisfaction. And your perception of these things should not be directly or indirectly influenced by your current degree of job satisfaction or of your degree of intent to quit the job. (This may be harder to establish.) Similarly, causes of degree of intent to quit cannot also be direct causes of degree of job satisfaction, if they are to serve as instruments of degree of intent to quit.

In Figure 14.14, I have entered as latent variables four instrumental variables each for each of the two variables in the reciprocal loop. These are in the path diagram at the left. I have also indicated the expected signs of the coefficients on the causal paths from these instruments to their target variables. Each latent instrumental variable also has four manifest indicators. We need multiple indicators to establish the objective validity of the instrumental variables, to show that there is a proportional influence of a common variable across all of the indicators.

In the diagram of Figure 14.14, I have not indicated correlations among the exogenous latent variables, the instrumental variables, simply to keep the diagram from being uselessly busy. They can all be intercorrelated. And one should free these correlations. I have also left the disturbances on the reciprocally caused variables uncorrelated. If the model does not fit with this specification, then the correlation between these disturbances should be tested with a Lagrange multiplier test, and freed if significant.

15

Model Evaluation

Introduction

After the free parameters of the confirmatory factor analysis model have been estimated, it is then possible to obtain a "reproduced" variance–covariance matrix $\hat{\boldsymbol{\Sigma}}_0 = \hat{\boldsymbol{\Lambda}}\hat{\boldsymbol{\Phi}}\hat{\boldsymbol{\Lambda}}' + \hat{\boldsymbol{\Psi}}^2$ among the observed variables based on the confirmatory factor analysis model's fixed, constrained, and estimated parameters. Or one may have estimated parameters of a SEM and used these to reproduce a model variance–covariance matrix given by

$$\hat{\boldsymbol{\Sigma}}_0 = \mathbf{G}\hat{\mathbf{B}}^{*-1}\hat{\boldsymbol{\Gamma}}^*\hat{\boldsymbol{\Phi}}\hat{\boldsymbol{\Gamma}}^{*\prime}\hat{\mathbf{B}}^{*\prime-1}\mathbf{G}'.$$

At this point, the researcher's natural inclination is to determine how well his or her model covariance matrix $\hat{\boldsymbol{\Sigma}}_0$ compares with the observed variance–covariance matrix $\mathbf{S}$, that is, how well the model covariance matrix "fits" the observed covariance matrix. But it is important to understand what the comparison means. Most discussions of model fit focus on how the fit of the model is a function of the estimated parameters of the model. In fact, the problem of parameter estimation so dominates these discussions that what is often lost sight of is what has been hypothesized and is to be tested in one's model. The hypothesis is about the fixed and constrained parameters in the framework of one's model. The "framework" is an *a priori* schema that one imposes upon the data. In our case it could be the common factor model equation or the SEM equation and the derived variance–covariance matrix—in the abstract, without values specified for its parameters, without a specified number of common factors. One seeks to interpret the observed covariance matrix in terms of this framework. And to formulate a hypothesis within this framework, one must specify the number of common factors or latent variables and values for some or all of the parameters of this model. If one specifies or constrains only some of the parameters, then the model is *incomplete*. If one specifies all of the parameters, then the model is *complete*. In incomplete models the estimated parameters are the unspecified portions of one's model.

Contrary to a popular misconception, estimated parameters do not represent hypothesized nonzero coefficients. Estimated parameters can take on any value between minus and plus infinity, including zero. They are the unknowns, what one is not able or willing to specify in one's hypothesized model. But the model cannot be evaluated without values for these unknown parameters to complete the model. So, there needs to be some filler in these unspecified places in the model. Parameter estimation provides this filler. But just not any values will do. The estimated parameters should not be ultimately responsible for lack of fit, for then an ambiguity will exist in interpreting the meaning of the fit. Is it due to the specified values and constraints in the framework of one's model, or is it due to the estimated parameter values, that were not in one's hypothesis? No, the parameters should be estimated in such a way that the model would fit as well as possible, except for values of the fixed and constrained parameters and their constraints on the estimated parameters. Only in that way can any lack of fit be attributed to the fixed and constrained values imposed on model parameters. That is why free (unspecified) parameters are estimated to optimize fit of the reproduced model variance–covariance matrix to the observed covariance matrix *conditional* on the fixed and constrained parameter values. So, the fit (or lack of fit) is ultimately about the fixed and constrained parameters in the model framework, for these are the only things specified in one's hypothesis before seeing the data.

Now, if the model fails to fit what can we infer about the model? Can we infer that it is due to just an improper specification for the specified parameters and/or constraints? Because so many things are included in one's hypothesis—for example, SEM framework, number of latent variables, fixed parameter values, constraints on parameters—one cannot logically draw any clear inference as to which of these is the source of lack of fit. It could be any and all of them. All lack of fit tells us is that something is wrong with the model. It may be that the framework of the common factor model itself imposed upon the data is inappropriate for the data, that no constraints on values in this framework would lead to a model that fits the data. Or one may just have the number of latent variables wrong, that there may be more additional latents one did not take into account in formulating one's model. Or the structural equation framework with the specified number of latents may be fine, but some or all of the specified values for the parameters are wrong. Or maybe something is wrong in our data collection and measurement technique. The presumed sampling distribution, for example, multivariate normality, may be wrong.

How the researcher will proceed after obtaining a lack of fit of his or her model to the data depends on how strongly he or she is committed to the model for the data in question. There are other structural models that might be considered besides the model in question. At this point, the researcher should reexamine the content of (not the covariances among) his or her variables and entertain the possibilities that they may be interrelated causally in other ways not considered. If one does so and concludes the model may not be appropriate, one exits the context of the model and takes up possibly another modeling framework. There may be even other mathematical models than a linear model that are more appropriate for the data.

But if one is still persuaded that the model holds, then one may wish to consider whether there are additional latent variables. Again, by examining the content of one's variables, one may be able to see the possibility for another latent variable and go back to the beginning with a respecified model that contains the additional latent variable and fit that model to the data.

There is also a way of obtaining clues as to possible additional latent variables by computing Lagrange multiplier (LM) tests for each of the zero covariances among the unique factors or disturbances. (More about these later.) These are single-degree-of-freedom chi-square tests that test whether a fixed parameter could be freed to produce significant improvement in fit. One can free the zero covariance with the largest LM test value and see how that improves one's fit by allowing it to take on nonzero values. One should also consider whether to do so would be theoretically justified in terms of substantive theory. One can go on freeing one unique factor covariance at a time and refitting until one feels one has found the basis for an additional common factor among the variables with correlated unique factors. Or one may be content with simply leaving the covariances among unique factors freed to this point as simply freed without trying to account for them in terms of a well-specified common factor. But regardless, one will pay a price for doing this, for one will lose degrees of freedom for each parameter freed to arrive at a better fitting model. (More about degrees of freedom later.) Or one may think the common factor model with the specified number of common factors is just fine. One can then also use LM tests to search among any spec- ified factor pattern loadings to see if they can be freed to improve model fit. Again the current wisdom is that one does this cautiously, justifying freeing each zero parameter value in terms of some substantive theoretical rationale and refusing to do so if one cannot.

On the other hand, what does good fit for a model imply? It does not nec- essarily imply that the model is correct for the data. In fact, there may not be any such thing as an "ultimately correct" model. There are just models that fit given sets of data within sampling and measurement error. But the theories that support them may be disconfirmed by extending them to other data and contexts thought to be covered by the theories in question and finding that the additional data and the original data do not cohere with the theoretical model in question. Or the theory on which the model is based may be disconfirmed in contexts external to the source of data in question. So any acceptable fit of a model to data may be regarded as only provisional support for any theory on which the hypothesis is based. That support may be undermined with additional data, larger samples with larger power to detect smaller discrepancies between model and data, or by disconfirmation of the theory in other contexts. For further discussion of these philosophical issues, see Mulaik and James (1995).

There are also numerous cases in which one mathematical model may be mathematically equivalent to another mathematical model in producing the same fit to the data, but have different interpretations with respect to the data. Some sets of data may not allow one to distinguish which of these models is the appropriate one for the data. This is a problem for all mathematical models. Other cases involve different mathematical models that fit exactly the same data, but one model requires estimating more parameters and has fewer degrees of freedom. One rule of thumb is to prefer (provisionally) the model with fewer estimated parameters (or conversely with more degrees of freedom). Mulaik and Quartetti (1997) showed that models with a first-order general factor and several group factors could also fit data generated by models with a second-order general factor and first-order group factors, although the latter would have more degrees of freedom if fit to the same data. Common factor models with a suitable number of factors are also known to be able to fit to high degrees of approximation data generated by simplex models with many more latent variables. So obtaining good fit is no guarantee of the truth of a model. Good fit only gives provisional support to a model.

There are some (Browne and Cudeck, 1993) who argue that "a null hypothesis that fits exactly in some population is known *a priori* to be false" (p. 137). I will not take such a strong position. To begin with we cannot know this *a priori*. To say that we do would be logically inconsistent with then conducting tests of *a priori* specified hypotheses. If anything, the assertion is not known *a priori* but is an inductive generalization based on finding that many models fail to fit data at some sample size. One can always overturn an inductive generalization with further experience. But there is no necessary reason that we will. Physicists certainly do not hold such a maxim because they have models in quantum physics based on precise ratio values for certain parameters that fit within measurement error and keep on testing them with larger and larger samples and in more and more situations. If they knew they would not fit *a priori*, they would not bother to go to all this trouble. There has to be an *a priori* possibility that a model will fit data to justify conducting tests of a model. Still we may never be able to determine precisely whether a model does fit exactly, and even if it does, that is no guarantee that it is "ultimately true." But it should be possible to "disconfirm" a model with lack of fit.

On the other hand, many models may not fit exactly but still be very good approximations. As good approximations they may allow for useful predictions and support provisionally certain lines of research that seek to improve the fit by adjusting parameters or making minor modifications of the framework of the model by including other variables. We will consider now the cases of testing for exact fit and the ways of measuring the degree of approximation of a model.

There are two complementary steps to take: (1) One begins first to test a hypothesis that the population variance–covariance matrix Σ is generated by a model with the specified parameters and parameter constraints and is as a consequence equal to Σ_0. For this, assuming the data have a multivariate normal distribution, a chi-square test based on a generalized likelihood ratio (LR) test is available. There are also confidence interval tests for testing hypotheses about individual parameters left freed in the model. (2) If the model fits to within sampling error in step (1), the researcher can provisionally accept the model as it is and draw substantive conclusions from this result. But if the model fails to fit to within sampling error by the chi-square test, the researcher should then examine the diagnostics of the computer output to gather clues as to the sources of lack of fit. The researcher may then find, especially if the lack of fit was large, that his or her approach to the data with the model was wrong, and then consider other possible models that would have to be tested with a new study and new data. But the researcher may wish to know how well $\hat{\Sigma}_0$, the estimated constrained model variance–covariance matrix estimated to fit the sample variance–covariance matrix $\mathbf{S}$, reproduces $\mathbf{S}$, that is, the degree to which it *approximates* $\mathbf{S}$. There have been numerous indices proposed to represent measures of approximate fit. We will consider the most important ones. But the researcher may then, on the basis that the fit was already a fairly high degree of approximation, consider freeing up certain parameters located by LM tests to see if the remaining constraints on the model then fit the data, either to within sampling error, and if not, to at least a very high degree of approximation.

Finally, the model must also be assessed from the point of view of its disconfirmability. Models that are not disconfirmable or have low degrees of disconfirmability are of no to little value compared with models that are highly disconfirmable from the point of view of establishing objective scientific knowledge. They may be useful from an exploratory point of view of providing values for parameters within the framework, say, of a factor analysis model with a specified number of factors that might be used as fixed values in future studies where these values are tested against independent data sets under other conditions. They might even be used in prediction studies where understanding the underlying causal mechanisms for the associations is not essential.

Next, when there is a lack of fit, the question will arise as to whether the model could be modified and reanalyzed with the same data set to obtain better fit or whether it is best to abandon the model altogether and rethink one's model. If one decides to retain the model, how should one make the modifications of the model? What penalties should one pay for these modifications after the fact?

Finally, the question will arise of whether there is a sequence in which one might proceed to test the assumptions and constraints of a model. We will consider each of the questions in the following sections.

Errors of Fit

Given two $p \times p$ covariance matrices $\mathbf{U}$ and $\mathbf{V}$, Browne (1982) proposed a scalar valued function $F(\mathbf{U}; \mathbf{V})$, known as a "discrepancy function," that has the following properties:

$F(\mathbf{U}; \mathbf{V}) \geq 0$.

$F(\mathbf{U}; \mathbf{V}) = 0$ if and only if $\mathbf{U}$ and $\mathbf{V}$ are equal.

$F(\mathbf{U}; \mathbf{V})$ is continuous in both $\mathbf{U}$ and $\mathbf{V}$.

The symbol $\mathbf{U}$ in the first position of the argument of the discrepancy function is usually a less restricted variance–covariance matrix to which a more restricted, constrained model variance–covariance matrix $\mathbf{V}$ in the second position is to be compared. There are a number of standard discrepancy functions, such as unweighted least squares, weighted least squares, and maximum likelihood's fit function. For example, the least-squares discrepancy function is

$$F_{LS} = \mathrm{tr}[(\mathbf{U} - \mathbf{V})(\mathbf{U} - \mathbf{V})]$$

and the maximum likelihood's fit function is

$$F_{ML} = \ln |\mathbf{V}| - \ln |\mathbf{U}| - \mathrm{tr}(\mathbf{U}\mathbf{V}^{-1}) - p,$$

which is a discrepancy function. Even chi-square defined as $\chi^2_{df} = (N-1)F_{ML}$ is a discrepancy function. Cudeck and Browne (1983) and Browne and Cudeck (1989, 1993) further developed the use of the discrepancy function and we will draw on their work here.

Using the concept of a discrepancy function, Cudeck and Henly (1991) introduced the concepts of different kinds of errors in comparison of the $n \times n$ matrices $\mathbf{\Sigma}$, $\tilde{\mathbf{\Sigma}}_0$, and $\hat{\mathbf{\Sigma}}_0$. $\mathbf{\Sigma}$ is the population variance–covariance matrix from which the sample variance–covariance matrix $\mathbf{S}$ has been drawn. It may or may not in fact conform to the model tested. $\tilde{\mathbf{\Sigma}}_0$ is the maximum-likelihood estimate of the model and its constraints fit to $\mathbf{\Sigma}$ instead of $\mathbf{S}$. $\hat{\mathbf{\Sigma}}_0$ is the sample-estimated constrained model variance–covariance matrix fit to $\mathbf{S}$. By *error of approximation*, we mean the value of the discrepancy function $F(\mathbf{\Sigma}; \tilde{\mathbf{\Sigma}}_0)$ that compares the population covariance matrix Σ to the reproduced covariance matrix $\tilde{\Sigma}_0$ based on fitting the hypothesized model to the population covariance matrix. Occasionally we will also call this the *population discrepancy*. This is a population parameter and cannot be directly observed, although it can be estimated. This source of discrepancy is due to the limitations of the model and contains no sampling error since both variance and covariance matrices are population matrices. It gives us the theoretical discrepancy between the model variance–covariance matrix estimated in the population and the population variance–covariance matrix. It can never take on a value less than zero, and will equal zero only when $\Sigma = \tilde{\Sigma}_0$. The error of approximation is the fundamental parameter that indices of fit should seek to estimate or be based upon.

On the other hand, $F(\tilde{\Sigma}_0; \hat{\Sigma}_0)$ represents the *error of estimation*, the discrepancy between sample estimate and population estimate for the model. It is not observable in the sample and can only be inferred and estimated. Because $\hat{\Sigma}_0$ varies as a function of S to which it has been fit, $F(\tilde{\Sigma}_0; \hat{\Sigma}_0)$ is a random variable and contains sampling error. It represents the sampling error due to the estimation of free parameters based on S. It usually does not contain all of the sampling error in S because the hypothesis does not estimate as many parameters as there are distinct parameters in S. Sampling error only concerns the estimated parameters and the portions of S that determine them. However, $\hat{\Sigma}_0$ converges in probability to $\tilde{\Sigma}_0$ as the sample size N increases indefinitely.

Another important discrepancy function value of interest is $F(\Sigma; \hat{\Sigma}_0)$, which represents the discrepancy between the population variance–covariance matrix that generates the data and the sample estimate of the model covariance matrix fit to the sample covariance matrix. This is a measure of the *overall discrepancy*. It is not directly observed in the sample and can only be estimated. Browne and Cudeck (1993) showed that

$$E[F(\Sigma; \hat{\Sigma}_0)] \approx F(\Sigma; \tilde{\Sigma}_0) + E[F(\tilde{\Sigma}_0; \hat{\Sigma}_0)], \tag{15.1}$$

which means that the overall error of fit of the sample-estimated variance–covariance matrix of the tested model to the population variance–covariance matrix equals approximately the sum of the error of approximation and the expected error of estimation. On the other hand,

$$E[F(\tilde{\Sigma}_0; \hat{\Sigma}_0)] \approx N^{-1}q, \tag{15.2}$$

where N is the sample size and q is the number of free parameters estimated in the model. In other words, the expected error of estimation (sampling error) is directly proportional to the number of estimated parameters and inversely proportional to the sample size. With more estimated parameters, the expected error of estimation goes up. But there is always an upper limit to this number at $q = n(n+1)/2$ because beyond that point the model will not be identified. On the other hand, as sample size increases, the expected error of estimation (sampling error) decreases, and eventually error of sampling is negligible. So, in summary,

$$E[F(\mathbf{\Sigma}; \hat{\mathbf{\Sigma}}_0)] \approx F(\mathbf{\Sigma}; \tilde{\mathbf{\Sigma}}_0) + N^{-1}q. \tag{15.3}$$

What this also means is that in larger and larger samples, fit becomes more and more dependent on the error of approximation. Nevertheless, a saturated model would ultimately have an error of approximation of zero and an expected error of sampling also of zero in an infinitely large sample. Thus on cross-validation to another same-sized sample from the population, where the reproduced $\hat{\mathbf{\Sigma}}_{01}$ from the first sample is then fit as a fixed matrix to the sample covariance matrix $\mathbf{S}_2$ from a second sample, an infinitely large sample would reproduce a saturated $\tilde{\mathbf{\Sigma}}_0$ perfectly. In small samples, however, where the sampling error of estimation can be larger with saturated models, the corresponding saturated model $\hat{\mathbf{\Sigma}}_0$ may always have zero discrepancy with $\mathbf{S}$, but may not be stable with respect to $\mathbf{\Sigma}$ and not cross-validate well.

At this point I must caution you that saturated models are not desirable in the context of hypothesis testing because they do not represent a testable hypothesis. They necessarily fit perfectly the variance–covariance matrix to which they are fit. There is no possibility of a test because there is no logical possibility of failing a test of fit. Furthermore, saturated models are not unique and many distinct saturated models are easily constructed for a given set of data. Whether indices of fit will lead one to reject saturated models despite their perfect fit to data is a question we will consider in the context of a discussion of parsimony and degrees of freedom.

Finally, $F(\mathbf{S}; \hat{\mathbf{\Sigma}}_0)$ represents the discrepancy between the sample variance–covariance matrix $\mathbf{S}$ and the sample-estimated constrained model variance–covariance matrix $\hat{\mathbf{\Sigma}}_0$. $F(\mathbf{S}; \hat{\mathbf{\Sigma}}_0)$ contains both error of approximation and sampling error. This value is directly observed.

Chi-Square Test of Fit

Karl Jöreskog (1969) described a chi-square test-of-fit statistic to be used in connection with testing a confirmatory factor analysis model. The chi-square test is based on a generalized LR statistic that is very powerful for testing composite hypotheses with many parameters. The generalized LR is given as

$$\Lambda = \frac{L(\hat{\omega})}{L(\hat{\Omega})},$$

where

$$L(\hat{\omega}) = \prod_{i=1}^{N} f(\mathbf{x}_i; \mathbf{g}, \hat{\mathbf{q}})$$

is the joint likelihood of the sample obtained as the product of the individual observation likelihoods $f(\mathbf{x}_i; \mathbf{g}, \hat{\mathbf{q}})$, with $\mathbf{x}_i$ an $n \times 1$ vector of observations on n observed random variables, γ an $n(n+1)/2 - q$ vector of fixed and constrained parameters that overidentify the model, and $\hat{\theta}$ a vector of q free parameters estimated by maximum likelihood (to maximize the joint likelihood). On the other hand, $L(\hat{\Omega})$ is given as $\prod_{i=1}^{N} f(\mathbf{x}_i; \hat{\mathbf{g}}, \hat{\mathbf{q}})$, where everything is as before except that $\hat{\gamma}$ and $\hat{\theta}$ are now both vectors of free parameters, making the model a saturated model with as many estimated parameters as the $n(n+1)/2$ distinct observed parameters in the sample variance–covariance matrix $\mathbf{S}$. The lowercase Greek letter ω denotes a constrained parameter space, with each dimension of ω corresponding to one of the m free parameters in $\hat{\theta}$. Maximum-likelihood estimation of the free parameters searches this restricted space to find that point in the space whose coordinates correspond to the values of each of the parameters that maximize the joint likelihood of the observations under the restricted model. The hat or $^\wedge$ over ω means that the likelihood is a function of those values in the parameter space ω that maximize the joint-likelihood function, in other words, the values in $\hat{\theta}$. Similarly, Ω is the less restricted parameter space of dimension $n(n+1)/2$ corresponding to the $n(n+1)/2$ free parameters in $\hat{\gamma}$ and $\hat{\theta}$. Again, the hat or $^\wedge$ over Ω means that $L(\hat{\Omega})$ is a function of those values for the parameters in the less restricted parameter space Ω that maximize the joint likelihood of the observation vectors.

Because maximum-likelihood estimation searches the spaces ω and Ω for the points whose coordinates are values for the parameters that maximize the likelihood function within their spaces, the likelihood $L(\hat{\omega})$ is never greater than the likelihood $L(\hat{\Omega})$. This is because ω is a subspace of Ω, and any point in Ω that represents a maximum likelihood over that space will correspond to a likelihood as large or larger than that found for the more restricted subspace ω. Thus L is a number <1, and one accepts the null hypothesis if $L > c$, some constant such that $P(\Lambda < c \mid H_0 \text{ is true}) \leq \alpha$, where α is the accepted hypothetical probability of rejecting the null hypothesis when it is false. However, L can be converted to a chi-square statistic by the following transformation:

$$\chi_{\text{df}}^2 = -2\ln \Lambda. \tag{15.4}$$

Although Jöreskog (1969) gave the formula for chi-square, Bollen (1989) showed how the generalized LR could be derived for any SEM, which includes confirmatory factor analysis as a special case:

$$\ln L(\hat{\omega}) = -\frac{(N-1)}{2} \left[\ln |\hat{\Sigma}_0| + \text{tr}(\hat{\Sigma}_0^{-1} \mathbf{S}) \right] \tag{15.5}$$

and

$$\ln L(\hat{\Omega}) = -\frac{(N-1)}{2} \left[\ln|\mathbf{S}| + \text{tr}(\mathbf{S}^{-1}\mathbf{S}) \right] = -\frac{(N-1)}{2} \left[\ln|\mathbf{S}| + n \right]. \tag{15.6}$$

Hence

$$U = -2\ln\frac{L(\hat{\omega})}{L(\hat{\Omega})} = -2\ln L(\hat{\omega}) + 2\ln L(\hat{\Omega})$$

$$= (N-1)\left[\ln|\hat{\Sigma}_0| + \text{tr}(\hat{\Sigma}_0^{-1}\mathbf{S})\right] - (N-1)(\ln|\mathbf{S}|+n) \qquad (15.7)$$

$$= (N-1)\left[\ln|\hat{\Sigma}_0| - \ln|\mathbf{S}| + \text{tr}(\hat{\Sigma}_0^{-1}\mathbf{S}) - n\right]$$

is a statistic that is distributed in increasingly larger samples as chi-square with df degrees of freedom when the model is correct and as the noncentral chi-square when the model is incorrect. The degrees of freedom of chi-square are df $= n(n+1)/2 - q$, where q is the number of free parameters in the model. The expression in brackets on the right-hand side of Equation 15.5 is the minimum value of the maximum-likelihood fit function,

$$F_{\text{ML}} = \ln\hat{\Sigma}_0 - \ln|\mathbf{S}| + \text{tr}(\hat{\Sigma}_0^{-1}\mathbf{S}) - n, \qquad (15.8)$$

that is minimized to obtain optimal estimates for the free parameters conditional on the constrained parameters of the model. This value is equivalent to $F(\mathbf{S}; \hat{\Sigma}_0)$. We can see that when $\mathbf{S}$ and $\hat{\Sigma}_0$ are equal, $\ln|\hat{\Sigma}_0| - \ln|\mathbf{S}| = 0$ and $\text{tr}(\hat{\Sigma}_0^{-1}\mathbf{S}) - n = 0$. When these two matrices are not equal, $F_{\text{ML}} > 0$ and can serve as a lack of fit measure in its own right.

The value $U = (N-1)F_{\text{ML}}$ is an expression for the chi-square statistic. This statistic is used to test the hypothesis that the constrained model covariance matrix is the population covariance matrix that generated the sample having sample covariance matrix $\mathbf{S}$. One rejects the null hypothesis that the model under the constraints generated the data represented by $\mathbf{S}$, when $\chi^2_{\text{df}} > c$, where c is some constant such that $P(\chi^2_{\text{df}} > c \mid H_0 \text{ is true}) \leq \alpha$.

However, the power of the chi-square statistic also increases with the increase in sample size. At the sample sizes where the statistic begins to have a good approximation to the chi-square statistic, it is also able to detect small discrepancies between the model and the data, leading to a rejection of the null hypothesis that the model and its constraints are correct. This does not mean that the null hypothesis is necessarily totally wrong. Quite possibly the discrepancy between data and model detected by the chi-square statistic may mean that the researcher has failed to account for some additional influence of a minor nature. For example, the assumption that a certain pair of unique factors is uncorrelated may be incorrect. Or some subjects in the sample may not conform to the model, violating the assumption of causal homogeneity for subjects in the study. That is, some subjects' responses may depend on different causes, or depend on the same causes with somewhat different magnitudes for the causal effects (e.g., have different factor loadings). Or there may be another minor common factor not accounted for in the model. But it could also mean that the common factor model is just not appropriate for the data. Some other data structure may underlie the data. The chi-square statistic cannot tell us which. The researcher must use his or her judgment to decide what to do next when the chi-square statistic indicates significant lack of fit. In many circumstances, the next thing one may consider is whether the model tested is nevertheless a good approximation to the data. Before we go on to discuss indices of approximation, we need to consider further properties of the chi-square distribution and another distribution related to it, the noncentral chi-square distribution.

We can also assert that the chi-square statistic is frequently undefined as the sample size becomes infinite because $\lim_{N \to \infty}(N-1)F_{ML}$ is undefined for all values of F_{ML} greater than zero.

Satorra–Bentler Corrected Chi-Square Statistic

When the population data are not distributed according to the multivariate normal distribution, then the usual chi-square statistic based on this distribution may be biased. Hu and Bentler (1995) report that Satorra and Bentler (1988a,b, 1994) developed a correction for the maximum-likelihood chi-square statistic using a scaling factor based on the model, the estimation method, and sample fourth-order moments. This statistic holds, then, regardless of the distribution of the variables. Hu and Bentler reported that it performed as well as the normal-theory methods under the condition of independence among latent variables. Although it performed well overall, it had a tendency, they say, to overreject true models at smaller sample sizes. However, they also report a recent follow-up study, Hu and Bentler (1995), that studied the Satorra–Bentler scaling correction applied to the generalized least-squares (GLS) chi-square statistic, and this scaled test statistic performed adequately even at smaller sample sizes. They regarded it as the most adequate statistic for testing model fit when sample size is small. With large samples both the maximum likelihood and GLS scaled statistics performed about as well.

Properties of Chi-Square and Noncentral Chi-Square

Statisticians define chi-square as a sum of squared, independent unit normal deviates. For example, suppose we have k independently distributed normal variables, $X_1, X_2, \ldots, X_k$, each with mean μ_i and variance σ_i^2. For each X_i construct a standardized variable, $Z_i = (X_i - \mu_i)/\sigma_i$, having a mean of 0 and a variance of 1. The Z_i also are normally distributed and are mutually independent. Mathematical statisticians define a chi-square distributed variable with k degrees of freedom as

$$\chi_k^2 = Z_1^2 + Z_2^2 + \cdots + Z_k^2. \tag{15.9}$$

Chi-square with k degrees of freedom is a sum of k independent squared unit normal deviates. Because $E(Z_i^2) = 1$,

$$E(\chi_k^2) = E(Z_1^2) + E(Z_2^2) + \cdots + E(Z_k^2) = k. \tag{15.10}$$

The mean of the chi-square distribution is equal to its degrees of freedom. The variance of the chi-square distribution in this case is $2k$.

The chi-square statistic can be used to test whether a model differs from the data in several dimensions. Each dimension may correspond to an observed parameter of the data, and the model generates a corresponding hypothetical set of values for the observed parameters. Differences between model and data are taken between observed values and hypothetical values, across each of the observed parameters. Standardizing the scores in each dimension puts them into a common metric. But there may be some loss of dimensions on which the comparison is to be made due to the process of parameter estimation in the model. The remaining number of dimensions will equal the degrees of freedom of the model. (More will be presented on this topic later.)

Suppose each Z_i represents a measure in one dimension of the difference between a model and the data. The variance of Z_i represents error of measurement of that difference in that dimension. The mean of each Z_i represents the actual difference between the model and the data in the ith dimension. Under the null hypothesis of no difference between the model variance–covariance matrix and the population variance–covariance from which the data have been sampled, one hypothesizes that each difference is equal to zero. The chi-square statistic represents a measure of the overall sum of squared differences, or the squared distance between model and data.

When the null hypothesis is false, think of each Z_i as having added to it a true difference d_i so that we may write each measure as $Z_i^* = Z_i + d_i$. Then the measure of discrepancy between model and data is given by

$$U^* = \sum_{i=1}^{k} Z_i^{*2}.$$

The expected value of U^* is then

$$E(U^*) = \sum_{i=1}^{k} E(Z_i^{*2}) = \sum_{i=1}^{k} E(Z_i^2 + 2Z_i d_i + d_i^2)$$

or

$$E(U^*) = \sum_{i=1}^{k} E(Z_i^2) + 2d_i \sum_{i=1}^{k} E(Z_i) + \sum_{i=1}^{k} d_i^2.$$

Because (a) the expected value of a squared unit normal deviate is 1 (its variance), (b) the expected value of a unit normal deviate is 0, and (c) the third term is a sum of constants, we may write this as

$$E(U^*) = k + \sum_{i=1}^{k} d_i^2 = k + \lambda, \tag{15.11}$$

where k is known as the degrees of freedom of U^* and λ is the squared distance between model and data known as the *noncentrality parameter*. In structural equation modeling and confirmatory factor analysis, λ is equal to $(N - 1)F(\mathbf{\Sigma}; \tilde{\mathbf{\Sigma}}_0)$. When the null hypothesis that $\lambda = 0$ is true, U^* is distributed as chi-square with k degrees of freedom. When $\lambda > 0$, then U^* is distributed according to the noncentral chi-square distribution. An unbiased estimate of λ is then

$$\hat{\lambda} = U^* - k. \tag{15.12}$$

The noncentrality parameter is going to assume an important role in the following discussion of measures of approximate fit and lack of fit. Steiger and Lind (1980) first suggested, in a paper given to the Psychometric Society, the use of the noncentrality parameter in evaluating SEMs. Steiger (1989) later described the use of the noncentrality parameter as well as a normalized noncentrality parameter that we are about to discuss. Bentler (1990), McDonald (1989), and McDonald and Marsh (1990) also described the use of the noncentrality parameter in measuring model lack of fit.

Closely related to the noncentrality parameter in Equation 15.9 is another parameter recommended by McDonald (1989), which is a normalized version of the noncentrality parameter, independent of sample size. In confirmatory factor analysis and structural equation modeling:

$$U^* = (N - 1)F_{\text{ML}}. \tag{15.13}$$

Let us substitute the expression on the right-hand side of Equation 15.12 in place of U^* and take the expectation:

$$E(\hat{\delta}^{*2}) = E[(N - 1)F_{\text{ML}}] - \text{df},$$
$$\lambda = (N - 1)E(F_{\text{ML}}) - \text{df}. \tag{15.14}$$

Here df represents the degrees of freedom of the model and F_{ML} represents the sample minimum fit function value for maximum-likelihood estimation under the constraints of the tested model. N is the sample size. We see now that as long as $E(F_{\text{ML}})$ is greater than zero, as the sample size N continues to increase, λ will tend to increase without bound. This contributes to the power of the chi-square test because a positive noncentrality parameter will, on average, increase with increases in sample size and make the difference to detect larger. But in infinitely large samples, λ, unless zero, is undefined because it is "infinite." If we divide $U^* - \mathrm{df}$ by $(N - 1)$, we then obtain a new index that converges in the limit to a finite value:

$$\hat{\delta} = \left(\frac{1}{N-1}\right)(U^* - \mathrm{df}) = \left(\frac{1}{N-1}\right)[(N-1)F_{\mathrm{ML}} - \mathrm{df}] = F_{\mathrm{ML}} - \frac{\mathrm{df}}{(N-1)}.$$

$$(15.15)$$

Browne and Cudeck (1992) argue that $\hat{\delta}$ is a less biased estimator of the population discrepancy $F(\mathbf{\Sigma}; \tilde{\mathbf{\Sigma}}_0)$ than is the raw $F(\mathbf{S}; \hat{\mathbf{\Sigma}}_0) = U^*/(N-1)$, which has for its expectation

$$E(F_{\mathrm{ML}}) = E[U^*/(N-1)] = (N-1)^{-1}E(U^*) = (N-1)^{-1}(\lambda + \mathrm{df})$$

$$= \frac{\lambda}{(N-1)} + \frac{\mathrm{df}}{(N-1)} = \frac{(N-1)F(\mathbf{\Sigma}; \tilde{\mathbf{\Sigma}}_0)}{(N-1)} + \frac{\mathrm{df}}{(N-1)} \qquad (15.16)$$

$$= F(\mathbf{\Sigma}; \tilde{\mathbf{\Sigma}}_0) + \frac{\mathrm{df}}{(N-1)}.$$

According to Bollen (1990), Browne (1982, 1984) stated that the average value of F_{ML} tended to be larger in smaller samples and smaller in larger samples. This is certainly borne out here because the second term on the right would be larger in smaller samples and smaller in larger ones. On the other hand, we also notice that on average F_{ML} tends to overestimate its population value by $\mathrm{df}/(N-1)$. If the model is correct, $F(\mathbf{\Sigma}; \tilde{\mathbf{\Sigma}}_0)$ is zero and the average value of F_{ML} is $\mathrm{df}/(N-1)$.

Combining Equations 15.15 and 15.16, the expected value of $\hat{\delta}$ is

$$E(\hat{\delta}) = E(F_{\mathrm{ML}}) - \frac{\mathrm{df}}{(N-1)}$$

$$= F(\mathbf{\Sigma}; \tilde{\mathbf{\Sigma}}_0) + \frac{\mathrm{df}}{(N-1)} - \frac{\mathrm{df}}{(N-1)} = F(\mathbf{\Sigma}; \tilde{\mathbf{\Sigma}}_0). \qquad (15.17)$$

So $\hat{\delta}$ is an (approximate) unbiased estimate of the population discrepancy, the approximation being due to the fact that U^* is only approximately distributed as noncentral chi-square in large samples when the null hypothesis is false. Finally, in the limit, as sample size increases without bound,

$$\lim_{N \to \infty} E(\hat{\delta}) = \lim_{N \to \infty} E(F_{\mathrm{ML}}) = \delta = F(\mathbf{\Sigma}; \tilde{\mathbf{\Sigma}}_0). \qquad (15.18)$$

Because $\hat{\delta}$ can take on negative values the index is often modified to

$$\hat{\delta}' = \mathrm{Max}\,(F_{\mathrm{ML}} - (N-1)^{-1}\hat{\delta}, 0) \qquad (15.19)$$

(Browne and Cudeck, 1993), but in doing so loses its unbiasedness. In Equation 15.18, $E(\hat{\delta}), E(F_{\text{ML}})$, and F_{ML} converge in the limit in probability as the sample size increases without bound to the population discrepancy function value of $F(\Sigma; \tilde{\Sigma}_0)$, where $\tilde{\Sigma}_0$ is the model-based reproduced variance–covariance matrix fitted to the population covariance matrix Σ.

Goodness-of-Fit Indices, CFI, and Others

Mulaik et al. (1989) introduced the term "goodness-of-fit index (GFI)" to refer to an index for assessing the degree of (approximate) fit of a model to data that range between zero and unity, with zero meaning "complete lack of fit" and unity indicating perfect fit. In contrast, indices such as the chi-square index range from zero to infinity and may be regarded as "lack-of-fit indices" or "badness-of-fit indices," with zero representing perfect fit and infinity, worst lack of fit. In this section we will examine a number of "GFIs."

GFIs of LISREL

Jöreskog and Sörbom (1981) proposed a family of GFIs to be used with the LISREL© program, an early program for confirmatory factor analysis and structural equation modeling. These indices were all variants of a single index, varying in terms of a scaling transformation matrix for evaluating the difference between the observed and the model variance–covariance matrix according to the method of estimation used to estimate parameters. The inspiration for these indices was likely the coefficient of determination (Wright, 1921),

$$R^2 = 1 - \frac{\text{Error variance}}{\text{Total variance}},$$

which estimates the proportion of total variance that is free of error variance. This index has its origins in Fisher's (1925) intraclass correlation. The GFI computes "error" as the sum of (weighted and possibly transformed) squared differences between the elements of the observed variance–covariance matrix **S** and those of the estimated model variance–covariance matrix $\hat{\Sigma}_0$ and compares this sum with the total sum of squares of the elements in **S**. The matrix $(\mathbf{S} - \hat{\Sigma}_0)$ is symmetric and produces the element-by-element differences between **S** and $\hat{\Sigma}_0$. **W** is a transformation matrix that weights and combines the elements of these matrices, depending on the method of estimation. Thus we have

$$\text{GFI} = 1 - \frac{\text{tr}[\mathbf{W}^{-1/2}(\mathbf{S} - \hat{\Sigma}_0)\mathbf{W}^{-1/2}][\mathbf{W}^{-1/2}(\mathbf{S} - \hat{\Sigma}_0)\mathbf{W}^{-1/2}]}{\text{tr}[\mathbf{W}^{-1/2}(\mathbf{S})\mathbf{W}^{-1/2}][\mathbf{W}^{-1/2}(\mathbf{S})\mathbf{W}^{-1/2}]},$$

where $\hat{\Sigma}_0$ is the model variance–covariance matrix, $\mathbf{S}$ is the unrestricted, sample variance–covariance matrix, and

$$\mathbf{W} = \begin{cases} \mathbf{I} - \text{Unweighted least squares} \\ \mathbf{S} - \text{Weighted least squares} \\ \hat{\Sigma}_0 - \text{Maximum likelihood} \end{cases}$$

How these error sum of squares and total sum of squares are obtained is most easily seen in the case of the unweighted least-squares version of GFI:

$$\text{GFI}_{\text{LS}} = 1 - \frac{\text{tr}[(\mathbf{S} - \hat{\Sigma}_0)(\mathbf{S} - \hat{\Sigma}_0)]}{\text{tr}(\mathbf{SS})}. \tag{15.20}$$

The trace of a square matrix is the sum of the elements on the principal diagonal of the matrix. In this case, the principal diagonal of $(\mathbf{S} - \hat{\Sigma}_0)(\mathbf{S} - \hat{\Sigma}_0)$ contains the sum of squares of the elements in each row, respectively, of $(\mathbf{S} - \hat{\Sigma}_0)$. Hence the sum of the principal diagonal of $(\mathbf{S} - \hat{\Sigma}_0)$ yields the total sum of squares of the elements of this matrix.

In GLS $\mathbf{W} = \mathbf{S} = (\mathbf{S}^{1/2}\mathbf{S}^{1/2})$. As a consequence, the numerator of Equation 15.19 is

$$\text{tr}[\mathbf{S}^{-1/2}(\mathbf{S} - \hat{\Sigma}_0)\mathbf{S}^{-1/2}\mathbf{S}^{-1/2}(\mathbf{S} - \hat{\Sigma}_0)\mathbf{S}^{-1/2}],$$

which, because of the invariance of traces under cyclic permutations of the matrices, becomes

$$\text{tr}[\mathbf{S}^{-1/2}\mathbf{S}^{-1/2}(\mathbf{S} - \hat{\Sigma}_0)\mathbf{S}^{-1/2}\mathbf{S}^{-1/2}(\mathbf{S} - \hat{\Sigma}_0)] == \text{tr}[(\mathbf{I} - \mathbf{S}^{-1}\hat{\Sigma}_0)(\mathbf{I} - \mathbf{S}^{-1}\hat{\Sigma}_0)]. \tag{15.21}$$

On the other hand, the denominator of Equation 15.19 in the case of GLS becomes $\text{tr}[(\mathbf{S}^{-1/2}\mathbf{SS}^{-1/2})(\mathbf{S}^{-1/2}\mathbf{SS}^{-1/2})] = \text{tr}(\mathbf{I}) = n$. Hence,

$$\text{GFI}_{\text{GLS}} = 1 - \frac{\text{tr}[(\mathbf{I} - \mathbf{S}^{-1}\hat{\Sigma}_0)(\mathbf{I} - \mathbf{S}^{-1}\hat{\Sigma}_0)]}{n}. \tag{15.22}$$

By similar reasoning,

$$\text{GFI}_{\text{ML}} = 1 - \frac{\text{tr}[(\mathbf{S}\hat{\Sigma}_0^{-1} - \mathbf{I})(\mathbf{S}\hat{\Sigma}_0^{-1} - \mathbf{I})]}{\text{tr}(\mathbf{S}\hat{\Sigma}_0^{-1}\mathbf{S}\hat{\Sigma}_0^{-1})}, \tag{15.23}$$

Equations 15.21 through 15.23 were originally worked out by Tanaka and Huba (1985). We might note that although the numerator of the formula for GFI$_{\text{ML}}$ in Equation 15.23 does not look like the discrepancy function F_{ML} of maximum-likelihood estimation, it nevertheless is essentially equivalent. Bentler (1989) discussed a variant of GLS estimation that is effectively equivalent to maximum-likelihood estimation, which he called "iteratively reweighted" GLS, which (following Steiger, 1995) is given by the discrepancy function

$$F_{\text{IRGLS}} = (\mathbf{S}, \hat{\boldsymbol{\Sigma}}_0 \mid \hat{\boldsymbol{\Sigma}}_0^{-1}) = \frac{1}{2}\text{tr}\left[(\mathbf{S}\hat{\boldsymbol{\Sigma}}_0^{-1} - \mathbf{I})(\mathbf{S}\hat{\boldsymbol{\Sigma}}_0^{-1} - \mathbf{I})\right]. \tag{15.24}$$

This discrepancy function was earlier described by Browne (1974) and designated as $F(\mathbf{S}, \hat{\boldsymbol{\Sigma}}_0 \mid \hat{\boldsymbol{\Sigma}}_0^{-1})$. Browne also provided a proof that it and maximum-likelihood estimation converge in the limit. Using Equation 15.24 as the function to minimize to yield estimates for free parameters will effectively produce the same estimates as maximum-likelihood estimation. What makes it iteratively reweighted is the fact that at each iteration, the weight matrix is revised based on the then current estimate of the model variance–covariance matrix based on the parameters estimated up to that point. So, it naturally appears in the numerator of the GFI for maximum likelihood. Since we know previously that in the limit $F(\mathbf{S}; \hat{\boldsymbol{\Sigma}}_0)$ converges to $F(\boldsymbol{\Sigma}; \tilde{\boldsymbol{\Sigma}}_0)$, we now know that $F(\mathbf{S}, \hat{\boldsymbol{\Sigma}}_0 \mid \hat{\boldsymbol{\Sigma}}_0^{-1})$ converges to $F(\boldsymbol{\Sigma}; \tilde{\boldsymbol{\Sigma}}_0)$ as well.

Some of the criticisms of the GFIs are that they tend to vary with sample size. Although Bollen (1989) notes that this is not due to the fact that N is involved explicitly in the formula for GFI, nevertheless Monte Carlo studies (Marsh, Balla, and McDonald, 1988) revealed that the average values of the GFI tend to increase with N.

Under certain fairly general circumstances an expression for a population value for the GFI may be obtained, for which we can then construct a confidence interval estimate. The idea of using a confidence interval estimate with a GFI was first proposed by Steiger and Lind (1980). Steiger described such a confidence interval estimate for the GFI population value in Steiger (1995). According to Browne (1974, proposition 8), there is a common case where we can derive an exact expression for the population value of the GFI. If $\boldsymbol{\Sigma}_0(\boldsymbol{\theta})$ is a model variance–covariance matrix that is a function of free parameters in θ, and if given an admissible estimate $\hat{\boldsymbol{\theta}}$ and any positive scalar α, there is an admissible $\boldsymbol{\theta}^*$ such that $\boldsymbol{\Sigma}_0(\boldsymbol{\theta}^*) = \alpha\boldsymbol{\Sigma}_0(\hat{\boldsymbol{\theta}})$ (the condition of invariance under a constant scaling function or ICSF), then, if $\hat{\boldsymbol{\theta}}$ is a maximum-likelihood estimate of the estimated parameters, $\text{tr}(\mathbf{S}\hat{\boldsymbol{\Sigma}}_0^{-1}) = n$. In this case

$$\begin{aligned}\text{GFI}_{\text{ML}} &= \frac{\text{tr}(\mathbf{S}\hat{\boldsymbol{\Sigma}}_0^{-1}\mathbf{S}\hat{\boldsymbol{\Sigma}}_0^{-1}) - \text{tr}[(\mathbf{S}\hat{\boldsymbol{\Sigma}}_0^{-1} - \mathbf{I})(\mathbf{S}\hat{\boldsymbol{\Sigma}}_0^{-1} - \mathbf{I})]}{\text{tr}(\mathbf{S}\hat{\boldsymbol{\Sigma}}_0^{-1}\mathbf{S}\hat{\boldsymbol{\Sigma}}_0^{-1})} \\ &= \frac{\text{tr}(\mathbf{S}\hat{\boldsymbol{\Sigma}}_0^{-1}\mathbf{S}\hat{\boldsymbol{\Sigma}}_0^{-1}) - \text{tr}[(\mathbf{S}\hat{\boldsymbol{\Sigma}}_0^{-1}\mathbf{S}\hat{\boldsymbol{\Sigma}}_0^{-1}) - \mathbf{S}\hat{\boldsymbol{\Sigma}}_0^{-1} - \mathbf{S}\hat{\boldsymbol{\Sigma}}_0^{-1} + \mathbf{I})]}{\text{tr}(\mathbf{S}\hat{\boldsymbol{\Sigma}}_0^{-1}\mathbf{S}\hat{\boldsymbol{\Sigma}}_0^{-1})}.\end{aligned}$$

If we next remove the brackets and substitute $\mathrm{tr}(\mathbf{S}\hat{\boldsymbol{\Sigma}}_0^{-1}) = n$, we have

$$\mathrm{GFI}_{\mathrm{ML}} = \frac{\mathrm{tr}(\mathbf{S}\hat{\boldsymbol{\Sigma}}_0^{-1}\mathbf{S}\hat{\boldsymbol{\Sigma}}_0^{-1}) - \mathrm{tr}(\mathbf{S}\hat{\boldsymbol{\Sigma}}_0^{-1}\mathbf{S}\hat{\boldsymbol{\Sigma}}_0^{-1}) + n + n - n}{\mathrm{tr}(\mathbf{S}\hat{\boldsymbol{\Sigma}}_0^{-1}\mathbf{S}\hat{\boldsymbol{\Sigma}}_0^{-1})}$$

or

$$\mathrm{GFI}_{\mathrm{ML}} = \frac{n}{\mathrm{tr}(\mathbf{S}\hat{\boldsymbol{\Sigma}}_0^{-1}\mathbf{S}\hat{\boldsymbol{\Sigma}}_0^{-1})}. \tag{15.25}$$

So, now, again given the special assumptions of ICSF, if we next substitute $\boldsymbol{\Sigma}$ for $\mathbf{S}$ and population $\tilde{\boldsymbol{\Sigma}}_0$ for sample $\hat{\boldsymbol{\Sigma}}_0$, we arrive at a population parameter

$$\Gamma_1 = \frac{n}{\mathrm{tr}(\boldsymbol{\Sigma}\tilde{\boldsymbol{\Sigma}}_0^{-1}\boldsymbol{\Sigma}\tilde{\boldsymbol{\Sigma}}_0^{-1})}, \tag{15.26}$$

which is a weighted population coefficient of determination for the multivariate (ICSF) model (Steiger, 1995, p. 3671).

Now, let us show how under the ICSF model we may express Γ_1 in terms of a population expression for the discrepancy

$$F_{\mathrm{IRGLS}}(\boldsymbol{\Sigma}; \tilde{\boldsymbol{\Sigma}}_0) = F_{\mathrm{ML}}(\boldsymbol{\Sigma}; \tilde{\boldsymbol{\Sigma}}_0) = \frac{1}{2}\mathrm{tr}[(\boldsymbol{\Sigma}\tilde{\boldsymbol{\Sigma}}_0^{-1} - \mathbf{I})(\boldsymbol{\Sigma}\tilde{\boldsymbol{\Sigma}}_0^{-1} - \mathbf{I})].$$

Note that just as we obtained Equation 15.26, we can expand the expression within brackets, take the trace, and substitute $\mathrm{tr}(\mathbf{S}\hat{\boldsymbol{\Sigma}}_0^{-1}) = n$ because the model is ICSF. If we do, we obtain

$$F_{\mathrm{IRGLS}}(\boldsymbol{\Sigma}; \tilde{\boldsymbol{\Sigma}}_0) = F_{\mathrm{ML}}(\boldsymbol{\Sigma}; \tilde{\boldsymbol{\Sigma}}_0) = \frac{1}{2}[\mathrm{tr}(\boldsymbol{\Sigma}\tilde{\boldsymbol{\Sigma}}_0^{-1}\boldsymbol{\Sigma}\tilde{\boldsymbol{\Sigma}}_0^{-1}) - n].$$

The denominator of Equation 15.26 is thus equal to $2F_{\mathrm{ML}}(\boldsymbol{\Sigma}; \tilde{\boldsymbol{\Sigma}}_0) + n$. Hence we can rewrite Equation 15.26 as

$$\Gamma_1 = \frac{n}{2F_{\mathrm{ML}}(\boldsymbol{\Sigma}; \tilde{\boldsymbol{\Sigma}}_0) + n}. \tag{15.27}$$

This will mean that if we have a consistent estimator of $F_{\mathrm{ML}}(\boldsymbol{\Sigma}; \tilde{\boldsymbol{\Sigma}}_0)$, we can use this in Equation 15.27 to obtain a consistent estimator of the (ICSF) model Γ_1. We have already shown that the sample $F_{\mathrm{ML}}(\mathbf{S}; \hat{\boldsymbol{\Sigma}}_0)$ is a consistent estimator of $F_{\mathrm{ML}}(\boldsymbol{\Sigma}; \tilde{\boldsymbol{\Sigma}}_0)$. Steiger (1989, 1995) points out that in this special case a consistent estimate of Γ_1 can be obtained from the sample $F_{\mathrm{ML}}(\mathbf{S}; \hat{\boldsymbol{\Sigma}}_0)$ and the number of estimated parameters by substituting $F_{\mathrm{ML}}(\mathbf{S}; \hat{\boldsymbol{\Sigma}}_0)$ in Equation 15.27 to obtain

$$\hat{\Gamma}_1 = \frac{n}{2F_{\mathrm{ML}}(\mathbf{S}; \hat{\boldsymbol{\Sigma}}_0) + n}. \tag{15.28}$$

We know from Equation 15.16 that $F_{ML}(S; \hat{\Sigma}_0)$ is a biased estimate of $F_{ML}(\Sigma; \tilde{\Sigma}_0)$. So, if we take the expectation of the estimator in Equation 15.28, we obtain

$$E(\hat{\Gamma}_1) \approx \frac{n}{2F_{ML}(\Sigma; \tilde{\Sigma}_0) + 2\text{df}/(N-1) + n}.$$

This result is consistent with the Monte Carlo results that suggested that as N increased, GFI tended to increase. We see that this would be so in the ICSF case, because on average the middle term in the denominator of $E(\hat{\Gamma}_1)$ becomes smaller as N increases. It also varies with the size of the number of degrees of freedom of the model, leading to greater underestimation of Γ_1 in smaller samples with many degrees of freedom.

We will now consider how a confidence interval test of Γ_1 may be obtained using the statistic $\hat{\Gamma}_1$ in Equation 15.28. Let the hypothesis to be tested be $H_0 : \Gamma_1 = \Gamma_0$ against $H_1 : \Gamma_1 \neq \Gamma_0$. We will construct a 95% confidence interval around the hypothesized value Γ_0. To find the upper and lower bounds of the confidence interval, we will use the noncentral chi-square distribution with a noncentrality of $\lambda = (N-1)F_{ML,0}$, where $F_{ML,0} = \text{df}(1 - \Gamma_0)/(2\Gamma_0)$. Let $\lambda_{.025}$ denote the value of the cumulative noncentrality distribution with df degrees of freedom and noncentrality λ, below which only 2.5% of the values of this distribution will be found. Similarly, let $\lambda_{.975}$ be the value below which 97.5% of the distribution is found. We may calculate these values using a procedure described by Browne and Cudeck (1993). Once these are found, we can calculate the values of Γ_1 that correspond to these values by first dividing them by $(N-1)$ and then substituting the resulting values for $F_{ML}(\Sigma; \tilde{\Sigma}_0)$ in Equation 15.27. If the estimated $\hat{\Gamma}_1$ is not found within these bounds, we reject the null hypothesis. It is also possible to construct confidence intervals around the estimate $\hat{\Gamma}_1$ by using that value in place of Γ_1 in the just-described procedure.

Again, we must stress that these results are known to apply if the model is ICSF.

CFI and Its Relatives

Normed Fit Index

Bentler and Bonett (1980) proposed a "normed fit index," the "NFI," as an index for assessing the fit of confirmatory factor analysis and SEMs. Like the GFI this index ranges from 0 to 1 and is a "GFI." This index is constructed from the values of chi-square for two models: M_k, the theoretical model to be tested, and M_{null}, a "null model." The null model hypothesizes that the variables of the study are all mutually unrelated, that is, have zero off-diagonal covariances between them. The population variance–covariance matrix for the null model is thus a diagonal matrix, Σ_{null}. The diagonal elements of

Σ_{null} are free parameters. The estimates of the diagonal elements of Σ_{null} are equal to the diagonal elements of Σ. Hence if there is any difference between Σ and Σ_{null}, this would be due to the off-diagonal elements of $(\Sigma - \Sigma_{null})$ being nonzero. So any lack of fit between the population null model and the true population model would be because there are nonzero covariances between the variables to be explained.

Now, Bentler and Bonett (1980) wanted to use the chi-square index value for the null model M_{null} as a kind of "norm" representing the worse possible fit you could obtain if there were any relationships between the variables. In other words, the norm represents the lack of fit due to covariation between variables when you hypothesize there is none. That represents information you are trying to account for with your model. You could compare this norm to the difference between the chi-square of the null model M_{null} and the chi-square of the tested model M_k to see how much of this information is accounted for by the reduction in lack of fit. This would produce an NFI:

$$\text{NFI} = \frac{(\chi^2_{null} - \chi^2_k)}{\chi^2_{null}}. \tag{15.29}$$

However, there is one other implicit assumption that must be recognized. The null model M_{null} must be "nested" within the model M_k. Now, a second model is nested within a first model if the value of each fixed parameter in the first model also remains with the same value as a fixed parameter within the second model, while the second model fixes or constrains some additional parameter values that correspond to free parameters in the first model. The null model will be nested within any confirmatory factor analysis model if the fixed values of the first model's factor pattern loadings are zeroes, other factor loadings are free parameters, factor variances are fixed to unity, and factor covariances are either free or fixed to zero. (We presume also that the covariances among the unique factors are fixed to zero, while the unique factor variances are free parameters.) The null model will be expressed as a factor analysis model in which all factor loadings are fixed to zero, it does not matter how the factor variances and covariances are specified, and the unique factor variance matrix is a diagonal matrix with free diagonal elements. So the null model's covariance matrix expressed as a factor analysis equation is given by

$$\hat{\Sigma}_{null} = \mathbf{0\Phi0}' + \mathbf{\Psi}^2 = \mathbf{\Psi}^2.$$

In this case the estimated model's covariance matrix will be a diagonal matrix, $\hat{\Sigma}_{null}$ with its diagonal elements equal respectively to the free variances of the unique factors, implying that the observed variances are entirely unique variances. The estimates of the unique variances turn out to be essentially the same respective values as in the principal diagonal of the unrestricted covariance matrix $\mathbf{S}$ for the observed variables. In other words, $\hat{\Sigma}_{null} = [\text{diag}\mathbf{S}]$. Then the chi-square for the null model is expressible as $\chi^2_{null} = (N-1)F(\mathbf{S}; \hat{\mathbf{\Sigma}}_{null})$ and may be computed from Equation 15.8 by substituting $\hat{\mathbf{\Sigma}}_{null}$ for $\hat{\mathbf{\Sigma}}_0$ and multiplying the result by $(N-1)$. Because n unique factor variances must be estimated, the degrees of freedom of the null model are $n(n-1)/2$.

The importance of being able to assume that the null model is nested within the hypothesized model is to insure that the chi-square of the null model is greater than or equal to the chi-square of the tested model. Otherwise, the NFI can take on a negative value in some rare instances. The reason this will be so when the null model is nested within the tested model is because of a general rule given by Bentler and Bonett (1980). The rule states that in a nested sequence of models, any model more constrained than models preceding it in the nested sequence will have a chi-square value greater than or equal to the chi-squares of the models preceding it. In general, if $M_1, M_2, \ldots, M_m$ are a nested sequence of models, each successive model more constrained than the model preceding (e.g., has additional fixed parameters), then $\chi^2_1 \le \chi^2_2 \le \cdots \le \chi^2_m$. Hence the tested model would never have a chi-square greater than the null model's chi-square. So, the difference between the null model chi-square and the model chi-square will always be greater than or equal to zero, and so, the NFI will always be zero or positive.

This requirement can be violated in some cases if the tested model M_k fixes factor pattern loadings to values other than zero. If the fixed loadings are generally considerably much greater than their true values, and the off-diagonal covariances of $\mathbf{\Sigma}$ are small, then they may possibly produce chi-squares that are greater than the null model chi-square. But this should be rare. So, this should not discourage the fixing of loadings to specific nonzero values if theory and past results suggest doing so. Fixing parameters to nonzero values on the basis of theory and past experience is an advance and yields more degrees of freedom, which, we will eventually see, correspond to distinct conditions by which a model could be disconfirmed by lack of fit, which is a desirable feature of scientific models. What this means then for the use of the NFI is that in some rare instances it can be less than zero, although negative NFIs would be readily interpretable as bad fits.

CFI

Marsh et al. (1988) conducted a series of Monte Carlo studies to assess the effect of sample size on a number of indices of fit. They reported that the NFI tended increasingly to underestimate its population value in smaller samples and did so nonnegligibly in samples smaller than 800. This form of bias would lead one to think that a model does not fit as well as it does in the population. Two years later, in the same issue of *Psychological Bulletin* MacDonald and Marsh (1990) and Bentler (1990) each presented, initially without awareness of the other's paper, a new index of fit that corrected for the bias in the NFI. Instead of using chi-squares for the null model M_{null} and the model to be tested M_k in the formula for the NFI, they recommended using unbiased estimates of the unnormalized noncentrality parameters of the null model and the model to be tested, respectively. Bentler (1990) gave this index the name "FI":

$$\text{FI} = \frac{(\hat{\delta}^*_{\text{null}} - \hat{\delta}^*_k)}{\hat{\delta}^*_{\text{null}}} = \frac{[(\chi^2_{\text{null}} - \text{df}_{\text{null}}) - (\chi^2_k - \text{df}_k)]}{(\chi^2_{\text{null}} - \text{df}_{\text{null}})}. \tag{15.30}$$

Bentler (1990) further corrected the FI to be 0 when it became negative, and to be 1 when it exceeded 1, just to keep it between the bounds of 0 and 1 in value. Out-of-bounds values could occur because in a sample a chi-square value could be less than its population mean—its degrees of freedom—so subtracting its degrees of freedom from the respective sample chi-square could produce a negative value as an estimate of the noncentrality parameter. He called the corrected FI the "CFI" for "comparative fit index", so,

$$\text{CFI} = \begin{cases} 0, & \text{FI} < 0 \\ \text{FI}, & 0 \leq \text{FI} \leq 1, \\ 1, & 1 < \text{FI} \end{cases}$$

We cannot establish by simple algebraic means that the uncorrected FI is an unbiased estimate of its asymptotic value (achieved when sample sizes grow to be infinitely large). The expected value of a ratio of random variables is not necessarily equal to the ratio of the expected values of the random variables. But Bentler (1990) argued that the FI is a consistent estimator of the asymptotic value. The CFI is also. Bentler (1990) also performed some sampling studies of the NFI, FI, and CFI. The NFI again showed the strongest tendency to underestimate its asymptotic value in small samples. The FI showed practically no bias at all sample sizes, and the mean value of the CFI was within a few thousandths of the mean value, from below, of the FI at almost all sample sizes. So, while the CFI shows a small bias in the direction of underestimating in small samples ($N = 50$), this bias is much smaller than that of the NFI and is generally negligible at moderate-to-large sample sizes.

As for a standard for the CFI that indicates a "good approximation," Bentler and Bonnet (1980) initially suggested using 0.90 for the NFI, and this standard was carried to the CFI by many. However, others (Carlson and Mulaik, 1993) frequently found that they could easily adjust their models by freeing just a few parameters based on LM tests to obtain fits of 0.95 or better, and so they recommended 0.95 as a standard. Hu and Bentler (1995) reviewed the goodness-of-fit literature and their own Monte Carlo studies and concluded that the 0.90 criterion was inadequate for "acceptable fit." Hu and Bentler (1999) suggested using a combination rule of rejecting models where CFI < 0.96 and standardized root mean residual (SRMR) > 0.09. Fan and Sivo (2005), however, questioned the conclusion that a combination index approach was needed, claiming that the finding by Hu and Bentler (1999) that the SRMR index was most sensitive to misspecified factor covariances, whereas the CFI and similar indices were most sensitive to misspecified factor loadings was not borne out in their Monte Carlo studies, but seemed to be the result of an artifact of the models studied by Hu and Bentler (1999). However, Fan and Sivo (2005) were reluctant to recommend the use of any particular cutoff for any particular fit index, considering the problem of fit complex and requiring further research.

My own opinion on this is that attempting to determine a cutoff value with Monte Carlo studies is beside the point. 0.95 is intuitively "close" as an approximation, regardless. The CFI and similar indices concern fit of the model-based reproduced covariance matrix to the empirical covariance matrix. Considering the wide variety of models that might be fitted, and the wide variety of conditions by which the data, on which the empirical covariances are based, may be generated, the aim is not to establish how the tested model is an approximation to the "real" model. There may not be a metric for such comparisons. The support for the hypothesized model is in how well the model-based reproduced covariance matrix fits the empirical covariance matrix. And this is only prima facie support, which may be overturned by producing a different model that fits as well or better—hopefully to within sampling error—with as many or more degrees of freedom.

McDonald's μ_k Index

In addition to what amounted to the FI, McDonald and Marsh (1990) proposed another GFI based on the normalized noncentrality parameter of a model $\delta = \delta^*/N$. The index is merely a monotonic transformation of the noncentrality parameter designed to guarantee that it ranges between 0 and 1. The formula for the population value of the index, denoted as μ_k, is

$$\mu_k = \exp\left[-\left(\frac{1}{2}\right)\delta_k\right]. \tag{15.31}$$

When the normalized noncentrality δ_k for model M_k is zero, μ_k is 1. As δ_k approaches infinity, μ_k approaches zero. A sample estimate of a model's μ_k is given as

$$\hat{\mu}_k = \exp[-(1/2)(\chi_k^2 - df_k)/N], \tag{15.32}$$

where χ_k^2 is the sample value of the chi-square statistic for the model, df_k is the degrees of freedom for the model, and N is the sample size.

McDonald's μ_k seems to drop off very rapidly from unity with small increases in lack of fit. A model with a chi-square of 1296.32 and 927 df for a sample of 280 had an estimated, normalized noncentrality parameter value of 1.319, which computed an estimated $\hat{\mu}_k$ of 0.517. But the CFI for this model was 0.969, while the GFI for the model was 0.835. Thus

McDonald's index may not produce values that are comparable to those of other fit indices. This index has not received much use in the literature. But there may be a place for variants of it and we will consider these later on in this chapter.

Tucker–Lewis (1973) Index

Bentler and Bonett (1980) popularized this index developed by Tucker and Lewis (1973) for factor analysis. The formula for Tucker–Lewis Index (TLI) is as follows:

$$\text{TLI} = \frac{\left(\chi^2_{\text{null}}/\text{df}_{\text{null}}\right) - \left(\chi^2_k/\text{df}_k\right)}{\left(\chi^2_{\text{null}}/\text{df}_{\text{null}}\right) - 1}, \tag{15.33a}$$

where χ^2_{null} is the chi-square value for the null model, df_{null} is the degrees of freedom for the null model, χ^2_k is the chi-square for the model being tested, and df_k is the degrees of freedom for the model being tested.

Tucker and Lewis (1973) patterned this index after a reliability index based on mean squares in components of variance analysis of analysis of variance (ANOVA). The minimum value of Jöreskog's maximum-likelihood fit function value F_m (see Equation 16.18) for a model m was treated like a sum of squares, and it was divided by the degrees of freedom for this model (because of the analogy with ANOVA) to yield a mean square M_m for error. In other words,

$$M_m = \frac{F_m}{\text{df}_m}. \tag{15.34}$$

What follows is a slightly simplified version of their derivation of this index.

What we seek is a ratio $\rho_m = \alpha_m/(\alpha_m + \delta_m)$, where ρ_m represents the "reliability" of model m, α_m represents a variance component due to a factor model with m factors, and δ_m represents a variance component involving a discrepancy between model m and the true model. In the analogy with ANOVA the mean square error

$$M_{\text{null}} = \frac{F_{\text{null}}}{\text{df}_{\text{null}}} \tag{15.35}$$

for a null model hypothesizing no relations (no common factors) among the observed variables (presumed nested within model M_k) has an expectation:

$$E(M_{\text{null}}) = \alpha_m + \delta_m + \varepsilon_m, \tag{15.36}$$

where ε_m is an additional error variance associated with random error of sampling. In other words, for the null model the total error consists of a component for variance due to the common factors not modeled, a component for error of fit of this model to the data at the population level, and an error due to sampling. The mean square error for model m is given by $M_m = F_m/df_m$ and has an expectation:

$$E(M_m) = \delta_m + \varepsilon_m. \tag{15.37}$$

The expected error for model m is due to error of approximation and sampling error.

From Equation 15.13 $(N-1)F_m = \chi_m^2$ is a chi-square variate with df_m degrees of freedom, with an expected value of df_m. Thus, we have from Equation 15.34

$$M_m = \frac{\chi_m^2}{(N-1)df_m}. \tag{15.38}$$

Hence, because the expected value of chi-square equals its degrees of freedom,

$$E(M_m) = \frac{df_m}{(N-1)df_m} = \frac{1}{(N-1)}.$$

In the case where model m is correctly specified, $\hat{\delta}_m = 0$, and Equation 15.37 yields

$$E(M_m) = \varepsilon_m = \frac{1}{(N-1)}.$$

Tucker and Lewis (1973) thus use this value as a value for the error variance component, which they now substitute into Equations 15.36 and 15.37 to obtain

$$E(M_{\text{null}}) = \alpha_m + \delta_m + \frac{1}{(N-1)} \tag{15.39}$$

and

$$E(M_m) = \delta_m + \frac{1}{(N-1)}. \tag{15.40}$$

We are now ready to formulate our "reliability coefficient":

$$\rho_m = \frac{\alpha_m}{\alpha_m + \delta_m}. \tag{15.41}$$

This they say will be analogous to an intraclass correlation. To obtain an approximate estimate of ρ_m, they recommend using as estimates for $E(M_{\text{null}})$ and $E(M_m)$, respectively, M_{null} and M_m. Hence $\hat{\alpha}_m = M_{\text{null}} - M_m$ and $\hat{\delta}_m = M_m - (1/(N-1))$, from which we can construct an approximate estimate as

$$\rho_m \approx \frac{M_{\text{null}} - M_m}{M_{\text{null}} - 1/(N-1)}. \tag{15.42}$$

If we multiply the numerator and denominator of Equation 15.41 by $(N-1)$, we obtain

$$\rho_m \approx \frac{(N-1)M_{null} - (N-1)M_m}{(N-1)M_{null} - 1}. \tag{15.43}$$

If we now substitute corresponding expressions for M_{null} and M_m from Equation 15.38 into Equation 15.43, we obtain

$$\rho_m \approx TLI = \frac{(\chi^2_{null}/df_{null}) - (\chi^2_m/df_m)}{(\chi^2_{null}/df_{null}) - 1}. \tag{15.33b}$$

To some extent the success of this index as an analogy depends on the degree to which one can show that modeling covariance matrices is the same as modeling sums of squares, mean squares, and degrees of freedom in ANOVA. However, Tucker and Lewis (1973) do not make explicit why degrees of freedom of ANOVA are equivalent to degrees of freedom of factor analysis. So, if we presume that $M_m = F_m/df_m$ because we make F_m play the role of a sum of squares and df_m the role of degrees of freedom in forming a mean square as in ANOVA, we need to ask, why is this reasonable or useful to do? One could just as easily have arrived at a "reliability index" based on the raw F_{LM} values without dividing them by degrees of freedom. In fact, the NFI would be the result. Is there any deeper mathematical reason why this correspondence should be made? Tucker and Lewis (1973) do not say. But the effect of dividing the chi-square by its degrees of freedom is to change the units of measurement of lack of fit from one model to the next. Differences are then computed between measures in these different units of measurement. To obtain an approximate form for the "reliability index," the derivation of the index is also based on the presumption that the error of approximation is zero in the case of the model chi-square, from which is derived an estimate for the sampling error as simply $1/(N-1)$. This will be presumed to be the same for any tested model. But the index is then going to be used where this assumption does not hold, with models that likely are not correct and that have nonzero errors of approximation. Thus it again must be emphasized that the index is an approximation of its intended coefficient and only works best when the model is correct.

Bentler (1990, p. 239) notes that "The degrees of freedom adjustment in the [TLI] index was designed to improve its performance near 1.0, not necessarily to permit the index to reflect other model features such as parsimony." He also notes that when $\chi^2_m = E(\chi^2_m) = df_m$, the TLI will equal 1.0. On the other hand, when $df_{null}/(\chi^2_m/df_m) > \chi^2_{null}$, as $\chi^2_{null} \geq df_{null}$ (in most cases), the index can become negative. If $\chi^2_m < df_m$, then the TLI can be larger than unity. Frequently in small samples, the index will be "anomalously small" and imply horrible fit, "when other indexes suggest an acceptable model fit" (Anderson and Gerbing, 1988). In sampling studies the TLI has a much larger variance than the NFI and other fit indices. Hu and Bentler (1995) also note that this index was not developed with a population parameter already defined, and then optimal sample estimates of this parameter were developed, as was the CFI and other indices yet to be considered in this chapter.

The Meaning of Degrees of Freedom

Degrees of freedom represent the number of dimensions in which data are free to differ from a model or curve with free parameters fit to the data as a result of constraints on some of the parameters of the model or curve.

To demonstrate this we will draw upon Mulaik (1990, 2001). He considered the general problem of fitting a function or model of several parameters to a series of observed parameters using least-squares estimation: Suppose we have p observed parameters $s_1, s_2, \ldots, s_p$, each modeled as a twice differentiable function $s_i = \sigma_i(\boldsymbol{\theta}) + e_i$, $i = 1, \ldots, p$ of m model parameters in the vector $\boldsymbol{\theta} = [\theta_1, \theta_2, \theta_3, \ldots, \theta_m]$, and e_i an error parameter. This allows us to write

$$s_1 = \sigma_1(\boldsymbol{\theta}) + e_1,$$

$$s_2 = \sigma_2(\boldsymbol{\theta}) + e_2,$$

$$\vdots$$

$$s_p = \sigma_p(\boldsymbol{\theta}) + e_p.$$

We may express these equations in vector form as

$$\begin{bmatrix} s_1 \\ s_2 \\ \vdots \\ s_p \end{bmatrix} = \begin{bmatrix} \sigma_1(\boldsymbol{\theta}) \\ \sigma_2(\boldsymbol{\theta}) \\ \vdots \\ \sigma_p(\boldsymbol{\theta}) \end{bmatrix} + \begin{bmatrix} e_1 \\ e_2 \\ \vdots \\ e_p \end{bmatrix}$$

and write simply $\mathbf{s} = \boldsymbol{\sigma}(\boldsymbol{\theta}) + \mathbf{e}$. We say then that the observed parameters are a vector function of parameters in $\boldsymbol{\theta}$ plus error. Each of the s_i in $\mathbf{s}$ is the coordinate of a point in p-dimensional space. Each of the $\sigma_i(\boldsymbol{\theta})$ is also a coordinate of the vector point $\boldsymbol{\sigma}(\boldsymbol{\theta})$ given a specific value for $\boldsymbol{\theta}$.

At the outset we will assume that the parameters in $\boldsymbol{\theta}$ are unknown and furthermore can be more in number than the number of observed parameters in $\mathbf{s}$; in other words, $m > p$. This means that we have more unknowns than knowns, and constraints on some of the model parameters must be introduced to achieve identification so that we can solve for the rest of the model parameters.

Once just-identifying constraints are introduced, we can then introduce from theory additional constraints on the model parameters to overidentify the model, which will allow for testing these constraints. In all, we will have k constraints on the model parameters. The reason we begin assuming the model parameters on the right can exceed in number the observed parameters is because this is typical for an SEM. Consider that a model equation for factor analysis is $\Sigma_{YY} = \begin{bmatrix} \Lambda & \Psi \end{bmatrix} \begin{bmatrix} \Phi_{XX} & \Phi_{XV} \\ \Phi_{VX} & I \end{bmatrix} \begin{bmatrix} \Lambda' \\ \Psi' \end{bmatrix}$, where Φ_{YY} is an $n \times n$ symmetric variance–covariance matrix with $p = n(n+1)/2$ distinct parameters (because of symmetry) that are to be determined from parameters in the matrices on the right, which correspond to the parameters in θ. Λ is an $n \times r$ common factor pattern matrix and Ψ is an $n \times n$ matrix of unique factor pattern loadings. Φ_{XX} is an $r \times r$ symmetric common factor variance–covariance matrix. Φ_{XE} and its transpose Φ_{EX} are $r \times n$ and $n \times r$ covariance matrices, respectively, representing covariation between common and unique factors.

As it stands, the model is underidentified with more parameters to estimate than there are observed parameters by which to determine their values: the total number of distinct model parameters is $m = 2nr + n(n + 1)/2 + r(r + 1)/2 > n(n + 1)/2 = p$. By fixing and constraining some of the model parameters, we may then be able to solve for the remaining free parameters. For example, by fixing the off-diagonal elements of Ψ to zero, we implement $n(n - 1)/2$ distinct constraints (because of symmetry of Ψ). By requiring the nr covariances between common factors and unique factors to be zero, we gain another nr constraints. By fixing various loadings to zero or other values in Λ, we implement further constraints. And if we fix the diagonal elements and some of the off-diagonal elements of Φ_{XX}, we will impose other constraints. Let k be the total number of constraints on the model parameters. Then this implies that $q = m - k$ is the number of free parameters. We will presume that $p - q > 0$ and that the model equation is overidentified, and write it as

$$\Sigma_{YY} = \begin{bmatrix} \Lambda & \Psi \end{bmatrix} \begin{bmatrix} \Phi_{XX} & 0 \\ 0 & I \end{bmatrix} \begin{bmatrix} \Lambda' \\ \Psi' \end{bmatrix}.$$

The $p = n(n + 1)/2$ distinct elements of Σ_{YY} can be rearranged systematically in the vector s, while for each element in s we can write an algebraic equation $\sigma_i(\theta)$ representing a function of the model parameters in Λ, Ψ, and Φ_{XX} (which will correspond to elements of θ) by which to derive the corresponding variance or covariance in Σ_{YY}, as the case may be, and place these in the vector $\sigma(\theta)$.

The problem of fitting the model or curve to the observed parameters by least squares is to find values for the free parameters in θ that yield an estimated vector $\sigma(\hat{\theta})$ that is minimally distant from s in the p-dimensional space spanned by s under k constraints placed on the parameters in θ by theory and requirements for identification of the estimated parameters. In other words, some of the parameters in θ are free to vary, and a particular combination of values for its free parameter values that produce a vector $\sigma(\hat{\theta})$ minimally distant from s is sought. To gauge "minimal distance" we will consider that the length of **e** represents that distance. However, this is also equivalent to minimizing $\mathbf{e'e} = \sum_{i=1}^{p} e_i^2$, that is, minimizing the sum of squared errors.

Mathematicians represent constraints on parameters in optimization studies by equations set equal to zero: Let $g_j(\boldsymbol{\theta}) = 0, j = 1, \ldots, k$ be the k constraints placed on the parameters in $\boldsymbol{\theta}$. The formulation of these equations of constraints can be based on substantive theory, but they should be sufficient in number so that $p + k > m$. But it is important that they be internally consistent because the surfaces defined by the equations should be able to intersect so that a consistent system is possible. Scientists, however, do not have many difficulties in formulating these constraints in simple ways, such as $\theta_5 = 0, \theta_{10} - 1 = 0, \theta_{22} - \theta_{33} = 0$, or $\theta_{11}^2 + \theta_{12}^2 + \cdots + \theta_{19}^2 - 1 = 0$.

Because the vector function $\boldsymbol{\sigma}(\boldsymbol{\theta})$ describes a point in p-dimensional space and furthermore varies as the values of $\boldsymbol{\theta}$ vary, this function is a mapping from a hypersurface G in m-dimensional model-parameter space, given by the equations of constraint, into a hypersurface S in the p-dimensional observed-parameter space. The hypersurface G of possible values for $\boldsymbol{\theta}$ is given by the equations of constraint

$$g_1(\boldsymbol{\theta}) = 0,$$
$$g_2(\boldsymbol{\theta}) = 0,$$
$$g_3(\boldsymbol{\theta}) = 0,$$
$$\vdots$$
$$g_k(\boldsymbol{\theta}) = 0$$

or

$$\mathbf{g}(\boldsymbol{\theta}) = \mathbf{0}.$$

The problem of least-squares parameter estimation becomes the problem of finding the value of a vector $\hat{\boldsymbol{\theta}}$ of model parameters on the hypersurface G of constrained parameters that maps via $\boldsymbol{\sigma}(\boldsymbol{\theta})$ to a point on the hypersurface S in p observed-parameter space that is minimally distant from the observed-parameter vector $\mathbf{s}$ in that same space.

The function to minimize is

$$L = \mathbf{e'e} + \boldsymbol{\lambda'}\mathbf{g}(\boldsymbol{\theta})$$

$$= \sum_{i=1}^{p} [s_i - \sigma_i(\boldsymbol{\theta})]^2 + \lambda_1 g_1(\boldsymbol{\theta}) + \cdots + \lambda_k g_k(\boldsymbol{\theta}), \qquad (15.44)$$

where $\boldsymbol{\lambda} = [\lambda_1, \lambda_2, \ldots, \lambda_k]$ is a $k \times 1$ vector of LMs whose values are to be solved in the process of minimizing this equation. (Do not confuse λ here with the noncentrality parameter discussed earlier. It is just that mathematicians like to designate LMs with the symbol λ.)

Mulaik (1990, 2001) argues that in an ε neighborhood of the point $\boldsymbol{\sigma}(\hat{\boldsymbol{\theta}})$, which is minimally distant from the observed parameter point $\mathbf{s}$, the function $\boldsymbol{\sigma}(\boldsymbol{\theta})$ may be said to vary in q dimensions. The proof of this is based

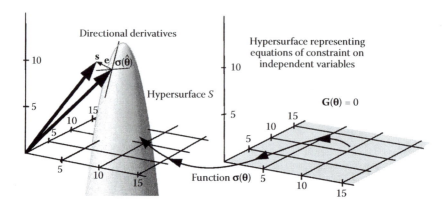

FIGURE 15.1 Graphical representation of the model fitting problem. Points in the constrained parameter space G are mapped by the model function $\sigma(\theta)$ into the hypersurface S in the observed parameter space. The point $\hat{\theta}$ produces a point $\sigma(\hat{\theta})$ on the hypersurface S that is minimally distant from the observed parameter vector $\mathbf{s}$. The error vector $\mathbf{e}$ is orthogonal to the hyperplane formed by the directional derivatives of $\sigma(\theta)$ with respect to the free parameters tangent to the hypersurface S at $\sigma(\hat{\theta})$. The vector $\mathbf{e}$ lies in the complementary space orthogonal to the tangent hyperplane and has for its length the distance between $\mathbf{s}$ and $\sigma(\hat{\theta})$.

on showing that in the neighborhood of the point $\hat{\theta}$ in model-parameter space G, $\sigma(\theta)$ is effectively a function of q parameters θ_q. By the implicit function theorem, the equation of constraint implicitly makes k parameters θ_k in the vector $\theta = \begin{bmatrix} \theta_q, \theta_k \end{bmatrix}$ a function $\mathbf{f}(\theta_q)$ of the remaining parameters so that $\mathbf{g}(\theta) = \mathbf{g}[\theta_q, \mathbf{f}(\theta_q)] = \mathbf{0}$. Next, by the affine approximation theorem of advanced calculus, we can construct an affine approximation $\mathbf{A}(\theta)$ to the image curve of $\sigma(\theta)$ at the point $\sigma(\hat{\theta})$ (Figure 15.1):

$$\mathbf{A}(\theta) = \sigma(\hat{\theta}) + \sigma'(\hat{\theta})(\theta - \hat{\theta}).$$

Here $\sigma'(\hat{\theta})$ is the derivative matrix of $\sigma(\theta)$ with respect to θ_q (which are the only varying parameters), evaluated at the value of $\hat{\theta}$ generating the point $\sigma(\hat{\theta})$ minimally distant from $\mathbf{s}$ (Williamson and Trotter, 1979, p. 226). The rank of the derivative matrix is q, which is also the number of dimensions of the vector space spanned by the linearly independent columns of the derivative matrix $\sigma'(\hat{\theta})$. The columns of $\sigma'(\hat{\theta})$ are vector partial derivatives and tangent vectors to the surface (generated by varying $\sigma(\theta)$ as a function of θ_q) at the point $\sigma(\hat{\theta})$. The line between $\mathbf{s}$ and $\sigma(\hat{\theta})$ is perpendicular to the tangent vectors at the point $\sigma(\hat{\theta})$, since that is the shortest distance between $\mathbf{s}$ and any point $\mathbf{A}(\theta)$ in the tangent space in the neighborhood of $\sigma(\hat{\theta})$ (because L, which is the squared length of the line from $\mathbf{s}$ to $\sigma(\hat{\theta})$, is minimized at $\sigma(\hat{\theta})$). The vector $\mathbf{e} = \mathbf{s} - \sigma(\hat{\theta})$, which may be made to originate at $\sigma(\hat{\theta})$, is collinear with the line between $\mathbf{s}$ and $\sigma(\hat{\theta})$ and is in the complement affine vector space of $p - q$

dimensions orthogonal to the q-dimensional space spanned by the tangent vectors originating at the point $\sigma(\hat{\theta})$.

In other words, the error vector **e** originating at the point on the surface generated by $\sigma(\theta)$ is orthogonal to each of the dimensions in which $\sigma(\theta)$ is varying at the point $\sigma(\hat{\theta})$. Because the dimensions in which $\sigma(\theta)$ varies is q in number, the error vector **e** must be in a space of at most $p - q$ dimensions orthogonal to the space in which $\sigma(\theta)$ varies in the vicinity of $\sigma(\hat{\theta})$. This space gives the maximum number of possible dimensions that **e** may span, or the number of dimensions in which the observed parameter vector **s** is free to differ from the reproduced model vector $\sigma(\hat{\theta})$. These correspond to the *degrees of freedom* of the model. Each degree of freedom corresponds to a condition by which the observed data vector **s** is free to differ from the reproduced model vector $\sigma(\hat{\theta})$. Thus the degrees of freedom of a model represent the number of distinct conditions by which the model can be disconfirmed by a lack of fit. Degrees of freedom are thus a quantitative measure of the disconfirmability of a model.

Generalization to Weighted Least-Squares Estimation

A more general form of least-squares estimation is weighted least-squares estimation. The above results also apply to this form of estimation. The GLS estimation criterion to minimize can be written as

$$L = [\sigma(\theta) - s]'W[\sigma(\theta) - s] + \Lambda'g(\theta). \tag{15.45}$$

Here **W** is a $p \times p$ nonsingular, Gramian, symmetric weight matrix. We can furthermore factor **W** as $W = HH'$. This allows us to rewrite Equation 15.45 as

$$L = [\sigma(\theta) - s]'HH'[\sigma(\theta) - s] + \Lambda'g(\theta).$$

Furthermore, we can now define $s^* = H's$ and $\sigma^*(\theta) = H'\sigma(\theta)$ and rewrite Equation 15.45 as

$$L = [\sigma^*(\theta) - s^*]'[\sigma^*(\theta) - s^*] + \Lambda'g(\theta). \tag{15.46}$$

This is simply a form of the least-squares criterion applied now to the transformed observations vector s^* and the model vector $\sigma^*(\theta)$.

Mulaik (1990) further considered the special problem where $s = \text{vecs}(S)$ is a vector of the $p = n(n + 1)/2$ nonduplicated elements of a sample variance–covariance matrix **S** and $\sigma(\theta) = \text{vec}(\Sigma(\theta))$ of a corresponding hypothetical variance–covariance matrix $\Sigma(\theta)$. He noted that Tanaka and Huba (1985) had shown that for structural equations modeling the different methods of estimation, (a) ordinary least squares, (b) GLS, and (c) maximum likelihood, may be treated as special cases of the general equation for weighted least squares. Let the elements of $\mathbf{s}$ and $\sigma(\theta)$ be indexed by the indices of the respective elements of $\mathbf{S}$ and $\Sigma(\theta)$, so that, for example, $\mathbf{s} = [s_{11}, s_{21}, s_{22}, s_{31}, s_{32}, s_{33}, \ldots, s_{n,n-1}, s_{nn}]$. Carrying this indexing to the elements of $\mathbf{W}$, $w_{ij,kl}$ is in the row corresponding to the ijth element of $\mathbf{s}$ and in the column corresponding to the klth element of $\mathbf{s}'$. Now, following Browne (1982, Equation 1.5.16, p. 83), let $\mathbf{W}$ be defined with elements $= w_{ij,kl} = [(2 - \delta_{ij})(2 - \delta_{kl})/4](v^{ik}v^{jl} + v^{il}v^{jk})$, where δ_{ij} and δ_{kl} are Kronecker deltas that equal unity when their left and right subscripts are the same and equal zero otherwise, and v^{ik} is the ikth element of the $n \times n$ symmetric matrix $\mathbf{V}^{-1}$, and so on. According to Tanaka and Huba (1985), for the case of ordinary least squares, $\mathbf{V} = \mathbf{I}_n$. For the case of GLS, $\mathbf{V} = \mathbf{S}$, the $n \times n$ sample covariance matrix. For the case of maximum-likelihood estimation, $\mathbf{V} = \Sigma = \Sigma(\hat{\theta})$. [Note that the equivalence with maximum likelihood is only approximate, with the approximation improving asymptotically (Browne, 1982, p. 84) as the sample size grows indefinitely.] But the fact that maximum-likelihood estimation at the population produces what is equivalent to a least-squares solution allows us to use the same interpretation here for degrees of freedom in the maximum-likelihood case.

Implications for Model Testing

Mulaik (1990, 2001) observed that we can now see how estimating more and more parameters until we have a saturated model will lead to ever-decreasing lack of fit, until we have a saturated model that fits perfectly. As we estimate more and more parameters, the number of dimensions in which the model vector $\sigma(\theta)$ is free to move around in the p-dimensional data space in search of a minimally distant point also increases. Usually a solution $\sigma(\hat{\theta})$ can be found that is closer and closer to $\mathbf{s}$. When we can search as many dimensions as there are observed parameters, we can find a model that passes through the observed parameter vector perfectly. This suggests that good fit that is obtained by freeing more and more parameters is suspect. We can also think of this as the estimating algorithm's "peeking at the data" to find estimates for parameters that best fit the observed parameters. This means that a model with good fit obtained by freeing more and more parameters is a model that has been formulated by adjusting it progressively to fit the very data that are to be a test of the model. This violates the often cited rule that one should not test a model against the same data used to formulate the model. A test requires a logical possibility of failing to pass the test. That logical possibility decreases with each estimated parameter until we get a saturated model that is impossible to reject for lack of fit. No test at all can be made with a saturated model.

On the other hand, we can interpret the lack of fit as concerning just the overidentifying constraints (in the context of some duly chosen just-identifying constraints) rather than testing all aspects of a model. As mentioned earlier, freeing parameters is not hypothesizing anything about them. The freed parameters are unknowns, and the estimating algorithms, unless constrained, will seek out any best-fitting value, which may be in the range from $-\infty$ to $+\infty$. Hence the test should not be about what is not asserted as a hypothesis *a priori*. A test is about the constraints hypothesized for parameters of the model.

"Badness-of-Fit" Indices, RMSEA, and ER

Up to now we have looked at "GFIs" that range generally between 0 and 1 to indicate degree of fit. We will now look at an index that is a "badness-of-fit index," where 0 indicates perfect fit and larger values indicate lack of fit. We have already indicated that the population error of approximation $F(\Sigma; \tilde{\Sigma}_0)$ is the desired measure of how well a model will fit the population variance–covariance matrix. An (approximately) unbiased sample estimate of this parameter is easily obtained by subtracting the degrees of freedom from the chi-square statistic and dividing by $(N-1)$ (*cf.* Equation 16.15). But the normalized noncentrality parameter concerns cumulative lack of fit over all the constraints placed on the model. When it comes to comparing models with different numbers of constraints placed on them, the noncentrality parameter for the model with more constraints can be larger than the model with fewer constraints. To place them on a common basis for comparison, it is recommended that we divide the noncentrality parameter by the number of degrees of freedom. The result is an index that yields the mean noncentrality per degree of freedom. Since the noncentrality parameter is a measure of squared distance, Steiger and Lind (1980) recommended taking the square root of this quantity, which they called the "root mean squared error of approximation" or the RMSEA index:

$$\text{RMSEA} = \sqrt{\frac{\chi^2_{\text{df}_k} - \text{df}_k}{(N-1)\text{df}_k}} \qquad (15.47)$$

Here $\chi^2_{\text{df}_k}$ is the chi-square index of fit with df_k degrees of freedom for model k, with N the sample size. As we have seen, the degrees of freedom are an index of the number of dimensions in which data are free to differ from a model with estimated parameters. Thus getting the average discrepancy per degree of freedom obtains an average measure of lack of fit per dimension of potential lack of fit. Browne and Cudeck (1993) reported some experience with typical values of this index for models they believed fit well and those that did not. On the basis of that they suggested using a value less than or equal to 0.05 for the RMSEA as a criterion for a model that has "good fit." Because the quantity under the radical sign can take on negative values in some instances, they further recommended setting any negative value under the radical to zero, otherwise retain any positive quantity as is. So, this index is currently defined for an arbitrary model k as

$$\text{RMSEA} = \sqrt{\text{Max}\left\{\left(\frac{\chi^2_{\text{df}_k} - \text{df}_k}{(N-1)\text{df}_k}\right), 0\right\}}. \qquad (15.48)$$

Browne and Cudeck (1993) indicate that a confidence interval estimate for the RMSEA is available to indicate the precision of the RMSEA estimate. To obtain this they begin by defining $G(\chi^2_{\text{df}_k} \mid \delta^*, \text{df})$ as the cumulative distribution function of the noncentral chi-square distribution with noncentrality parameter value δ^* and df degrees of freedom. Given a specific value for $\chi^2_{\text{df}_k}$ derived from fitting by maximum-likelihood estimation a model to the sample covariance matrix **S** and values for δ^* and df_k for a model, one can calculate the cumulative probability of obtaining such a value for the chi-square statistic.

$$P = G(\chi^2_{\text{df}_k} \mid \delta^*, \text{df}_k).$$

Furthermore, if δ^* is not given and instead the probability is given, one can solve for the value of δ^* that would define a distribution in which $\chi^2_{\text{df}_k}$ occurs with a specified cumulative probability of P. For a 90% confidence interval on $\chi^2_{\text{df}_k}$, we need to find values for $\delta^* = \delta^*_U$ such that $P = G(\chi^2_{\text{df}_k} \mid \delta^*_U, \text{df}_k) = 0.05$, which will be the upper limit, and $\delta^* = \delta^*_L$ such that $P = G(\chi^2_{\text{df}_k} \mid \delta^*_L, \text{df}_k) = 0.95$, which will be the lower limit. Then a confidence interval on the RMSEA is given as $\left(\sqrt{\delta^*_L/[(N-1)\text{df}_k]}, \sqrt{\delta^*_U/[(N-1)\text{df}_k]}\right)$. Browne and Cudeck (1993) refer to several algorithms available for calculating these values of the noncentrality parameter, but we will not go into them here. However, they note that if the lower limit is zero, the test of the null hypothesis based on $\chi^2_{\text{df}_k}$ would not reject the null hypothesis that $\delta^* = 0$ at the 5% level. Thus the interval provides the corresponding information given in the chi-square test. Of course, in this case, if the upper limit is nonzero, one cannot be certain that the model is correct, since nonzero values would also be consistent within this interval. In any case, the confidence interval gives a measure of precision for the estimated RMSEA.

There may now be some evidence (Olsson, Foss, and Breivik, 2004) that the noncentral chi-distribution may not be appropriate for establishing these confidence intervals. What Olsson et al. (2004) have shown with Monte Carlo studies is that the noncentral chi-square distribution is a good approximation to the empirical distribution in models with a relatively small number of variables and relatively small noncentrality parameters. But they also found that the empirical distribution, while still centered very close to the mean of the noncentral chi-square distribution, was more spread out in the tails than in the noncentral chi-square distribution, and, in fact, seemed more closely approximated by a normal distribution. Yuan (2008) argues mathematically that chi-square of the likelihood ratio statistic follows asymptotically a normal distribution. So, caution is signaled in using these confidence intervals. [See later in the discussion of the Akaiki information criterion (AIC) index.]

Some have suggested that the RMSEA index takes "parsimony" of the model into account by dividing by the degrees of freedom. But this is not so, as we will see shortly in a discussion of parsimony in model evaluation. One could have a model with a modest number of degrees of freedom and a certain RMSEA value and begin to add further constraints. The normalized noncentrality will likely increase. But each additional constraint may in some cases add only the same degree of lack of fit, and the average discrepancy per degree of freedom will not change. So, the RMSEA does not necessarily favor models with more degrees of freedom. Like the chi-square index, it should be interpreted primarily in terms of the fit of the constraints in the model rather than of the "whole model."

Exponentialized Negative RMSEA Index

In Equation 16.32, we saw how McDonald formulated a GFI by obtaining the exponential of minus one half of the normalized noncentrality obtained by dividing the noncentrality parameter by $(N - 1)$. Over nonnegative values for the noncentrality parameter from 0 to infinity, the index ranges from 1 to 0. However, we noted that the index falls off rather rapidly from 1 with small increases in the normalized noncentrality and yields values that are much lower than those of the GFIs or CFIs for models considered to be very good approximations. This seemed to be an undesirable feature of this index. But a variant of this index may be made to behave if we obtain the exponential of the negative RMSEA (ER) index:

$$ ER = \exp(-RMSEA) = \exp\left(-\sqrt{Max\left\{\left(\frac{\chi^2_{df_k} - df_k}{(N-1)df_k}\right), 0\right\}}\right). \quad (15.49) $$

(Thanks to Paul Dudgeon for suggesting "ER" for the name of this index.) When the RMSEA index is 0.05, the value of ER is 0.951. When the RMSEA is 0.10, ER is 0.904. When the RMSEA is 0.20, ER is 0.818. So, this behaves very much like other GFIs in the region of acceptable approximate fit. We can regard a model with an ER of 0.95 or better as an acceptable approximation, much as we would with other indices that range between 0 and 1. And because we can construct confidence intervals for the RMSEA, we can do the same with the ER, using the lower and upper bounds of the interval for the RMSEA to construct a corresponding interval for the ER. There will be other advantages of this index when we consider parsimony.

Parsimony

Parsimony is an important concept in the evaluation of models. We should note here that the dictionary definition of *parsimony* is that it is synonymous with "extreme economy or frugality." Because parsimony has played an important role in the history of science, we will begin our discussion of it by reviewing its history. (We will draw upon the account of this history given by Mulaik et al. (1989).) In the fourteenth century, the nominalist philosopher and theologian William of Occam proposed a principle that is taken by many to be fundamental in science, known as "Occam's razor": *Entities are not to be multiplied unnecessarily.* This came to be understood as requiring that theories should be as simple as possible. Immanuel Kant (1787/1996) recognized this as a regulative principle of reason impelling us to unify experience as much as possible by means of the smallest number of concepts. But Kant also warned that the principle should not be applied uncritically. Against it there is another principle, that the varieties of things should not be rashly diminished, if we are to capture their individuality and distinctness. Toward the end of the nineteenth century, the Austrian physicist and Kantian philosopher of science Heinrich Hertz (1894) advanced the view that our theories are not merely summary descriptions of that which is given to us in experience, but constructs or models actively imposed by us onto experience. So, for any given phenomenon there are many models that we might construct with respect to it. But given a number of models to consider, we need criteria by which to choose the best from among them. He therefore suggested these criteria: the model should be (1) logically and/or formally consistent, (2) empirically adequate (meaning, be related to real-world data), (3) be able to represent more of the essential relations of the objects of the phenomenon (good fit), and (4) the simplest (Janik and Toulmin, 1973). Hertz's stress on simplicity had a considerable impact on physicists who followed him. Thus parsimony in explanation involved frugality or fewness of concepts in an explanation of something.

By the 1930s and 1940s, simplicity of theories was often cited as a fundamental principle by scientists. For example, L. L. Thurstone (1947, p. 52) in his text on factor analysis stated that, "The criterion by which a new construct in science is accepted or rejected is the degree to which it facilitates the comprehension of a class of phenomena which can be thought of as examples of a single construct rather than as individualized events." He then asserted, "But in order for this reduction may be accepted as science, it must be demonstrated, either explicitly or by implication, that the number of degrees of freedom of the construct is smaller than the number of degrees of freedom of the phenomena that the reduction is expected to subsume" (Thurstone, 1947, p. 52). Here I believe he used the term "degrees of freedom" differently from the way we did in the previous section. He meant number of free parameters. He cited an example: Suppose we propose a rational equation as the law governing the relation between two variables. We obtain three observations. If the equation has three independent parameters, then the number of degrees of freedom of the phenomena is the same as the number of degrees of freedom of the equation, and the relation is undemonstrated. Here, I believe, he was thinking of a just-identified equation. There are as many unknown free parameters in the equation as there are observations to fit the equation's curve to. Since in these situations scientists use the data to estimate free parameters in such a way as to optimize fit to the data conditional on any identifying or overidentifying constraints, the fit of the three-parameter curve to the three observations would be perfect but nothing would be demonstrated. The perfect fit is a mathematical necessity regardless of the empirical setting. But then he says, suppose we have 100 observations and an equation with three parameters. Such an equation is of scientific interest. He concluded by saying "The convincingness of a hypothesis can be gauged inversely by the ratio of its number of degrees of freedom to that of the phenomena which it has demonstrably covered" (p. 52). In other words, the ratio of the number of free parameters to the number of observations is inversely related to the convincingness of a hypothesis.

And we have to use parsimony as a way of choosing the best hypothesis. Thurstone concluded by saying, "It is in the nature of science that no scientific law can ever be proved to be right. It can only be shown to be plausible. The laws of science are not immutable. They are only human efforts toward parsimony in the comprehension of nature" (p. 52). Parsimony then became a key idea in his use of the method of factor analysis. It played a role in his ideas of minimum rank, the overdetermination of factors by observed variables, and simple structure.

In the mid-1930s, the Austrian philosopher of science who later became a British subject and was knighted, Sir Karl Popper, believed that the simplicity or parsimony of a hypothesis is essential to evaluating the merits of a hypothesis before and after it is subjected to empirical tests. "The epistemological questions which arise in connection with the concept of simplicity," he said, "can all be answered if we equate this concept with *degree of falsifiability*" (Popper, 1934/1961, p. 140). He seemed to grasp the meaning of how, in connection with a given set of observations, a hypothesis with few freely estimated parameters may be subjected to more tests of possible disconfirmation than a hypothesis containing numerous freely estimated parameters. However, he seemed unable to formulate explicitly the principle of why estimating fewer parameters yields more ways to disconfirm a model.

However, we have just done so in connection with developing the concept of degrees of freedom. We saw how one begins with a number of observed parameters and seeks to formulate an equation by which to determine these points. If the equation has no free parameters, so all parameters have values by hypothesis, then each observed parameter is a value against which the predicted values of the theoretical equation can be compared. There are as many degrees of freedom, dimensions, as there are observed parameters by which the observed parameters are free to differ from their predicted values. But if out of ignorance for them, we free certain model parameters and estimate them to optimize fit to the observations conditional on the constrained model parameters, then, for each freed parameter, we lose a degree of freedom, a dimension, by which the observations may differ from the predicted values. So, models are disconfirmable to the extent that there are dimensions by which they may differ from the observations to be modeled. And this is inversely related to the number of free parameters relative to the number of observations.

Parsimony is the fewness of free parameters relative to the number of observations. This is a ratio, as Thurstone (1947) seemed to understand. Parsimony is not degrees of freedom, which is the difference between the number of observations and the number of free parameters in the (identified) model. But parsimony is related to degrees of freedom. Both concepts are derived from the same information. But degrees of freedom do not convey information about the number of observations to account for with the model. Degrees of freedom can be large in number, but in a model with many thousands of observed data points to fit, the degrees of freedom may still be small relative to the number of observations, because proportionately there are still many free parameters in a complex model. So, the size of the degrees of freedom does not indicate the degree to which the observations are accounted for by the hypothesized parameters of the model as opposed to the estimating of parameters. But a ratio can convey this idea.

James et al. (1982) argued in connection with the NFI of Equation 15.29 that "The parsimony of a model is indicated by the ratio of the degrees of freedom of the model to the number of degrees of freedom available in the data as indicated by the number of degrees of freedom available of the null model for those data." They reasoned that the null model, a covariance matrix whose off-diagonal covariances were all zero, and whose variances in the diagonal were all freely estimated, provided a norm which only concerned the lack of fit of the null model to the off-diagonal elements of the observed covariance matrix. In the formula for the NFI, $\text{NFI} = (\chi^2_{\text{null}} - \chi^2_t)/\chi^2_{\text{null}}$, the difference in lack of fit between the null model and the tested model t is compared with the lack of fit of the null model. Since the lack of fit of the null model only concerns its fit with the $n(n-1)/2$ off-diagonal elements of the covariance matrix $\mathbf{S}$, the maximum number of degrees of freedom is therefore only $n(n-1)/2$. So they recommended that one multiply the ratio

$$\text{PR} = \frac{\text{df(model)}}{\text{df(maximum)}} = \frac{\text{df(model)}}{n(n-1)/2} \qquad (15.50)$$

with the value of the NFI to obtain a "parsimonious fit index." Later when the NFI was replaced by the CFI of Equation (15.30), Mulaik et al. (1989) recommended multiplying this value for the PR with the CFI for the model. The resulting quantity represented a combining of information about the disconfirmability of a model with its fit to the data to provide an index of the overall quality of the model. Perhaps the PR ratio could also be called the "disconfirmability ratio." An ideal model would be one that had perfect fit to the observed covariance matrix and a PR of 1, meaning every parameter of the model was fixed by hypothesis. The resulting index PR*CFI would then equal unity. The model would be the most disconfirmable and at the same time best fitting. On the other hand, a perfectly fitting saturated model would obtain a combined PR*GFI or PR*CFI of 0, since the PR would be zero. Mulaik et al. (1989) also recommended that a different ratio be used for PR in connection with the GFI. In this case, since the GFI concerned the fit to all of the elements of the covariance matrix, the maximum degrees of freedom is $n(n + 1)/2$, so

$$PR = \frac{df(model)}{df(maximum)} = \frac{df(model)}{n(n + 1)/2}. \tag{15.51}$$

Then one would obtain the value of PR*GFI and interpret this similarly as an index of model quality. Carlson and Mulaik (1993) added further interpretations to the use of parsimony-adjusted GFIs. They said,

> One would not need to obtain the parsimony adjusted goodness of fit value PCFI = (df_k/df_0)CFI if one used the CFI index with *large* samples simply as an index of the approximate correctness of the *overidentified conditions* placed on the tested model, in a manner analogous to the way the generalized likelihood ratio chi square test is used to test the same overidentified conditions as an exact hypothesis using the sampling distribution of chi square with corresponding degrees of freedom as a basis for evaluating the magnitude of the chi square statistic. In this case the CFI is used because large samples make sampling distribution theory less relevant to the assessment of the usefulness of the *overidentifying constraints* as an approximation. But for various reasons, researchers have tended to use chisquare and goodness of fit indices to evaluate not simply the overidentified conditions of the model but the *whole model* ... as if the indices provided a test of the whole model, to be used in the comparison and selection of models. It is for such applications that the parsimony adjustment of goodness of fit indices has been introduced. (Carlson and Mulaik, 1993, p. 130)

It is quite possible to use the PR ratio alone as a measure of model disconfirmability when considering the fit of a model. This serves to remind one of how much has been tested with one's model. Values of the PR*CFI or the PR*GFI above 0.85 represent models with both good fit and high disconfirmability. (This is not a hard and fast rule!) Models with values below this could use improvement, either in fit or in specifying more parameters. The PR ratio can also be used with the ER index, since it also ranges between 0 and 1 and is interpretable in a manner similar to parsimony adjustments of the other indices that range between 0 and 1.

Information Theoretic Measures of Model Discrepancy

AIC Index

The Japanese statistician Hirotugu Akaiki (Akaiki, 1973, 1987; Sakamoto, Ishiguro, and Kitagawa, 1986; Bozdogan, 1987) has argued that a measure of the discrepancy in fit of a (fully specified) model to the true model is given by the difference between the expected log likelihood of the hypothesized model and the expected log likelihood of the true model, with the expectations taken in both cases with respect to the true distribution generating the data. In other words, let X be a random variable with $f(x)$ its true (but possibly unknown) distribution function. Let $h(x)$ be a hypothesized distribution for X. Then assuming counterfactually that we know $f(x)$, the measure of the discrepancy between the true distribution and the hypothesized distribution would be

$$I(f;h) = E_X[\log f(x)] - E_X[\log h(x)]$$
$$= \int_{-\infty}^{\infty} f(x) \log f(x)\, dx - \int_{-\infty}^{\infty} f(x) \log h(x)\, dx,$$

where $I(f;h)$ is known as the Kullback–Leibler information quantity. This quantity measures the amount of information on average with an observation of the random variable that would allow one to discriminate between the true distribution and the hypothesized distribution. The integrals are just the expressions for the expected values of continuously distributed variables. It is important here to see that the likelihood of a value x, $h(x)$, represents a transformation of that value, and that $\log h(x)$, which is a monotonic transformation of $h(x)$, does also. So both of these function like scores. So we compute the average $\log f(x)$ and compare that with the average $\log h(x)$ and obtain their difference.

When applied to a sample value x_i of the random variable X, the likelihood $h(x_i)$ represents a measure of the support of the observed value for the hypothesis. If the observed value x_i has very low likelihood $h(x_i)$, this is weak support for the hypothesis. If its likelihood is high, this is strong support. We can conceive of sampling a single observation x of the random variable X and transforming it by applying the likelihood function $h(x)$ to it. This produces a new random variable $\log(h(X))$. The expected value of the resulting quantity with respect to the true distribution thus represents a population measure of fit of the hypothesized distribution to the actual distribution of the random variable. When the hypothesized distribution $h(x)$ equals the true distribution $f(x)$, then $E_X[\log h(x)]$ attains its maximum value, which is equal to $E_X[\log f(x)]$, and $I(f;h) = 0$, implying no discrepancy between hypothesized distribution and actual distribution. When $h(x)$ does not equal the true distribution, then $E_X[\log h(x)]$ will be less than $E_X[\log f(x)]$ and $I(f;h) > 0$, indicating a discrepancy between the true distribution and the hypothesized distribution.

Note, however, that over different hypothesized models, only the right-most term $E_X[\log h(x)]$ in the expression for $I(f;h)$ varies; the left-hand term $E_X[\log f(x)]$ is the same for all hypothesized models for the same variable X. So, the rightmost term is the only important term for comparisons of models. Over a range of models compared, the model with the largest value for $E_X[\log h(x)]$ will be the closest to the true model.

In practice, we ordinarily do not know the true distribution $f(x)$. Some think we can get around this problem if we estimate the quantity $E_X[\log h(x)]$ by obtaining a large sample of observations of $X, x_1, x_2, \ldots, x_N$ and computing the sample mean log likelihood $1/N \sum \log h(x_i)$. Closely related to this is the quantity $\sum \log h(x_i) = \log \Pi h(x_i)$, which is the log likelihood for the sample under the hypothesized model. Its expected value (under the assumption that the observations are independently and identically distributed) is

$$E_X\left[\sum_{i=1}^{N} \log h(x_i)\right] = N \cdot E_X[\log h(x)].$$

AIC Index

Akaiki (1973, 1987) considered the case where the distribution is known and all that differed between models were the parameters. So, given a true model $f(x \mid \boldsymbol{\theta}^*)$ with true parameters $\boldsymbol{\theta}^*$, other models with different values for the parameters would be designated as $f(x \mid \boldsymbol{\theta})$. Furthermore, he generalized the likelihood concept used as a measure of support for a model with completely specified parameters to consider models with estimated parameters. In doing so he showed that

$-2(\text{maximum log likelihood of a model}) - (\text{number of free parameters})$

(15.52)

is an asymptotically unbiased estimator of a population parameter he called the "mean expected log likelihood" (Sakamoto et al., 1986). This parameter is analogous to $N \cdot E_X[\log h(x)]$ in the above discussion. However, if we simply took $\hat{\boldsymbol{\theta}}$, the parameter vector with sample maximum-likelihood estimates for the free parameters for some sample, and considered $N \cdot E_Z[\log f(z \mid \hat{\boldsymbol{\theta}})]$ for some other random variable Z that is identically but independently distributed with respect to X, this would not yield the desired expectation analogous to $N \cdot E_X[\log h(x)]$. This is because the free parameters in the parameter vector $\hat{\boldsymbol{\theta}}$ are random variables that are functions of some vector of values in a random sample $x = [x_1, x_2, \ldots, x_N]$ of the random variable X. We need to consider finding the mean of $N \cdot E_Z[\log f(z \mid \hat{\boldsymbol{\theta}})]$ over all random samples x. This we can do if we find

$$E_X\{N \cdot E_Z[\log f(z \mid \theta(\mathbf{X}))]\} = \int_{R^N} N \cdot E_Z[\log f(z \mid \theta(\xi))] \prod_{i=1}^{N} f(x_i \mid \theta^*) \, d\xi.$$

The expectation on the left is taken with respect to a random vector of N identically and independently distributed random variables $\mathbf{X} = [X_1, X_2, \ldots, X_N]$, which represents a random sample of observations of the random variable X having the distribution function $f(x)$. A realization of $\mathbf{X}$ is denoted by the vector x. The free parameters in $\theta(\mathbf{X})$ are functions of the random vector of observations $\mathbf{X}$. The integral on the right is a multiple integral in N-dimensional real-valued space R^N, whereas $\prod_{i=1}^{N} f(x_i \mid \theta^*)$ is the joint density of sample values, each having the true distribution $f(x \mid \theta^*)$. Because this appears to be an expected value of an expected value, this parameter is called the "mean expected log likelihood" by Akaiki. As an estimate of this parameter, Akaiki proposed an index known as the *Akaiki information criterion* (AIC), which is

$$\text{AIC} = -2(\text{maximum log likelihood of model})$$
$$+ 2(\text{number of free parameters of model}).$$

When a number of models are to be compared, the model with the smallest AIC value is considered the best. However, this is not a test statistic since no accept–reject criterion is provided. Given a set of models, one may have the lowest AIC value, but still it may not fit the data very well.

Variants of the AIC

Another index that has frequently been called the AIC is the index

$$\text{AIC}_2 = \chi^2_{df} - 2df. \tag{15.53}$$

Here χ^2_{df} is the chi-square statistic with df degrees of freedom for the model. This index differs from the true AIC in that it represents the difference

$$\text{AIC}(t) - \text{AIC}(s) = \chi^2_{df} - 2df$$

between the AIC of the hypothesized model t and the AIC(s) of a saturated model s. Since the AIC(s) of the saturated model would be subtracted from the AIC of each of a series of models, it does not change the relative ordering of the models, so this index can be used for selecting the model with the lowest value. A derivation of this index for the factor analysis model is as follows.

From the Wishart distribution we know that the natural logarithm of the likelihood function for modeling the covariance matrix $\Sigma(\theta)$ as a function of other parameters, given the estimated sample covariance matrix $\mathbf{S}$ is

$$\ln l(\boldsymbol{\theta}) = -K - \frac{1}{2}(N-1)\ln\left|\boldsymbol{\Sigma}(\hat{\boldsymbol{\theta}})\right| - \frac{1}{2}(N-1)\mathrm{tr}[\mathbf{S}\boldsymbol{\Sigma}^{-1}(\hat{\boldsymbol{\theta}})]$$

(Jöreskog, 1969), where K is a function only of sample size N and number of variables n, but not of model parameters. Subsequently,

$$\mathrm{AIC}(t) = 2K + (N-1)\ln\left|\boldsymbol{\Sigma}(\hat{\boldsymbol{\theta}})\right| + (N-1)\mathrm{tr}[\mathbf{S}\boldsymbol{\Sigma}^{-1}(\hat{\boldsymbol{\theta}})] + 2q$$

is the AIC value for model t, with q the number of free parameters. On the other hand,

$$\mathrm{AIC}(s) = 2K + (N-1)\ln|\mathbf{S}| + (N-1)\mathrm{tr}[\mathbf{S}\mathbf{S}^{-1}] + 2n(n+1)/2,$$

which reduces to

$$\mathrm{AIC}(s) = 2K + (N-1)\ln|\mathbf{S}| + (N-1)n + 2n(n+1)/2.$$

Hence

$$\mathrm{AIC}(t) - \mathrm{AIC}(s) = 2K + (N-1)\ln\left|\boldsymbol{\Sigma}(\hat{\boldsymbol{\theta}})\right| + (N-1)\mathrm{tr}[\mathbf{S}\boldsymbol{\Sigma}^{-1}(\hat{\boldsymbol{\theta}})] + 2q$$
$$- 2K - (N-1)\ln|\mathbf{S}| - (N-1)n - 2n(n+1)/2$$
$$= (N-1)\left[\ln\left|\boldsymbol{\Sigma}(\hat{\boldsymbol{\theta}})\right| - \ln|\mathbf{S}| + \mathrm{tr}[\mathbf{S}\boldsymbol{\Sigma}^{-1}(\hat{\boldsymbol{\theta}})] - n\right]$$
$$- 2\left[n(n+1)/2 - q\right].$$

From Equation 15.7 we see that the first expression on the right is $\chi^2_{df_t}$, whereas the second is simply $-2df$; hence

$$\mathrm{AIC}(t) - \mathrm{AIC}(s) = \chi^2_{df_t} - 2df.$$

Another variant of the AIC is

$$\mathrm{AIC}_3 = \chi^2_{df_t} + 2q, \tag{15.54}$$

where q is the number of free parameters in the model. The justification for equating this index with the AIC is that it represents the result of adding another constant, $2n(n+1)/2$, to AIC_2, so it too should maintain the same ordering among models as the AIC. While the AIC and AIC_2 may occasionally take on negative values, the AIC_3 takes on positive values only.

AIC Does Not Correct for Parsimony

Although frequently regarded as an index that "corrects for complexity" or the estimation of parameters, it does not really do so in the sense that the PR does. Parsimony is not regarded in the AIC as a ratio of free parameters to observed parameters. Rather the term $2q$ is a technical correction for the small sample bias in estimating the mean expected log likelihood with a model that has q free parameters. And, as we are about to demonstrate, the effect of this correction diminishes to vanishing as the sample size gets increasingly large.

The AIC has received much attention from professional statisticians, perhaps because of the sophisticated rationale behind it and a dubious belief that the task of scientists is to construct numerous models and select from among them the best fitting. The problem with this is an excessive reliance on fit alone in judging the quality of a model. Akaiki originally developed the index in the context of regression models, where the data points to be fitted are the individual observations. Rarely do researchers formulate saturated or near-saturated regression models, especially when the sample size is huge. But the factor analysis model (and SEMs as well) are concerned with fitting a model covariance matrix to an observed covariance matrix. The number of distinct data points to which the model is fit is $n(n + 1)/2$, and this remains the same regardless of sample size for the observations. It is well within reach in both exploratory and confirmatory factor analysis to free up so many parameters as to obtain near-saturated and even saturated models. So, among other things, we need to see the behavior of the AIC in models that approach being saturated. We also need to look at the effect of sample size on the AIC.

To do this we will focus on AIC_2, since the theory associated with the chi-square term in it is well-developed. Given $AIC_2 = \chi^2_{df_t} - 2df$, we know that the expected value of AIC_2 is

$$E(AIC_2) = E(\chi^2_{df}) - 2df = (\lambda + df) - 2df = (N - 1)\delta - df, \qquad (15.55)$$

where $\lambda = (N - 1)\delta$ is the unnormalized noncentrality parameter and δ the normalized noncentrality parameter. Recall from the discussion following Equation 16.11 that $\lambda = (N - 1)F(\Sigma; \tilde{\Sigma}_0)$, whereas from Equation 15.18 $\delta = F(\Sigma; \tilde{\Sigma}_0)$, the population error of approximation.

The following is taken from Mulaik (2001): Let us now consider three models M_1, M_2, and M_3, with expected AIC_2 values of

$$E[AIC_2(M_1)] = (N - 1)\delta_1 - df_1,$$

$$E[AIC_2(M_2)] = (N - 1)\delta_2 - df_2,$$

$$E[AIC_2(M_3)] = 0 - 0 = 0.$$

Model M_3 is a saturated model with, of mathematical necessity, zero lack of fit and zero degrees of freedom, regardless of sample size. Now, under what conditions will it be the case that

$$(N - 1)\delta_1 - \mathrm{df}_1 < (N - 1)\delta_2 - \mathrm{df}_2$$

so that the first model will have a smaller expected value for AIC_2 than the second model? A little algebra reveals that this will happen when

$$(\delta_1 - \delta_2) < \frac{(\mathrm{df}_1 - \mathrm{df}_2)}{(N - 1)}. \tag{15.56}$$

When the difference between the normalized noncentrality values δ_1 and δ_2 for models M_1 and M_2, respectively, is less than the difference between their degrees of freedom divided by $(N-1)$, model M_1 will be preferred on average over model M_2.

According to proponents of the AIC, when $\delta_1 = 0$ and $\mathrm{df}_1 > \mathrm{df}_2$, then model M_1 will be favored over any model M_2, including the saturated model M_3. But for nonzero values of δ there is an important exception. Recall that $\lim_{N \to \infty} N\delta$ is undefined for any positive δ. As N approaches infinity, with positive values for δ_1 and δ_2, the $E(\mathrm{AIC}_2)$ becomes undefined and cannot distinguish between any pair of such models. To control for increasing sample size, McDonald (1989) suggested for the chi-square index and the AIC_2 that one divide them, respectively, by $(N-1)$. Then

$$E\left(\frac{\mathrm{AIC}_2}{(N - 1)}\right) = \delta - \frac{\mathrm{df}}{(N - 1)}. \tag{15.57}$$

Although this shows that the $\mathrm{AIC}_2/(N–1)$ on average underestimates the population error of approximation, the bias diminishes with increasing sample size. But we see that this will not help to distinguish between models with positive degrees of freedom and a zero δ and saturated models that necessarily have zero δ and zero degrees of freedom. For as N approaches infinity, $\mathrm{df}/(N - 1)$ approaches zero, and this means that Equation 15.56 will not hold, since it will become the equality $0 = 0$. So, at the population level the *normalized* AIC will not distinguish, on average, between a perfect fitting model with positive degrees of freedom and a perfect fitting saturated model with zero degrees of freedom.

Kieseppä's Critique and a Rejoinder

Kieseppä (2003) has taken exception to the argument that the raw AIC must be replaced by the normalized AIC. Although he does not quarrel with the results based on expected values, he argues that they do not consider the question of whether or not in the limit there is still a probability greater than 0.5 for favoring, say, a model with zero δ and $\mathrm{df} > 0$, over a saturated model with zero δ and df $= 0$. Using well-developed theory (Hogg and Craig, 1965, pp. 318–320), he states that

$$E(\text{AIC}(k)) \approx (N-1)\delta - \text{df} \qquad (15.58a)$$

is the expected value of the AIC for some model k, whereas

$$D^2(\text{AIC}(k)) \approx 4(N-1)\delta + 2\text{df} \qquad (15.59a)$$

is the variance of the AIC.

Kieseppä then notes that if model M_1 is a correct model with $\delta_1 = 0$ and $\text{df}_1 > 0$, while M_3 is the saturated model with $\delta = 0$ and df $= 0$, then model M_1 should be preferred to model M_3 if

$$\text{AIC}(M_1) < \text{AIC}(M_3).$$

He is not concerned with expected values but the probability that we will observe cases where the AIC for M_1 will be less than the AIC for M_3. He next reasons $\text{AIC}(M_3) = 0$ with probability 1. This is reasonable since it is a mathematical necessity that the AIC of a saturated model will equal 0 and have 0 df. It should have no variance at all, regardless of sample size. So, let us now turn to the distribution of M_1. Suppose we now assume that model M_1 has a normalized noncentrality $\delta = 0$. He assumes in this case that in large but not necessarily infinite samples that the distribution of $\text{AIC}(M_1)$ is approximately normal with mean and variance given by Equations 15.58a and 15.59a. So, to determine what the probability is under this distribution that the $\text{AIC}(M_1)$ is less than zero, he computes the cumulative probability

$$\Pr[\text{AIC}(M_1) < 0 \mid \mathbf{s}_N] \approx F\left(\frac{0 - E(\text{AIC}(M_1))}{\sqrt{D^2(\text{AIC}(M_1))}}\right).$$

Let us focus on the expression inside the parenthesis of the cumulative normal probability function on the right. We will now substitute Equations 15.58a and 15.59a for $E(\text{AIC}(M_1))$ and $D^2(\text{AIC}(M_1))$, and then we will immediately multiply the result by the ratio $[1/(N-1)/(1/(N-1))]$.

$$\frac{-(N-1)\delta + \text{df}}{\sqrt{4(N-1)\delta + 2\text{df}}} = \left(\frac{1/(N-1)}{1/(N-1)}\right)\frac{-(N-1)\delta + \text{df}}{\sqrt{4(N-1)\delta + 2\text{df}}}.$$

We will further carry out the obvious multiplications involving $1/(N-1)$ and will obtain the following, which we can further simplify as

$$= \frac{-\delta + (\text{df}/(N-1))}{\sqrt{(4(N-1)\delta + 2\text{df}/(N-1)^2)}} = \frac{-\delta + (\text{df}/(N-1))}{\sqrt{(4\delta/(N-1)) + (2\text{df}/(N-1)^2)}}.$$

Now, let us impose our assumption that $\delta = 0$. We get:

$$\frac{df/(N-1)}{\sqrt{2df/(N-1)^2}} = \frac{df}{\sqrt{2df}} = \frac{df\sqrt{2df}}{2df} = \frac{\sqrt{2df}}{2} = \frac{\sqrt{2}\sqrt{df}}{\sqrt{2}\sqrt{2}} = \sqrt{df/2}.$$

The expressions involving $(N-1)$ cancel in the numerator and denominator, producing a resulting ratio that does not depend on sample size. Consequently he says that we can write

$$P = \lim_{N\to\infty} \Pr[\text{AIC}(M_1) < 0 \mid s_N] \approx F\left(\frac{0 - E(\text{AIC}(M_1))}{\sqrt{D^2(\text{AIC}(M_1))}}\right)$$

$$= F\left(\frac{df_1}{\sqrt{2df_1}}\right) = F(\sqrt{df_1/2}). \tag{15.60}$$

Here s_N denotes a sample of size N, $F()$ is the cumulative normal distribution function, and everything else is as before. Since $\sqrt{df/2}$ is a positive quantity, it must be greater than the mean and so P must be greater than 0.5. In fact, the probability that $\text{AIC}(M_1)$ is less than the AIC for the saturated model increases beyond 0.5 with the size of the degrees of freedom, but is 0.5 when $df = 0$. So, Kieseppä's result shows that a "correct model" ($\delta = 0$) with more degrees of freedom will be preferred over a "correct model" with fewer degrees of freedom and even over a saturated model with 0 df.

However, paradoxically, the distinguishability between models with $\delta = 0$ and different degrees of freedom or even zero degrees of freedom does not occur with normalized AIC, $\text{AIC}/(N-1)$, which is designed to avoid the effect of sample size. On the other hand, we can distinguish between models with $\delta > 0$, so there is a trade-off in advantages and disadvantages. We have

$$E\left[\frac{\text{AIC}}{(N-1)}\right] = \delta - \frac{df}{(N-1)}, \tag{15.58b}$$

$$D^2\left[\frac{\text{AIC}}{(N-1)}\right] = \frac{1}{(N-1)^2}[4(N-1)\delta + 2df] = \frac{4\delta}{(N-1)} + \frac{2df}{(N-1)^2}. \tag{15.59b}$$

In the limit

$$\lim_{N\to\infty}\left[E\left(\frac{\text{AIC}}{(N-1)}\right)\right] = \lim_{N\to\infty}\left[\delta - \frac{df}{(N-1)}\right] = \delta \tag{15.61}$$

and

$$\lim_{N\to\infty}\left[D^2\left(\frac{\text{AIC}}{(N-1)}\right)\right] = \lim_{N\to\infty}\left[\frac{4\delta}{(N-1)} + \frac{2df}{(N-1)^2}\right] = 0. \tag{15.62}$$

So, AIC/$(N-1)$ converges to δ with zero variance (or with probability 1) and is defined in the limit. All other values of AIC/$(N-1)$ have zero probability. However, it does not have an approximately normal distribution because it has a zero variance. We can derive with the AIC/$(N-1)$ the same ratio that Kieseppä obtains with the AIC in Equation 15.59b. Nevertheless, when $\delta = 0$, in the limit, the probability is 1 that AIC/$(N-1) = 0$ regardless of the degrees of freedom, and thus a saturated model would be indistinguishable from a model with $\delta = 0$ and df > 0 on the basis of AIC/$(N-1)$. The ratio in Equation 15.59b of $\sqrt{df_1/2}$ is the number of standard deviations away from the mean. But with the standard deviation equal to zero in the limit, this is $\sqrt{df_1/2}$ times zero away from the mean of zero. Kieseppä would argue that this is a defect of the normalized AIC, even though it would allow us to distinguish between models with different nonzero normalized noncentralities, which we cannot do in the limit with the AIC. So, there are trade-offs in the advantages and disadvantages of these indices.

But as I put it in Mulaik (2001), critics of the AIC also point out that in practice almost no model fits data perfectly, especially at large sample sizes, so comparisons are generally to be made between models with positive δ's. From Equation 15.56 we know that on average that a model M_1 with a larger normalized noncentrality parameter than a second model M_2 will be preferred if $(\delta_1 - \delta_2) < (df_1 - df_2)/(N-1)$. However, for any positive $(df_1 - df_2)$ there will always be a large-enough sample size N at which a model M_2 with fewer degrees of freedom and smaller δ will be favored on average. If in Equation 15.56 $\delta_1 > \delta_2$ and $df_1 > df_2$, then there should be a sample size N where $\delta_1 - \delta_2 > (df_1 - df_2)/(N-1)$, meaning the model with smaller normalized noncentrality and fewer degrees of freedom will be *preferred on average* to the model with somewhat larger normalized noncentrality and more degrees of freedom.

This point was first made by McDonald (1989) and McDonald and Marsh (1990), which they demonstrated both mathematically and empirically. McDonald and Marsh (1990) took a table published by Cudeck and Browne (1983) of the values of various fit indices, including the AIC, computed for nine common factor models with increasing numbers of factors from 1 to 9 fitted to 18×18 sample covariance matrices based on subsamples of the same data of varying size from 75, through 200, to 1338 observations. McDonald and Marsh (1990) further provided additional computed fit indices. They noted that "As already shown by Cudeck and Browne (1983, p. 253), the complexity [having more free parameters] of the model selected by the AIC increases with sample size, and by extrapolation one may see that with a sufficiently large sample size the saturated model would be selected." In increasingly larger samples, the AIC selected models *on average* with larger numbers of factors having increasing numbers of free factor loadings and preferred a saturated model to the model with small nonzero normalized noncentrality and positive degrees of freedom. So, in these cases the AIC tended in larger samples to give too much weight to better fit achieved by estimating more parameters. On the other hand, in small samples the AIC would favor the model with small δ and more degrees of freedom over models with equally small δ and fewer degrees of freedom.

Kieseppä (2003) and Haughton, Oud, and Jansen (1997), however, would still object that given two models t and u, where t is more constrained (has more degrees of freedom) than u, it is not sufficient to show that $E[AIC(t)] - E[AIC(u)]$ is negative to demonstrate that the AIC typically prefers the more constrained model. In practice, with results from a single sample, we would accept model t over model u if $AIC(t) < AIC(u)$, or, equivalently, if $AIC(t) - AIC(u)$ is negative. What we need to show in general is that $\Pr[(AIC(t) - AIC(u)) < 0] > 0.5$. In other words, the proper way to show that the AIC generally prefers the more constrained model over the less constrained model is to show that it prefers the more constrained model more than 50% of the time. The AIC for finite sample sizes is not symmetric but positively skewed by nonzero noncentrality, and furthermore the distribution is shifted in the negative direction by increasing degrees of freedom. So, while the mean value of $E[AIC(t)] - E[AIC(u)]$ may be positive (favoring the less constrained model), the median of $(AIC(t) - AIC(u))$ may still be negative, favoring the more constrained model more than 50% of the time.

The AIC may indeed favor more constrained models over less constrained models at relatively small sample sizes. But this does not characterize the large sample or even the "population" behavior of the AIC. I think, however, that inferences as to the large sample behavior of the random variable $AIC(t)$-$AIC(u)$ can be obtained by applying the mean–median–mode inequality that states that in any positively skewed distribution, $\mu \geq m \geq M$, where μ is the mean, m is the median, and M is the mode of the distribution, and with this inequality we can place bounds on the median and thereby determine conditions under which it will be sufficient that the random variable will prefer one or the other model more than 50% of the time.

I will argue that $AIC(t) - AIC(u)$ is distributed as a translated noncentral chi-square distribution and from this we can establish these bounds.

If we substitute the definition for the AIC in Equation 15.53 into $AIC(t) - AIC(u)$, we obtain

$$AIC(t) - AIC(u) = \chi^2_{df_t} - 2df_t - \chi^2_{df_u} + 2df_u. \tag{15.63}$$

Steiger, Shaprio, and Browne (1985) proved that for nested sequences of models estimated by the maximum-likelihood fit function, $\chi^2_{df_t} - \chi^2_{df_u}$ is distributed asymptotically as a noncentral chi-square distribution with $df_t - df_u$ degrees of freedom and noncentrality parameter $\lambda = \lambda^*_t - \lambda^*_u$. Hence Equation 15.63 is distributed as a noncentral chi-square distribution with $p = df_t - df_u$ degrees of freedom and noncentrality parameter $\lambda = \lambda^*_t - \lambda^*_u$, with translation of the distribution by $-2(df_t - df_u)$. Thus the mean, median, and mode of $\chi^2_{df_t} - \chi^2_{df_u}$ will be their usual values minus $2p = 2(df_t - df_u)$.

Sen (1989) notes that Pearson (1895) conjectured on the basis of empirical experience that for a positively skewed unimodal distribution the following inequality holds:

$$M \le m \le \mu,$$

where M is the mode, m is the median, and μ is the mean. It is known that for the central chi-square distribution with p degrees of freedom that the mean is given as

$$\mu(p,0) = p,$$

the mode is given by

$$M(p,0) = (p-2),$$

and the median is a value $m(p,0)$ such that

$$(p-1) < m(p,0) < p.$$

This satisfies Pearson's conjectured inequality.

Sen (1989) gives the mean, median, and mode of a noncentral chi-square distribution with p degrees of freedom and noncentrality parameter λ as follows:

$$\mu(p,\lambda) = p + \lambda = \mu(p,0) + \lambda \qquad (15.64)$$

is the mean, where $\mu(p,0)$ is the mean of the central chi-square distribution with p degrees of freedom and zero noncentrality. The mode of this distribution has the properties, for all $p \ge 2$ and $\lambda \ge 0$,

$$M(p,\lambda) \le p - 2 + \lambda, \qquad (15.65)$$

$$M(p,\lambda) \ge p - 2 + p^{-1}(p-2)\lambda \qquad (15.66)$$

and

$$M(p-2,\lambda) \le M(p,\lambda). \qquad (15.67)$$

The median of this distribution has the property

$$M(p,\lambda) \le p - 2 + \lambda < m(p,\lambda) \le p + \lambda = \mu(p,\lambda). \qquad (15.68)$$

From these results, we can now infer that

$$E[\text{AIC}(t) - \text{AIC}(u)] = \mu(p,\lambda) - 2p = \lambda + p - 2p = \lambda - p. \qquad (15.69)$$

The mode of the distribution of $\text{AIC}(t) - \text{AIC}(u)$ has the properties

$$M(\text{AIC}(t) - \text{AIC}(u)) \le p - 2 + \lambda - 2p = \lambda - p - 2.$$
$$M(\text{AIC}(t) - \text{AIC}(u)) \ge p^{-1}(p-2)\lambda - p - 2. \qquad (15.70)$$

The median $m(\text{AIC}(t) - \text{AIC}(u))$ of the distribution of $\text{AIC}(t) - \text{AIC}(u)$ equals $m(p, \lambda) - 2p$ and satisfies the inequality

$$M(\text{AIC}(t) - \text{AIC}(u)) \leq \lambda - p - 2 < m(p, \lambda) - 2p \leq \lambda - p. \qquad (15.71)$$

We are now in a position to determine when $\text{AIC}(t) - \text{AIC}(u)$ favors the more constrained or the less constrained model. If $p^{-1}(p - 2)\lambda - p - 2$ is positive, then $\text{AIC}(t) - \text{AIC}(u)$ favors the less constrained model, for both the median and the mean of this variable will be greater than this value, because the mode is always greater than or equal to this value. An even stronger bound is given when λ and $\lambda - p - 2$ are positive, for the median will be greater than this, and $\text{AIC}(t) - \text{AIC}(u)$ will be positive greater than 50% of the time, implying that the less constrained model will be favored more than 50% of the time. Because λ and p are positive, $\lambda - p - 2$ will be positive iff $\lambda - 2 > p$.

I will argue that, when λ is positive, there is always a finite sample size, beyond which $\lambda - 2 > p$. To show this, let us recall that $\lambda = \lambda_t^* - \lambda_u^*$ and $p = (\text{df}_t - \text{df}_u)$. Now, McDonald and Marsh (1990) defined the noncentrality parameter for a sample of size N in terms of a *rescaled* or *normalized* noncentrality parameter δ as $\lambda^* = (N - 1)\delta$. The *rescaled* or *normalized* noncentrality parameter is invariant with respect to sample size and is a more appropriate population parameter for characterizing the lack of fit of a model to the data. An unbiased estimate of δ is given by

$$\hat{\delta} = \frac{\chi_{\text{df}}^2 - \text{df}}{N - 1}. \qquad (15.72)$$

Now, we may express λ in terms of normalized noncentrality parameters as

$$\lambda = (N - 1)(\delta_t - \delta_u) \qquad (15.73)$$

and p in terms of the original degrees of freedom of the two respective models. The expression $(\delta_t - \delta_u)$ is necessarily greater than or equal to zero, since the noncentrality parameter of a less constrained model in a nested sequence of models is always less than or equal to the noncentrality parameter of the more constrained model (Bentler and Bonnet, 1980). Thus a sufficient condition that $\lambda - 2 > p$ is that

$$(N - 1)(\delta_t - \delta_u) - 2 > (\text{df}_t - \text{df}_u). \qquad (15.74)$$

If we make Equation 15.74 an equality, and consider only cases in which $\delta_t > 0$, we may solve for the value of $(N - 1)$ which makes this an equality as

$$(N - 1) = \frac{(\text{df}_t - \text{df}_u) + 2}{(\delta_t - \delta_u)}. \qquad (15.75)$$

Any value for $(N-1)$ greater than that given by the right-hand expression of Equation 15.75 will be sufficient to make $\text{AIC}(t) - \text{AIC}(u)$ favor the less constrained model, which is what I sought to prove. However, if $\delta_t = 0$, then the expression on the left of Equation 15.74 will always be negative and less than the expression on the right, implying that the more constrained model will be favored, only when the more constrained model fits perfectly, which almost never occurs in practice.

Is the Noncentral Chi-Square Distribution Appropriate?

Olsson et al. (2004) produced Monte Carlo evidence that the noncentral chi-square distribution is not always a good approximation to the chi-square statistic of goodness of fit in structural equation modeling and factor analysis. However, their demonstration principally showed that the tails of the empirical distribution for the chi-square statistic for a misspecified model were more extreme than that of the corresponding theoretical noncentral chi-square distribution. The mean of the empirical distribution was still close to the mean of the theoretical distribution. In fact in large samples, while the noncentral chi-square distribution seemed to fit less well with large noncentrality parameter values, a normal distribution seemed more appropriate, albeit with a larger variance than that of the corresponding theoretical noncentral chi-square distribution. These results may weaken the use of confidence intervals for the RMSEA index based on the theoretical noncentral chi-square distribution, but I believe it is reasonable to conjecture that these findings should not affect the essential conclusions involving the means and the medians argued here.

What Olsson et al. (2004) have shown is that the noncentral chi-square distribution is a good approximation to the empirical distribution in models with a relatively small number of variables and relatively small noncentrality parameters. So, our conclusions, that the AIC and similar indices will tend to favor less constrained models over more constrained but slightly ill-fitting models beyond some sample size, still hold. But they point out that for models with many variables and larger noncentrality parameter values, the empirical distributions tend to become more normal (symmetric) in shape while at the same time retaining mean values very close to the theoretical noncentrality parameter values, but with larger variances. The implication is that the median of the empirical distribution will tend to be closer to the mean in the empirical distribution in these cases than they are in the theoretical noncentral chi-square distribution. The empirical distribution of the statistic $\text{AIC}(t) - \text{AIC}(u)$ should also have a closer approximation to the normal distribution, while retaining approximately the same expected value as that of the corresponding distribution based on the noncentral chi-square distribution. And its median should be closer to the mean of the empirical distribution than it is in the distribution based on the noncentral chi-square. Hence the lower bound on the median of the noncentral chi-square distribution, which is less than that in the empirical distribution, should apply as well in the empirical case. It is not necessary for the argument that we be able to specify exactly the sample size at which the better fitting less constrained model becomes preferred to the more constrained poorer fitting model. It is sufficient to show that it is reasonable to believe that at some sufficiently large but possibly unknown sample size close to the size given for the noncentral chi-square distribution, this will occur.

The method of arriving at these results may be extended to other indices of model fit of the type

$$\text{GOF} = \chi^2_{\text{df},\lambda} + C, \tag{15.76}$$

where $\chi^2_{\text{df},\lambda}$ is distributed asymptotically as a noncentral chi-square distribution and C is a constant representing a penalty for parameter estimation. In other words, if k is the number of parameters estimated, then $C > k$. In parallel with Equation 15.63 we will then have the difference variable

$$\text{GOF}(t) - \text{GOF}(u) = \chi^2_{\text{df}_t,\lambda_t} - \chi^2_{\text{df}_u,\lambda_u} + C_t - C_u. \tag{15.77}$$

The rule then is to accept the more constrained model t if $\text{GOF}(t)$ is smaller than $\text{GOF}(u)$, implying that Equation 15.77 is negative. We may again regard this variable as equivalent to

$$\text{GOF}(t) - \text{GOF}(u) = \chi^2_{p,\lambda} + (C_t - C_u), \tag{15.78}$$

which implies that $\text{GOF}(t) - \text{GOF}(u)$ is distributed asymptotically as a noncentral chi-square distribution with $p = (\text{df}_t - \text{df}_u)$ degrees of freedom and noncentrality parameter $\lambda = (\lambda_t - \lambda_u)$ that has been translated by the added constant $(C_t - C_u)$.

The mean of the distribution of $\text{GOF}(t) - \text{GOF}(u)$ is

$$\mu[\text{GOF}(t) - \text{GOF}(u)] = p + \lambda + (C_t - C_u).$$

The mode of $\text{GOF}(t) - \text{GOF}(u)$ obeys the inequalities

$$M[\text{GOF}(t) - \text{GOF}(u)] \leq p - 2 + \lambda + (C_t - C_u)$$

and

$$M[\text{GOF}(t) - \text{GOF}(u)] \geq p - 2 + p^{-1}(p - 2)\lambda + (C_t - C_u).$$

The median of $m(p, \lambda) + (C_t - C_u)$ of $\text{GOF}(t) - \text{GOF}(u)$ obeys the inequality

$$p - 2 + \lambda + (C_t - C_u) < m(p, \lambda) + (C_t - C_u) \leq p + \lambda + (C_t - C_u).$$

Thus the median will be greater than zero whenever $p - 2 + \lambda + (C_t - C_u)$ is positive, and the model preferred will be the less constrained model.

Under what conditions will $p - 2 + \lambda + (C_t - C_u)$ be positive? Note that it must be that $C_t < C_u$, for the less constrained model must receive a larger positive penalty for parameter estimation than the more constrained model. Hence $(C_t - C_u)$ is negative. Now, if p and λ are positive, implying that λ_t of $\lambda = \lambda_t - \lambda_u$ is positive, then for $p - 2 + \lambda + (C_t - C_u)$ to be positive, it must be that $p - 2 + \lambda > (C_u - C_t)$ (note that C_t and C_u are reversed in position). But if $p - 2 > (C_u - C_t)$, then it does not matter what positive value λ has, for then the less constrained model will always be selected regardless of fit. But this does not happen if $(C_u - C_t)$ is some multiple of p greater than p, which is the case if the penalties are multiples of the number of parameters estimated, so that $(C_u - C_t)$ is a multiple $w(k_u - k_t)$, $w > 1$, of the difference in parameters estimated $(k_u - k_t)$, and this difference also necessarily equals the difference in degrees of freedom p. This happens to be the case with the AIC and the Bayes information criterion (BIC), so we can consider the more likely case that $p - 2 \leq (C_u - C_t)$. Consequently, for $p - 2 + \lambda + (C_t - C_u)$ to be positive under this additional constraint, it must be the case that $\lambda > (C_u - C_t) - p + 2$. Expressing this inequality in terms of normalized noncentrality parameters and differences in degrees of freedom, we require that

$$(N - 1)(\delta_t - \delta_u) > (C_u - C_t) - (\text{df}_t - \text{df}_u) + 2. \tag{15.79}$$

Assuming then too that δ_t is positive but smaller than the right-hand expression of Equation 15.79, there exists a large enough sample size N such that for all samples of this size or greater

$$(N - 1) > \frac{(C_u - C_t) - (\text{df}_t - \text{df}_u) + 2}{(\delta_t - \delta_u)}, \tag{15.80}$$

meaning that the less constrained model would be preferred in that case. So, any GFI that penalizes a model's noncentrality chi-square with an additive constant that increases monotonically with the number of parameters estimated will prefer the less constrained model at some sufficiently large sample size, as long as the normalized noncentrality of the model is greater than 0, no matter how small.

BIC

A special case of the just-described kind of index that has been offered as an improvement over the AIC is the BIC (Schwartz, 1978; Raftery, 1986, 1993). The formula for the BIC is

$$\text{BIC} = \chi^2_{\text{df},\lambda} + k \cdot \ln(N),$$

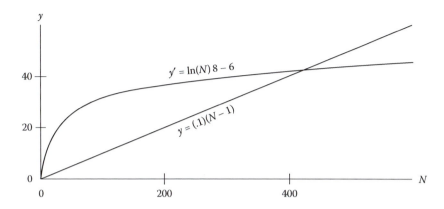

FIGURE 15.2 Superimposed graphs of $y = (N-1)(0.1)$ (straight line) and $y' = \ln(N)8 - 6$ (logarithmic curve), illustrating how there is a value (about 425) of N beyond which the expression on the left-hand side of the inequality in Equation 15.81 exceeds the value of the expression on the right, implying that a less constrained model would be preferred.

where k is the number of estimated parameters of the model. This index increases the penalty for the number of parameters estimated as the logarithm of the sample size. Beyond certain values of N, this rate of increase in the penalty is not as rapid as the increase in the nonzero raw noncentrality parameter with increase in sample size. In this case, the corresponding form of Equation 15.79 is

$$(N-1)(\delta_t - \delta_u) > \ln(N)(k_u - k_t) - (df_t - df_u) + 2 \qquad (15.81)$$

and the less constrained model is accepted if the inequality holds. There is no closed-form solution for N that satisfies the inequality. However, we can see in Figure 15.2 of a graph of a typical case where $(\delta_t - \delta_u) = 0.1$, $(k_u - k_t) = 8$, and $-(df_t - df_u) + 2 = -6$, how the left hand of the inequality rises as a straight line, $y = (N-1)(0.1)$, whereas the right hand follows the form of a logarithmic function, $y' = \ln(N)(8) - 8 + 2$. Because the logarithmic function rises much more slowly with larger values of N than does the left-hand expression, the curves generated by each cross at around $N = 425$, and the value of the curve on the left eventually exceeds the value of the curve on the right after that.

So, the BIC also shows the tendency to prefer less constrained models after samples reach a certain large magnitude.

Discussion

The motivation for the current investigation has been to demonstrate that GFIs based on penalty weights *added* to a noncentral chi-square variable, such as the AIC and the BIC, are not good models of the concept of par-simony and correction for parameter estimation. They have the defect that when the noncentrality parameter is positive, there will exist a sufficiently large sample size at which a less constrained and less parsimonious model is preferred to the more constrained and more parsimonious model, even if the noncentrality parameter is only slightly above zero. In theory, in these cases, a saturated model with no degrees of freedom, estimating as many param-eters as there are distinct elements of the data to fit, and having necessarily a zero noncentrality, will be preferred, when the sample size is sufficiently large. However, saturated models are not unique. And in the parts of models having estimated parameters there are often nonunique alternative equiv-alent model structures, and the same number of free parameters that have equivalent fit to the data (MacCallum et al., 1993). This fact undermines, I believe, an implicit assumption made by those who formulate information theoretic indices such as the AIC and the BIC, that the model framework or structure is essentially correct, and that penalties for parameter estimation are only for the purpose of correcting for small sample bias in the estimates. This would imply then that in larger samples the estimated parameters are con-sidered to be less biased, so the penalty corrections can be relatively weaker, and asymptotically, the parameter estimates should be unbiased and so no penalty corrections should then be applied. Hence, one could accept a sat-urated model over any other model with a nonzero noncentrality, since the saturated model would then have correct estimates of the parameters. But if the saturated model and near-saturated models are not unique structurally, but equivalent in fit, there is no reason to believe one has obtained (nearly) correct values for estimated parameters in very large samples, merely because one has near-perfect or perfect fit.

In contrast, I believe that instead of a model-selection approach that relies principally on model comparison among models tested against a given set of data to select "optimal models," a hypothesis-testing approach seeks to establish the objective validity of a given model by demonstrating that its parameters are invariants independent of the observer's methods and per-spective (Mulaik, 2004). By asserting fixed values for parameters, one asserts values of the parameters that are supposed to be invariant. By testing one's hypothesized assertions about the invariant parameters with data not used in the formulation of the model and its fixed parameters, one makes the degree of fit independent of the researcher's prior experience and manner of fram-ing the model of the hypothesis. Since hypothesized fixed parameters may be based on prior experience with other forms of data, testing against a new set of data is a test of their invariance. Estimated parameters are not part of the hypothesis, but simply filler, estimated to optimize fit conditional on the fixed or constrained parameters, so that any lack of fit can be attributed to the fixed and constrained parameters.

Establishing objective results also does not depend on a given set of data, but is a continuing process, where parameters estimated at one stage with one set of data may be used as fixed parameters at another stage, with other sets of data in new contexts, to establish their invariance and hence "objective" status. However, objectivity does not imply incorrigibility. What is deemed objective at one phase of an investigation may be deemed not objective at a later phase as one formulates tests of hypotheses against new sets of data and finds that what one thought would be an invariant is not. Objectivity also implies the modularity of model components: that certain causal structures between variables in one model may be extracted from one model and incorporated in another model with other variables and causal structures, and the causal parameters in those transplanted structures remain invariant. So, the manner of establishing objective results via hypothesis testing across a series of data sets is a quite different process from a model-comparison approach that focuses on selecting one model from many that fits a given set of data optimally. Objectivity also seeks a conception of the model that is independent of sample size, which is a factor relative to the researcher. Hence model fit should be a population parameter, and it should be possible to conceive of what the model would look like if the sample were infinite and the model were fitted to an unrestricted covariance matrix that is essentially the population unrestricted covariance matrix. Parsimony of the model, implying fewness of parameters estimated, relative to the number of data points to fit, should be a property of the model independent of sample size. In fact, a free parameter is a dimension in which the model is free to adjust itself to better fit the data. We saw when we considered equivalent models how parameter estimation is the reason for the existence of equivalent models that fit the same regardless of the data, and how this creates ambiguities in interpreting the meaning of good fit for a model.

Cross-Validation Index

We may get some further insight into why the AIC behaves as it does if we look at a related index, the cross-validation index (CVI) (Cudeck and Browne, 1983; Browne and Cudeck, 1993). Suppose we have two random samples from the same population. Let C denote the "calibration" sample, N_C its size, and S_C its sample variance–covariance matrix. Let V denote the "validation sample," N_V its size, and S_V its sample variance–covariance matrix. When in the calibration sample the model is fit to the sample covariance matrix S_C, the resulting model produces the estimated variance–covariance matrix $\hat{\Sigma}_C$. Now, if we use $\hat{\Sigma}_C$ as a matrix of fixed parameters to see how well it would fit the validation sample variance–covariance matrix S_V, we would assess the discrepancy between them by

$$\text{CVI} = F(S_V, \hat{\Sigma}_C). \tag{15.82}$$

The fit of $\hat{\boldsymbol{\Sigma}}_C$ to $\mathbf{S}_V$ should in probability be poorer than the fit of $\hat{\boldsymbol{\Sigma}}_C$ to $\mathbf{S}_C$, simply because of sampling error that generally would make $\mathbf{S}_V$ differ from $\mathbf{S}_C$ to which $\hat{\boldsymbol{\Sigma}}_C$ was designed to fit optimally. However, as sample size increases, the difference between $\mathbf{S}_V$ and $\mathbf{S}_C$ should diminish in probability because plim $\mathbf{S}_V = \boldsymbol{\Sigma}$ and plim $\mathbf{S}_C = \boldsymbol{\Sigma}$. Consequently as sample size increases without bound

$$\text{plim } [F(\mathbf{S}_V, \hat{\boldsymbol{\Sigma}}_C) - F(\mathbf{S}_C, \hat{\boldsymbol{\Sigma}}_C)] = 0. \qquad (15.83)$$

The expression "plim" means the probability limit, that is, the point on which a sampling distribution collapses as the sample size increases without bound (Goldberger, 1964).

Browne and Cudeck (1993) give the conditional expectation of the CVI "over validation samples holding the calibration sample fixed" (p. 147) as approximately

$$E_V\text{CVI} = E_V[F(\mathbf{S}_V, \hat{\boldsymbol{\Sigma}}_C) \mid \hat{\boldsymbol{\Sigma}}_C] \approx F(\boldsymbol{\Sigma}_0, \hat{\boldsymbol{\Sigma}}_C) + (N_V - 1)^{-1}[n(n+1)/2].$$
$$(15.84)$$

This formula corresponds to the total error plus a factor that is directly proportional to the degrees of freedom and inversely to the sample size -1 of the validation sample. In this case $n(n+1)/2$ is the degrees of freedom since the calibration variance–covariance matrix is a completely fixed matrix with no estimated parameters when applied to the validation sample variance–covariance matrix. Consequently, the CVI is a biased estimator of the total error for the estimate of the model in the calibration sample. Although it would be possible to easily correct for the bias, Browne and Cudeck (1993) advise against it, because to do so would produce at times inadmissible negative values and it is hardly worth doing since the bias term $(N_V - 1)^{-1}[n(n+1)/2]$ would be the same for all competing models fitted to the calibration sample. This would not change the rank ordering of competing models in a model-comparison context.

Although it is also possible to use the validation sample as a calibration sample and validate against the original calibration sample (reversing the roles of the two samples), Browne and Cudeck advise against this also. We have in the two samples the basis for one full-sized sample, from which we can obtain more exact estimates. But first they devised an estimate of the expected CVI (ECVI) taking the expectation over both samples where $N = N_C = N_V$.

Consider that if we take the expectation of the CVI over both calibration and validation sampling, by substituting Equation 15.2 into Equation 15.83, we obtain

$$\text{ECVI} = E_C E_V F(\mathbf{S}_V, \hat{\boldsymbol{\Sigma}}_C) \approx F(\boldsymbol{\Sigma}, \tilde{\boldsymbol{\Sigma}}) + (N - 1)^{-1}[n(n+1)/2 + q] \qquad (15.85)$$

with N the sample size of both calibration and validation samples, n the number of variables, and q the number of free parameters (Browne and Cudeck, 1993, p. 148).

Now, let us consider a single combined sample of size $N = N_C + N_V$. Let **S** for this combined sample correspond to $\mathbf{S}_C$ and let the estimated model variance–covariance matrix be $\hat{\boldsymbol{\Sigma}} = \hat{\boldsymbol{\Sigma}}_C$. Then from Equation 15.16, because $E[F(\mathbf{S}; \hat{\boldsymbol{\Sigma}})] = F(\boldsymbol{\Sigma}; \tilde{\boldsymbol{\Sigma}}) + \mathrm{df}/(N-1)$ and $\mathrm{df} = n(n+1)/2 - q$, the expression $F(\mathbf{S}; \hat{\boldsymbol{\Sigma}}) + (N-1)^{-1}2q$ has for its expectation

$$E[F(\mathbf{S}; \hat{\boldsymbol{\Sigma}}) + (N-1)^{-1}2q] = F(\boldsymbol{\Sigma}, \tilde{\boldsymbol{\Sigma}}) + (N-1)^{-1}[n(n+1)/2 - q + 2q]$$

or

$$E[F(\mathbf{S}; \hat{\boldsymbol{\Sigma}}) + (N-1)^{-1}2q] = F(\boldsymbol{\Sigma}, \tilde{\boldsymbol{\Sigma}}) + (N-1)^{-1}[n(n+1)/2 + q] = \mathrm{ECVI}.$$

So, an unbiased estimator of the ECVI estimated from a single sample is

$$c = F(\mathbf{S}; \hat{\boldsymbol{\Sigma}}) + (N-1)^{-1}2q. \tag{15.86}$$

But we should note that this is equivalent to $\mathrm{AIC}_3/(N-1) = (\chi^2_{\mathrm{df}} + 2q)/(N-1)$. So the single-sample estimate of ECVI produces values that would maintain the same ordering of models as given by $\mathrm{AIC}_1, \mathrm{AIC}_2$, or $\mathrm{AIC}_2/(N-1)$. We should note that the statistic c is always positive, whereas with the AIC_2 its values can be negative.

Now, the expected value of c is also given as

$$E(c) = E\left(\frac{\chi^2_{\mathrm{df}} + 2q}{(N-1)}\right) = \delta + \frac{\mathrm{df} + 2q}{(N-1)}.$$

Being equivalent to a variant of the $\mathrm{AIC}/(N-1)$, we should expect the estimate of the ECVI in the single sample to have the same general properties as the $\mathrm{AIC}/(N-1)$. For example, we can derive the case where $\mathrm{ECVI}(1) < \mathrm{ECVI}(2)$ for two models M_1 and M_2, respectively, to show that it will occur analogously to Equation 15.56 when

$$\delta_1 - \delta_2 < \frac{q_2 - q_1}{(N-1)}, \tag{15.87}$$

where δ_1 and δ_2 are the normalized noncentrality parameters and q_1 and q_2 are the number of estimated parameters, respectively, of models M_1 and M_2. So, suppose $\delta_1 > \delta_2$ but $q_2 > q_1$, implying that M_2 has better fit achieved by estimating more parameters and thereby having fewer degrees of freedom than in model M_1. There will be a sample size N sufficiently large to make the model with more estimated parameters be necessarily preferred on average to the model with fewer estimated parameters but larger normalized noncentrality. And if model M_2 is a saturated model with $\delta_2 = 0$ and $q_2 = p(p+1)/2$, there is always a sample size large enough that it will be preferred to any model M_1 with normalized noncentrality δ_1 greater than zero and nonzero degrees of freedom.

In larger and larger samples, both the estimate c of the ECVI and the variants of the AIC tend to favor models with smaller normalized noncentrality parameter values achieved by estimating more and more parameters. In the context of cross-validation, consider that the model-based estimate of the population variance–covariance matrix $\hat{\Sigma}_C$ derived in a calibration sample is used as a matrix with all its elements fixed to be compared with a second validation sample variance–covariance matrix S_V. Estimates of elements in $\hat{\Sigma}_C$ based on estimated factor loadings and factor covariances no longer are treated as estimates but as fixed values. Certainly the error in cross-validating will be the sampling error implicit in the elements of $\hat{\Sigma}_C$ and those of S_V. But that error diminishes as sample sizes increase because plim $S_V = \Sigma$ and plim $S_C = \Sigma$. In huge samples it would be expected that $\hat{\Sigma}_C$ would differ little from a corresponding $\hat{\Sigma}_V$ obtained by fitting the model to S_V. So, in very large samples the c statistic (and the variants of the AIC) does not give much weight to penalizing estimation of parameters and, in fact, seems to downplay the fact that some if not all of the small-to-zero lack of fit may be achieved by estimating parameters; hence only the relative size of the normalized noncentrality parameter becomes paramount.

The idea of cross-validation to another random sample from the same population also tends to be confused with the idea of a rigorous test against an independent data set not used in the formulation of the hypothesis, obtained under other conditions. In large samples, in estimating free parameters, one has extracted much of the information about the population Σ via the calibration sample S_C, and the resulting $\hat{\Sigma}_C$ will necessarily differ hardly at all from a corresponding $\hat{\Sigma}_V$ based on S_V. So, the cross-validation is not against an independent data set not used in the formulation of the hypothesis. It should not be thought of or used as a test of the hypothesis. Its only value is to give information about the sampling stability of the estimated model. And in huge samples this becomes a trivial concern.

Confusion of "Likelihoods" in the AIC

Part of the problem with the AIC is that it is based on a conflation of two senses of likelihood. On the one hand, the conception of likelihood as support for a hypothesis about a parameter or model is based on the model's being fully specified. When there are unspecified parameters and one uses *maximum likelihood* to estimate the unspecified parameters, the overall likelihood for the model is euphemistically called a *generalized likelihood*, but it is neither purely one nor the other kind of likelihood, since some of the likelihood is due to the prespecified, constrained parameters, and some of the likelihood is due to obtaining estimates of parameters that have a maximum likelihood under the constraints. When the model is fully prespecified, the likelihood of the model is the basis for a test of the model: how likely are the data given the model? But when the model does not prespecify all its parameters and estimates unspecified parameters by maximum likelihood, the "likelihood" of the model in support of the model is contaminated by maximizing aspects of that likelihood by estimating parameters in such a way as to make the model fit the data maximally conditional on the constraints. So, if one uses the overall "likelihood" for the model as a measure of support for the model as a whole, one will increasingly prefer models that estimate more and more parameters. Although this tendency is also built into the chi-square statistic, it is countered by requiring us to refer the chi-square statistic to a distribution of the chi-square distribution having for its degrees of freedom those of the model. And because the chi-square statistic is based on an LR comparing the likelihood of a more constrained model to the likelihood of a less constrained model in which the more constrained model is nested, the chi-square statistic is to be interpreted as concerning just the difference in constraints. The chi-square statistic obtained for a given constrained model effectively compares the constrained model to the saturated model.

The AIC seemingly corrects for the number of free parameters, but the formula is misleading. The effect of the correction diminishes with increasing sample size. The chi-square part of the formula with nonzero noncentrality increases with increasing sample size, whereas the degrees of freedom part remains constant. At some large sample size, models with more free parameters and resulting smaller noncentrality will be preferred to models with fewer free parameters and somewhat larger noncentrality.

The AIC is an estimate of a population parameter, the "mean expected log likelihood," corrected for bias resulting from estimating parameters. Users of the AIC presume that the parts of the model involving estimated parameters are unproblematic and that with ever larger samples we improve our estimates of these parameters' true values, since the correction for bias due to their estimation diminishes and their standard errors diminish. There is no concern that the parts of the model involving unspecified estimated parameters may be structurally unsound or indistinguishable from alternative structures with similarly free parameters that generate equivalent covariance matrices for the model (Stelzl, 1986; Lee and Hershberger, 1990; MacCallum et al., 1993). This is the question of the objective validity of the model. How is the model a representation of something in the world as opposed to a statistical artifact obtained by using the observed data to provide values for the unknown parameters of the problematic and untested structures of the model? This is a problem often overlooked by statisticians who favor the AIC.

This question is understood most clearly in the case of the saturated model. Saturated models fit the data perfectly by mathematical necessity. But they are not unique. Countless saturated models based on widely varying mathematical structures can be formulated and fit perfectly to any given set of data. That different researchers may be advocates for different such structures (e.g., factor analysis versus simplex models) makes the issue of their subjectivity apparent, for they become researcher-dependent. And in the case of saturated versions of these models, they are indistinguishable in their ability to fit the data. They all do so perfectly.

Other Information Theoretic Indices, ICOMP

Bozdogan (2000) has reviewed the developments in connection with the AIC and has offered his own index based on what he calls "information complexity" (ICOMP). He notes that the term "complexity" has different meanings in different contexts. For example, it is often associated with the number of free parameters in a model. However, he has a different concept of complexity: "Complexity of a system (of any type) is a measure of the degree of interdependency between the whole system and a simple enumerative composition of its subsystems or parts" (p. 72). He then says that by this definition we should be interested in the amount by which a whole system, say, *S*, differs from the composition of its components. He then designates *C(S)* as a real-valued measure of the amount of the difference between the whole system and its parts. In information theoretic terms he defines this amount as "the discrimination information of the joint distribution of the probability model at hand against the product of its marginal distributions. Discrimination information is equal to zero if the distributions are identical and is positive otherwise" (p. 72).

Complexity

For a vector **X** of random variables, he declares that its complexity is measured by the interaction or dependency between its components. So, the informational measure of dependence among the random variables $x_1, x_2, \ldots, x_n$ is given by

$$I(\mathbf{X}) = I(x_1, x_2, \ldots, x_n) = E_f \left[\log \frac{f(x_1, x_2, \ldots, x_n)}{f(x_1)f(x_2) \cdots f(x_n)} \right]$$

$$= \int_{-\infty}^{\infty} \cdots \int_{-\infty}^{\infty} f(x_1, x_2, \ldots, x_n) \log \frac{f(x_1, x_2, \ldots, x_n)}{f(x_1)f(x_2) \cdots f(x_n)} \, dx_1 \cdots dx_n.$$

What is compared here is the joint distribution $f(x_1, x_2, \ldots, x_n)$ of the random variables to the product of their marginal distributions $f(x_1)f(x_2) \cdots f(x_n)$ under the assumption that they are independently distributed. To the extent that the joint distribution differs from the distribution of the variables under the assumption that they are independent, this is a measure of the interdependence among the variables. Note that if the two distributions are the same (the variables jointly are then independent), the ratio of numerator to denominator will equal 1, and the log will be zero and the information measure will be equal to zero. In other words, there is zero complexity. Otherwise there is positive complexity.

Given a multivariate normal distribution for the n variables in $\mathbf{X}$, distributed $N_n(\boldsymbol{\mu}, \boldsymbol{\Sigma})$, the interdependency among the variables is given by their covariance matrix $\boldsymbol{\Sigma}$. To measure the complexity implied by this interdependency, we will compare this distribution with the distribution for the variables under the assumption that they are independently distributed, but with the same variances as in $\boldsymbol{\Sigma}$. Let $\mathbf{D_x} = [\text{diag } \boldsymbol{\Sigma}]$ be a diagonal variance–covariance matrix whose diagonal elements are the same as the diagonal elements of $\boldsymbol{\Sigma}$. Under the independence assumption then $\mathbf{x}$ is distributed $N_n(\boldsymbol{\mu}, \mathbf{D_x})$. Bozdogan (2000) then shows that the informational measure of complexity for the random variables amounts to

$$C_0(\boldsymbol{\Sigma}) = \frac{1}{2} \log |\mathbf{D_x}| - \frac{1}{2} \log |\boldsymbol{\Sigma}| = \frac{1}{2} \sum_{j=1}^{n} \log \sigma_j^2 - \frac{1}{2} \log |\boldsymbol{\Sigma}|. \qquad (15.88)$$

This is analogous to the lack of fit for the null model when fitted to the $\mathbf{S}$ matrix that is used in the CFI as a norm.

Bozdogan (2000), however, indicates that the measure of complexity in Equation 15.88 is not independent of the coordinates of the original variables. He thus seeks the maximum amount of complexity in $\boldsymbol{\Sigma}$ under orthogonal transformations $\mathbf{Y} = \mathbf{T}'\mathbf{X}$ of the variables, where $\mathbf{T}'\mathbf{T} = \mathbf{I}$ and $\mathbf{T}$ is $n \times n$, as a reasonable measure of complexity. He thus derives a maximal information theoretic measure of complexity of a covariance matrix $\boldsymbol{\Sigma}$ of a multivariate normal distribution to be

$$C_1(\boldsymbol{\Sigma}) = \max_{\mathbf{T}} C_0(\boldsymbol{\Sigma}) = \frac{n}{2} \log \left[\frac{\text{tr}(\boldsymbol{\Sigma})}{n} \right] - \frac{1}{2} \log |\boldsymbol{\Sigma}|. \qquad (15.89)$$

How he arrives at this is interesting. We first note that $|\boldsymbol{\Sigma}|$ is invariant under similarity transformations with orthonormal transformation matrices $\mathbf{T}$, that is, $|\boldsymbol{\Sigma}| = |\mathbf{T}'\boldsymbol{\Sigma}\mathbf{T}|$ under the constraint that $\mathbf{T}'\mathbf{T} = \mathbf{I}$. This is because $|\mathbf{T}| = |\mathbf{T}'|$ and $|\mathbf{T}'\boldsymbol{\Sigma}\mathbf{T}| = |\mathbf{T}'| |\boldsymbol{\Sigma}| |\mathbf{T}| = |\mathbf{T}'| |\mathbf{T}| |\boldsymbol{\Sigma}| = |\mathbf{T}'\mathbf{T}| |\boldsymbol{\Sigma}| = |\mathbf{I}| |\boldsymbol{\Sigma}| = |\boldsymbol{\Sigma}|$. So, the right-hand term in Equation 15.88 will be invariant under orthonormal similarity transformations of $\boldsymbol{\Sigma}$, which correspond to orthogonal transformations $\mathbf{T}'\mathbf{X}$.

We also need to note that $\text{tr}(\mathbf{T}'\boldsymbol{\Sigma}\mathbf{T}) = \text{tr}(\boldsymbol{\Sigma}\mathbf{T}\mathbf{T}') = \text{tr}(\boldsymbol{\Sigma}\mathbf{I}) = \text{tr}(\boldsymbol{\Sigma})$ by reason of the invariance of the trace under cyclic permutations of the matrices and the fact that for an orthonormal matrix $\mathbf{T}' = \mathbf{T}^{-1}$.

So, if we seek a maximum $\max_{\mathbf{T}} C_0(\boldsymbol{\Sigma})$, it will result from finding the similarity transformation of $\boldsymbol{\Sigma}$, $\mathbf{T}'\boldsymbol{\Sigma}\mathbf{T}$, such that $\log |[\text{diag}\mathbf{T}'\boldsymbol{\Sigma}\mathbf{T}]|$ is a maximum. Actually, we do not need to find the matrix $\mathbf{T}$ that accomplishes this because our task is simply to find an expression for the maximum value for $\log |[\text{diag}\mathbf{T}'\boldsymbol{\Sigma}\mathbf{T}]|$ under the constraint that $\text{tr}(\mathbf{T}'\boldsymbol{\Sigma}\mathbf{T}) = \text{tr}(\boldsymbol{\Sigma})$, which incorporates both the orthogonality property for $\mathbf{T}$ and the invariance of the trace under orthonormal similarity transformations.

Although at this point finding the expression for the maximum seems formidable, it is actually very easy to find. Consider that $\log \left| [\text{diag} \mathbf{T}' \boldsymbol{\Sigma} \mathbf{T}] \right|$ concerns the logarithm of the determinant of a diagonal matrix, which we know is simply the logarithm of the product of the diagonal elements of the diagonal matrix. Let these diagonal elements be designated as $\sigma_{11}, \sigma_{22}, \ldots, \sigma_{nn}$. Then their product $\prod_{j=1}^{n} \sigma_{jj}$ is to be a maximum under the constraint that $\sum_{j=1}^{n} \sigma_{jj} = \text{tr}(\boldsymbol{\Sigma})$. To find the solution, Bozdogan (1990) uses two means, the arithmetic mean and the geometric mean of the diagonal elements $\sigma_{11}, \sigma_{22}, \ldots, \sigma_{nn}$ in $[\text{diag} \mathbf{T}' \boldsymbol{\Sigma} \mathbf{T}]$. According to Cauchy's theorem the geometric mean is always less than or equal to the arithmetic mean of a series of numbers. Hence

$$\left(\prod_{j=1}^{n} \sigma_{jj} \right)^{1/n} \leq \bar{\sigma} = \frac{1}{n} \sum_{j=1}^{n} \sigma_{jj} \tag{15.90}$$

with equality when $\sigma_{11} = \cdots = \sigma_{nn}$. The product of the diagonal elements will be a maximum under the constraints when their geometric mean is equal to the arithmetic mean of the diagonal elements of $[\text{diag} \mathbf{T}' \boldsymbol{\Sigma} \mathbf{T}]$ and this when all the diagonal elements are equal. Bozdogan (1990) notes that "Van Emden (1971, p. 66) showed that there always exists an orthogonal similarity transformation that transforms any covariance matrix to have equal diagonal elements." So, the desired transformation matrix $\mathbf{T}$ is one that produces a matrix $\mathbf{T}' \boldsymbol{\Sigma} \mathbf{T}$ such that $[\text{diag} \mathbf{T}' \boldsymbol{\Sigma} \mathbf{T}] = \sigma \mathbf{I}$ and $\text{tr}[\text{diag} \mathbf{T}' \boldsymbol{\Sigma} \mathbf{T}] = \text{tr}(\sigma \mathbf{I}) = \text{tr}(\boldsymbol{\Sigma})$. We are now able to solve for σ. We know almost immediately that $\text{tr}(\sigma \mathbf{I}) = n\sigma$, so our solution for σ is given by solving the equation $n\sigma = \text{tr}(\boldsymbol{\Sigma})$, which we see is

$$\sigma = \bar{\sigma} = \frac{\text{tr}(\boldsymbol{\Sigma})}{n}.$$

So, our criterion becomes

$$C_1(\boldsymbol{\Sigma}) = \max_{\mathbf{T}} C_0(\boldsymbol{\Sigma}) = \frac{1}{2} \log |\bar{\sigma} \mathbf{I}| - \frac{1}{2} \log |\boldsymbol{\Sigma}| \tag{15.91}$$

or

$$C_1(\boldsymbol{\Sigma}) = \frac{1}{2} \log(\bar{\sigma}^n) - \frac{1}{2} \log |\boldsymbol{\Sigma}| = \frac{n}{2} \log \left(\frac{\text{tr}(\boldsymbol{\Sigma})}{n} \right) - \frac{1}{2} \log |\boldsymbol{\Sigma}|,$$

which was to be demonstrated.

Bozdogan (2000) then shows that the complexity of a square symmetric variance–covariance matrix $\boldsymbol{\Sigma}$ can be expressed exclusively in terms of its eigenvalues. Recall that the trace of a square symmetric matrix equals the sum of its eigenvalues, whereas the determinant of the matrix equals the product of its eigenvalues. Thus the expression $\log(\text{tr}(\boldsymbol{\Sigma})/n) = \log \left((1/n) \sum_{i=1}^{n} \lambda_i \right) =$

$\log \bar{\lambda}_a$ is the logarithm of the arithmetic mean of the eigenvalues of $\boldsymbol{\Sigma}$. On the other hand,

$$\log |\boldsymbol{\Sigma}| = \log \left(\prod_{i=1}^{n} \lambda_i \right) = n \log \left(\prod_{i=1}^{n} \lambda_i \right)^{1/n} = n \log \bar{\lambda}_g,$$

where $\bar{\lambda}_g$ is the geometric mean of the eigenvalues. Thus we can rewrite Equation 15.89 as

$$C_1(\boldsymbol{\Sigma}) = \frac{n}{2} \log \left(\frac{\bar{\lambda}_a}{\bar{\lambda}_g} \right). \tag{15.92}$$

Thus the maximal information theoretic measure of the complexity of a covariance matrix is given by $n/2$ times the logarithm of the ratio of the arithmetic mean $\bar{\lambda}_a$ to the geometric mean $\bar{\lambda}_g$ of the eigenvalues of the matrix. As Bozdogan notes, "It measures how unequal the eigenvalues of $\boldsymbol{\Sigma}$ are, and it incorporates two simplest scalar measures of multivariate scatter, namely the trace and the determinant into one single function" (Bozdogan, 2000, p. 75). He then goes on to say, "In general large values of complexity indicate a high interaction between the variables, and a low degree of complexity represents less interaction between the variables. The minimum of $C_1(\boldsymbol{\Sigma})$ corresponds to the *least complex* structure. In other words, $C_1(\boldsymbol{\Sigma}) \to 0$ as $\boldsymbol{\Sigma} \to \mathbf{I}$ or $\boldsymbol{\Sigma} \to s\mathbf{I}$, where s is a scalar.

ICOMP

Bozdogan next seeks to combine the complexity measure $C_1(\boldsymbol{\Sigma})$ with an informational measure of fit to get an overall measure of model quality in an index he calls ICOMP, for "information and complexity." He suggests several versions of this index. For a multivariate linear (or nonlinear) structural model, the maximal information-theoretic overall measure-of-complexity ICOMP using the estimated inverse of Fisher's information matrix (IFIM) $\hat{F}^{-1}$—which gives the estimated variances and covariances among the estimated parameters—is defined by

$$\text{ICOMP}[\hat{F}^{-1}(\hat{\boldsymbol{\theta}})] = C_1(\hat{\boldsymbol{\Sigma}}(\hat{\boldsymbol{\theta}})) - 2 \log L(\hat{\boldsymbol{\theta}}_q)$$

$$= \frac{\dim \hat{F}^{-1}}{2} \log \left[\frac{\text{tr}\hat{F}^{-1}}{\dim \hat{F}^{-1}} \right] - \frac{1}{2} \log \left| \hat{F}^{-1} \right| - 2 \log L(\hat{\boldsymbol{\theta}}_q),$$

$$\tag{15.93}$$

where $\hat{\boldsymbol{\Sigma}}(\hat{\boldsymbol{\theta}})$ denotes the variance–covariance matrix among the estimated parameters of the model (not the estimated model covariance matrix among the variables), $-2 \log L(\hat{\theta}_q)$ denotes -2 times the (natural) logarithm of the joint likelihood of the model in the sample, $\dim \hat{F}^{-1} = \operatorname{rank} \hat{F}^{-1}$, and q is the number of estimated parameters in the model.

The matrix $\hat{F}^{-1}$ is given by Bollen (1989) for a confirmatory factor analysis or structural equation modeling algorithm as

$$\hat{F}^{-1} = \left\{ -E \left[\frac{\partial^2 F_{\text{ML}}}{\partial \theta_i \partial \theta_j} \right]_{\hat{\theta}} \right\}^{-1}, \tag{15.94}$$

where the expression in brackets is the $q \times q$ matrix of second derivatives of the maximum-likelihood fit function with respect to the free parameters evaluated using values of the estimated parameters $\hat{\theta}$ when the fit function has attained a minimum, and θ_i and θ_j are any two free parameters. Jöreskog (1979) has given expressions for obtaining the elements of the second derivative matrix in descriptions of his programs for confirmatory factor analysis and structural equation modeling. The diagonal elements of $\hat{F}^{-1}$ contain the estimated variances or squared standard errors of the estimated parameters, while the off-diagonals of this matrix contain their covariances.

Bollen (1989, p. 133) gives the expression for the logarithm of the joint likelihood of an SEM with n variables distributed according to a multivariate normal distribution, which appears in the right-hand expression of Equation 15.93, as

$$\log L(\hat{\theta}) = - \left(\frac{N}{2} \right) \left(n \log(2\pi) + \log \left| \hat{\Sigma} \right| + \operatorname{tr}(S^* \hat{\Sigma}^{-1}) \right), \tag{15.95}$$

where $\hat{\Sigma}$ is the $n \times n$ estimated variance–covariance matrix for the observed variables according to the model, S^* is the sample maximum-likelihood estimate (not the unbiased estimate) of the population variance–covariance matrix, n is the number of observed variables, and N is the sample size.

Using these results we may rewrite the ICOMP formula as

$$\text{ICOMP} = C_1(\hat{F}^{-1}) + N \left(n \log(2\pi) + \log \left| \hat{\Sigma} \right| + \operatorname{tr}(S^* \hat{\Sigma}^{-1}) \right). \tag{15.96}$$

Bozdogan describes other approaches to constructing ICOMP formulas, but they will not be discussed here.

We now can see that like the AIC, ICOMP combines a penalty (complexity) with the log likelihood of the model. Complexity also concerns the estimated parameters. The idea with both the AIC and the ICOMP is to select the model with the smallest value for the respective index. There is no absolute criterion of model quality. I would expect then that some of the same kinds of criticisms directed to the AIC also apply to ICOMP. For example, because the expression in the second term of Equation 15.96 is multiplied by the sample size N, this expression increases with increases in sample size, while the complexity term $C_1(\hat{F}^{-1})$ converges toward a finite value and is bounded above by the maximal information complexity of a saturated model. So, ICOMP tends to infinity with unbounded increases in sample size. In the meantime, the relative impact of the complexity penalty on the overall value of ICOMP also diminishes with increasing sample size. Thus we should expect that there will be a sample size at which a model with more complexity but smaller lack of fit will be preferred on average to a model with less complexity and small lack of fit. Thus in large samples estimated parameters tend to be treated as if they are prespecified fixed parameters because their standard errors are vanishing. But again the problem with this is that numerous equivalent models may be formulated in portions of the model containing free parameters while holding constant the model structure of the fixed parameters (Stelzl, 1986; Lee and Hershberger, 1990; MacCallum et al., 1993). An objective test of these estimated parts of the model is not yet made, since nothing about them is asserted as a part of one's hypothesis. One is in effect peeking at the data and adjusting the free parameters to optimally fit conditional on the fixed parameters and then declaring the resulting model fit to be evidence for the correctness of the full model. To prevent falling into such traps, it is essential to keep in mind the distinction between conducting an objective test of an asserted invariant (a hypothesis) in an incompletely specified model and estimating parameters of a subjectively formulated structural model so as to get optimal fit to a set of data, which involves the generation of a hypothesis. There may be alternative structures that have not yet been ruled out, including, possibly, the researcher's.

LM, Wald, and LR Tests

Many of the statistical tests in structural equation modeling were developed by econometricians. They have established that most tests are based on the Wald, LR, or LM principle (Engle, 1984). Given a *null* hypothesis and its *alternative*, the Wald test begins with the alternative hypothesis and seeks to determine whether the null would be more appropriate. The LR test compares the null hypothesis with the alternative hypothesis, treating each on an equal basis. The LM test begins with the null hypothesis and seeks to determine whether the alternative would be a significant improvement in fit (Engle, 1984). In structural equation modeling, the Wald test would begin with estimated free parameters and test whether they could be replaced with specified fixed parameters without significantly increasing lack of fit. An LR test would compare two models, such as a constrained model versus the saturated model, as in the usual chi-square goodness-of-fit test. An LM test begins with one or more fixed parameters and tests whether they could be freed to improve fit significantly. The symmetry among these tests suggests that they are equivalent, and, indeed, they are *asymptotically* equivalent.

To give a general outline of these tests consider a simple case, and here we follow Engle (1984). Let $\mathbf{y}_j$ be a $p \times 1$ observation vector of a random vector $\mathbf{Y}$, whose density function $f(\mathbf{y} \mid \theta)$ depends on the $k \times 1$ parameter vector θ. Assume we have a sample of N independently and identically distributed observations $\mathbf{y}_j$ of $\mathbf{Y}$. The joint likelihood of θ, given the N observations, is

$$L(\theta \mid \mathbf{y}_1, \ldots, \mathbf{y}_N) = \prod_{j=1}^{N} f(\mathbf{y}_j \mid \theta),$$

whereas the log of the joint likelihood is

$$\ell(\theta \mid \mathbf{y}_1, \ldots, \mathbf{y}_N) = \sum_{j=1}^{N} \ln f(\mathbf{y}_j \mid \theta).$$

For the sake of simplicity, we will hereafter write $\ell(\theta)$ for $\ell(\theta \mid \mathbf{y}_1, \ldots, \mathbf{y}_N)$. Suppose further that the likelihood is maximized at the value $\hat{\theta}$, which satisfies the condition that $\frac{\partial \ell(\hat{\theta})}{\partial \theta_1} = 0, \frac{\partial \ell(\hat{\theta})}{\partial \theta_2} = 0, \ldots, \frac{\partial \ell(\hat{\theta})}{\partial \theta_k} = 0$.

In other words, each of the partial derivatives of the log-likelihood function with respect to each one of its parameters evaluated with the values of the estimated parameters in $\hat{\theta}$ is zero. This occurs when the derivatives are evaluated at the values that maximize the function.

Now for some notation: given a differentiable single-valued function f, the function ∇f is defined as

$$\nabla f(\mathbf{x}) = \left(\frac{\partial f}{\partial x_1}(\mathbf{x}), \ldots, \frac{\partial f}{\partial x_n}(\mathbf{x}) \right)' \tag{15.97}$$

and is known as the $n \times 1$ gradient vector of f (Williamson and Trotter, 1979). In other words, it is the vector of partial derivatives of the function with respect to each of the parameters (in $\mathbf{x}$) of the function. It indicates the slope of the function in the direction of each of the parameters x_i in parameter space.

On the other hand, by $\nabla^2 f$ we mean the $n \times n$ matrix of second partial derivatives of the function f defined as

$$\nabla^2 f(\mathbf{x}) = \begin{bmatrix} \dfrac{\partial^2 f(\mathbf{x})}{\partial x_1^2} & \dfrac{\partial f(\mathbf{x})}{\partial x_1 \partial x_2} & \cdots & \dfrac{\partial f(\mathbf{x})}{\partial x_1 \partial x_n} \\ \dfrac{\partial f(\mathbf{x})}{\partial x_2 \partial x_1} & \dfrac{\partial^2 f(\mathbf{x})}{\partial x_2^2} & \cdots & \dfrac{\partial f(\mathbf{x})}{\partial x_2 \partial x_n} \\ \vdots & \vdots & \ddots & \vdots \\ \dfrac{\partial f(\mathbf{x})}{\partial x_n \partial x_1} & \dfrac{\partial f(\mathbf{x})}{\partial x_n \partial x_2} & \cdots & \dfrac{\partial^2 f(\mathbf{x})}{\partial x_n^2} \end{bmatrix}. \tag{15.98}$$

This is also known as the Hessian matrix, which is often denoted as $\mathbf{H}(\mathbf{x})$. If $\nabla f = \mathbf{0}$ and the Hessian matrix $\mathbf{H}(\mathbf{x})$ is continuous and positive (negative) definite at a point $\mathbf{x}$, then $\mathbf{x}$ is a local minimum (maximum) of the function (Monahan, 2001).

The Newton–Raphson iterative method of solving for the values of the parameters in $\mathbf{x}$ that maximize (or minimize) the function is then given by

$$\mathbf{x}_{(j+1)} = \mathbf{x}_{(j)} - \mathbf{H}(\mathbf{x}_{(j)})\nabla f(\mathbf{x}_{(j)}), \tag{15.99}$$

where the subscripts in parentheses, (j) and $(j+1)$, denote the values used or to be used in the jth or $j+1$st iteration, respectively; $\mathbf{H}(\mathbf{x}_{(j)})$ is the Hessian matrix evaluated using the values of the parameters in the jth iteration, whereas $\nabla f(\mathbf{x}_{(j)})$ is the gradient vector of partial derivatives evaluated with the values of the parameters in the jth iteration. The iterations are started with a value $\mathbf{x}_{(0)}$ chosen as a reasonable but possibly crude approximation to the expected solution. And the iterations continue until the difference between successive estimates is very small in absolute value.

Now, returning to the problem of testing a hypothesis with parameters estimated by maximum likelihood, let us first partition the $k \times 1$ parameter vector $\boldsymbol{\theta}$ into $\boldsymbol{\theta} = [\boldsymbol{\theta}_1, \boldsymbol{\theta}_2]$, where $\boldsymbol{\theta}_1$ and $\boldsymbol{\theta}_2$ are $m \times 1$ and $[(k-m) \times 1]$ subvectors of $\boldsymbol{\theta}$. Let $H_0 : \boldsymbol{\theta}_1 = \boldsymbol{\theta}_1^{(0)}$ be the null hypothesis and $H_1 : \boldsymbol{\theta}_1 \neq \boldsymbol{\theta}_1^{(0)}$, the alternative hypothesis. This compares a simple hypothesis against a composite alternative.

Under the null hypothesis, with fixed $\boldsymbol{\theta}_1^{(0)}$ the maximum-likelihood estimate of $\boldsymbol{\theta}$ is denoted as $\tilde{\boldsymbol{\theta}} = [\boldsymbol{\theta}_1^{(0)}, \tilde{\boldsymbol{\theta}}_2]$, where $\boldsymbol{\theta}_1^{(0)}$ is a subvector of fixed parameters, whereas $\tilde{\boldsymbol{\theta}}_2$ is a subvector of estimated free parameters. Under the alternative hypothesis, the maximum-likelihood estimate of $\boldsymbol{\theta}$ is $\hat{\boldsymbol{\theta}} = [\hat{\boldsymbol{\theta}}_1, \hat{\boldsymbol{\theta}}_2]$, with all free parameters. Before going on, we need to define several important matrices.

Fisher's information matrix is defined as

$$\mathbf{I}(\boldsymbol{\theta}) = \frac{-E\left[\partial^2 \ell(\theta)/\partial\theta\partial\theta'\right]}{N} = \frac{-E[\mathbf{H}(\boldsymbol{\theta})]}{N}, \tag{15.100}$$

which is the expected value of the negative of the matrix of second derivatives of the log-likelihood function with respect to the parameters, divided by sample size N. (*Note*: $\mathbf{I}(\boldsymbol{\theta})$ is not an identity matrix.) But $\mathbf{I}(\boldsymbol{\theta})$ is also often given as

$$\mathbf{I}(\boldsymbol{\theta}) = \frac{-E\left[\partial^2 \ell(\theta)/\partial\theta\partial\theta'\right]}{N} = \frac{-E[\nabla\ell(\boldsymbol{\theta})\nabla\ell(\boldsymbol{\theta})']}{N} = \text{cov}(\nabla\ell(\boldsymbol{\theta})). \tag{15.101}$$

To obtain this version of the information matrix, one must algebraically carry out the multiplication of the mathematical expression for the gradient vector times its transpose, and then take the expectation of whatever elements in the resulting expression are random. One then replaces the expected values by the sample estimates of them. Sometimes this is easier than working out the Hessian matrix mathematically.

We may partition the information matrix to conform to the partitioning of our vector of parameters:

$$\mathbf{I}(\theta) = \begin{bmatrix} \mathbf{I}_{11} & \mathbf{I}_{12} \\ \mathbf{I}_{21} & \mathbf{I}_{22} \end{bmatrix}.$$

Then the asymptotic covariance matrix among the k parameter estimates is given by the inverse of the partitioned information matrix as (Graybill, 1969)

$$\text{cov}(\hat{\theta}) = \mathbf{I}(\hat{\theta})^{-1} = \begin{bmatrix} [\mathbf{I}_{11} - \mathbf{I}_{12}\mathbf{I}_{22}^{-1}\mathbf{I}_{21}]^{-1} & -\mathbf{I}_{11}^{-1}\mathbf{I}_{12}[\mathbf{I}_{22} - \mathbf{I}_{21}\mathbf{I}_{11}^{-1}\mathbf{I}_{12}]^{-1} \\ -\mathbf{I}_{22}^{-1}\mathbf{I}_{21}[\mathbf{I}_{11} - \mathbf{I}_{12}\mathbf{I}_{22}^{-1}\mathbf{I}_{21}]^{-1} & [\mathbf{I}_{22} - \mathbf{I}_{21}\mathbf{I}_{11}^{-1}\mathbf{I}_{12}]^{-1} \end{bmatrix}_{\theta=\hat{\theta}},$$

which is evaluated at the estimated values $\hat{\theta}$ for θ.

However, according to Monahan (2001), because the likelihood function is approximately quadratic in the region of the maximum, then the quadratic nature of the observed likelihood and not the expected likelihood determines the accuracy of the maximum-likelihood estimates. Thus he says that Efron and Hinkley (1978, p. 206) "... convincingly argued the superiority of the observed information $\nabla^2 \ell_n$ over the expected information $\mathbf{J}_n(\theta)$" in estimating variances and covariances among the estimated parameters. This suggests using the Hessian in place of the information matrix. Nevertheless, the expected information matrix often is easier to obtain and still is a close approximation to the Hessian, so that the convergence of the iterations in estimating θ using the information matrix is almost as quick as using the Hessian. On the other hand, it is also possible to numerically and iteratively produce a very close approximation to the Hessian at the minimum by using the quasi-Newton secant method of Broydon, Fletcher, Goldfarb, and Shanno (Monahan, 2001).

Let

$$\mathbf{d}_{(j+1)} = \nabla f(\mathbf{x}_{(j+1)}) - \nabla f(\mathbf{x}_{(j)}) = \mathbf{H}(\mathbf{x}_{(j)})(\mathbf{x}_{(j+1)} - \mathbf{x}_{(j)}) = \mathbf{H}(\mathbf{x}_{(j)})\mathbf{s}_{(j+1)};$$

then the approximate Hessian is given for the $j + 1$st iteration as

$$\mathbf{H}_{(j+1)} = \mathbf{H}_{(j)} + \frac{\mathbf{d}_{(j+1)}\mathbf{d}'_{(j+1)}}{\mathbf{d}'_{(j+1)}\mathbf{s}_{(j+1)}} - \frac{\mathbf{H}_{(j)}\mathbf{s}_{(j+1)}\mathbf{s}'_{(j+1)}\mathbf{H}_{(j)}}{\mathbf{s}'_{(j+1)}\mathbf{H}_{(j)}\mathbf{s}_{(j+1)}}$$

(Monahan, 2001).

Then $\mathbf{H}_{(j+1)}$ is substituted for $\mathbf{H}_{(j)}$ for the next iteration of the Newton–Raphson equation. Monahan (2001) refers to Goldfarb (1976) for details of the update of this equation. An initial approximation to the Hessian in the zeroth iteration could be obtained with numerical differentiation, examples of which are described by Monahan (2001, pp. 184–187).

Returning to our problem of maximizing the log likelihood, we need to simplify notation. Let

$$\boldsymbol{\Omega} = \begin{bmatrix} \boldsymbol{\Omega}_{11} & \boldsymbol{\Omega}_{12} \\ \boldsymbol{\Omega}_{21} & \boldsymbol{\Omega}_{22} \end{bmatrix} \equiv \mathrm{cov}(\hat{\boldsymbol{\theta}}) = \mathbf{I}(\hat{\boldsymbol{\theta}})^{-1}$$

be the covariance matrix among the estimated parameters or the inverse of the information matrix. Then if we can assume that $\hat{\boldsymbol{\theta}}$ has an asymptotic multivariate normal distribution and $\mathbf{I}(\theta)$ is consistently estimated by $\mathbf{I}(\hat{\boldsymbol{\theta}})$,

$$W = N(\hat{\boldsymbol{\theta}}_1 - \boldsymbol{\theta}_1^{(0)})' \boldsymbol{\Omega}_{11}^{-1}(\hat{\boldsymbol{\theta}}_1 - \boldsymbol{\theta}_1^{(0)}) = \chi_m^2 \qquad (15.102)$$

is known as the Wald test statistic (Engle, 1984). This is formally equivalent to the multivariate Z^2 test (Rencher, 2002) of the null hypothesis on the mean vector, $H_0 : \boldsymbol{\mu} = \boldsymbol{\mu}_0$ against $H_1 : \boldsymbol{\mu} \neq \boldsymbol{\mu}_0$:

$$Z^2 = N(\bar{\mathbf{y}} - \boldsymbol{\mu}_0)' \boldsymbol{\Sigma}^{-1}(\bar{\mathbf{y}} - \boldsymbol{\mu}_0).$$

The Wald test is based on the difference between $\hat{\boldsymbol{\theta}}_1$ and $\boldsymbol{\theta}_1^{(0)}$.

The LR test

$$\mathrm{LR} = -2(\ell(\boldsymbol{\theta}_1^{(0)}, \tilde{\boldsymbol{\theta}}_2 \mid \mathbf{y}) - \ell(\hat{\boldsymbol{\theta}}_1, \hat{\boldsymbol{\theta}}_2 \mid \mathbf{y}))$$

compares the difference between the log likelihoods and has an asymptotic chi-square distribution under the null hypothesis (Engle, 1984).

The LM test is based on the principle of constrained maximization using LMs. If we seek to maximize the log likelihood under the constraint that $\boldsymbol{\theta}_1 = \boldsymbol{\theta}_1^{(0)}$, then setting up the problem with LMs multiplied by each equation of constraint, we have the function

$$H = \ell(\boldsymbol{\theta} \mid \mathbf{y}) - \boldsymbol{\lambda}'(\boldsymbol{\theta}_1 - \boldsymbol{\theta}_1^{(0)}),$$

where $\boldsymbol{\lambda}$ is an $m \times 1$ vector of LMs. Taking partial derivatives we obtain

$$\frac{\partial H}{\partial \boldsymbol{\theta}} = \left[\frac{\partial \ell(\boldsymbol{\theta})}{\partial \theta_i}\right]_{k \times 1} - \left[\frac{\partial(\boldsymbol{\theta}_1 - \boldsymbol{\theta}_1^{(0)})}{\partial \theta_i}\right]_{k \times m} \boldsymbol{\lambda} = \left[\frac{\partial \ell(\boldsymbol{\theta})}{\partial \theta_i}\right]_{k \times 1} - \begin{bmatrix} \mathbf{I}_m \\ \mathbf{0} \end{bmatrix}_{k \times m} \boldsymbol{\lambda}$$

$$= \mathbf{D} + \mathbf{L}\boldsymbol{\lambda}$$

$$\frac{\partial H}{\partial l} = \left[\frac{\partial \ell(q)}{\partial \lambda_h}\right]_{m \times 1} - \left[\frac{\partial l'(q_1 - q_1^{(0)})}{\partial \lambda_h}\right]_{m \times 1} = -(q_1 - q_1^{(0)}).$$

First, to simplify our notation further, let us write ℓ for $\ell(\boldsymbol{\theta})$. Then setting each of these partial derivatives equal to null vectors and solving for $\boldsymbol{\lambda}$ and $\boldsymbol{\theta}_1$, we obtain

$$
\begin{bmatrix} \dfrac{\partial \ell}{\partial \boldsymbol{\theta}_1} \\[2mm] \dfrac{\partial \ell}{\partial \boldsymbol{\theta}_2} \end{bmatrix} = \begin{bmatrix} \boldsymbol{\lambda} \\ \mathbf{0} \end{bmatrix}.
$$

Hence

$$
\frac{\partial \ell}{\partial \boldsymbol{\theta}_1} = \boldsymbol{\lambda}; \quad \frac{\partial \ell}{\partial \boldsymbol{\theta}_2} = \mathbf{0}_{(k-m)\times 1}, \quad \boldsymbol{\theta}_1 = \boldsymbol{\theta}_1^{(0)}.
$$

Hence at the maximum (minimum)

$$
\begin{bmatrix} \dfrac{\partial \tilde{\ell}}{\partial \boldsymbol{\theta}_i} \end{bmatrix}_{k\times 1} = \begin{bmatrix} \tilde{\boldsymbol{\lambda}} \\ \mathbf{0} \end{bmatrix} = \begin{bmatrix} \dfrac{\partial \tilde{\ell}}{\partial \boldsymbol{\theta}_1} \\ \mathbf{0} \end{bmatrix} = \tilde{\mathbf{D}}.
$$

These results will hold at the maximum (minimum) and to a close approximation at convergence of the iterations to be performed.

The LM statistic is given in two forms (Breusch and Pagan, 1980), with tildes over symbols indicating values evaluated at the maximum (minimum):

$$
\text{LM} = \tilde{\mathbf{D}} \mathbf{I}(\tilde{\boldsymbol{\theta}})^{-1} \tilde{\mathbf{D}} = \tilde{\boldsymbol{\lambda}}' \tilde{\mathbf{L}}' \mathbf{I}(\tilde{\boldsymbol{\theta}})^{-1} \tilde{\mathbf{L}} \tilde{\boldsymbol{\lambda}}.
$$

According to Breusch and Pagan (1980), the term $\tilde{\mathbf{D}} \mathbf{I}(\tilde{\boldsymbol{\theta}})^{-1} \tilde{\mathbf{D}}$ is known as the "score" statistic (Rao, 1973), whereas $\tilde{\boldsymbol{\lambda}}' \tilde{\mathbf{L}}' \mathbf{I}(\tilde{\boldsymbol{\theta}})^{-1} \tilde{\mathbf{L}} \tilde{\boldsymbol{\lambda}}$ is traditionally known as the LM statistic. The two statistics are identical and so the choice of which form to use reduces to the one that is most convenient, and this is the score test (Breusch and Pagan, 1980); however, the name "Lagrange multiplier test" or "LM test" remains.

We will now define a quantity known as "the score," which is none other than the gradient of the log-likelihood function evaluated at $\tilde{\boldsymbol{\theta}} = [\boldsymbol{\theta}_1^{(0)}, \tilde{\boldsymbol{\theta}}_2]$:

$$
\mathbf{s}(\tilde{\boldsymbol{\theta}} \mid \mathbf{y}) = \tilde{\mathbf{D}} = \begin{bmatrix} \dfrac{\partial \tilde{\ell}}{\partial \boldsymbol{\theta}_1} \\ \mathbf{0} \end{bmatrix} = \nabla \ell(\tilde{\boldsymbol{\theta}}).
$$

Then the LM test statistic is given by

$$
\text{LM} = \frac{\mathbf{s}(\tilde{\boldsymbol{\theta}} \mid \mathbf{y})' \mathbf{I}(\tilde{\boldsymbol{\theta}})^{-1} \mathbf{s}(\tilde{\boldsymbol{\theta}} \mid \mathbf{y})}{N} = \frac{\mathbf{s}(\tilde{\boldsymbol{\theta}} \mid \mathbf{y})' \boldsymbol{\Omega}(\tilde{\boldsymbol{\theta}}) \mathbf{s}(\tilde{\boldsymbol{\theta}} \mid \mathbf{y})}{N}.
$$

To show the equivalence to the Wald statistic, let us write the information matrix in partitioned form:

$$
\mathbf{I}(\boldsymbol{\theta}) = \begin{bmatrix} \mathbf{I}_{11} & \mathbf{I}_{12} \\ \mathbf{I}_{21} & \mathbf{I}_{22} \end{bmatrix}.
$$

Then

$$
\mathrm{LM} = \left(\frac{\partial \ell}{\partial \boldsymbol{\theta}_1}', \mathbf{0}' \right) \begin{bmatrix} \mathbf{I}_{11} & \mathbf{I}_{12} \\ \mathbf{I}_{21} & \mathbf{I}_{22} \end{bmatrix}^{-1} \begin{bmatrix} \frac{\partial \ell}{\partial \boldsymbol{\theta}_1} \\ \mathbf{0} \end{bmatrix} = \left(\frac{\partial \ell}{\partial \boldsymbol{\theta}_1} \right)' (\mathbf{I}_{11} - \mathbf{I}_{12} \mathbf{I}_{22}^{-1} \mathbf{I}_{21})^{-1} \left(\frac{\partial \ell}{\partial \boldsymbol{\theta}_1} \right),
$$

but this is the same as

$$
\mathrm{LM} = \left(\frac{\partial \ell}{\partial \boldsymbol{\theta}_1} \right)' \mathbf{I}^{11} \left(\frac{\partial \ell}{\partial \boldsymbol{\theta}_1} \right) = \left(\frac{\partial \ell}{\partial \boldsymbol{\theta}_1} \right)' \boldsymbol{\Omega}_{11} \left(\frac{\partial \ell}{\partial \boldsymbol{\theta}_1} \right) = \chi_m^2,
$$

where $\mathbf{I}^{11}$ denotes the $(1, 1)$ $m \times m$ partition block of $\mathbf{I}(\boldsymbol{\theta})^{-1}$ that corresponds to $\boldsymbol{\Omega}_{11}$, the covariances among estimated parameters in $\boldsymbol{\theta}_1$. As noted, LM is distributed as chi-squared with m degrees of freedom.

Now suppose that after our iterative algorithm converges, we run another iteration, by "the method of scoring." We have

$$
\hat{\boldsymbol{\theta}} = \tilde{\boldsymbol{\theta}} + \mathbf{I}(\tilde{\boldsymbol{\theta}})^{-1} \left(\frac{\partial \tilde{\ell}}{\partial \boldsymbol{\theta}} \right).
$$

Next, subtract $\tilde{\boldsymbol{\theta}}$ from both sides; then premultiply both sides by $\mathbf{I}(\boldsymbol{\theta})$ to yield

$$
\left(\frac{\partial \tilde{\ell}}{\partial \boldsymbol{\theta}} \right) = \mathbf{I}(\tilde{\boldsymbol{\theta}})(\hat{\boldsymbol{\theta}} - \tilde{\boldsymbol{\theta}}),
$$

which we may substitute for $\left(\partial \tilde{\ell} / \partial \boldsymbol{\theta} \right)$ to obtain

$$
(\hat{\boldsymbol{\theta}} - \tilde{\boldsymbol{\theta}})' \mathbf{I}(\tilde{\boldsymbol{\theta}}) \mathbf{I}(\tilde{\boldsymbol{\theta}})^{-1} \mathbf{I}(\tilde{\boldsymbol{\theta}})(\hat{\boldsymbol{\theta}} - \tilde{\boldsymbol{\theta}}) = (\hat{\boldsymbol{\theta}} - \tilde{\boldsymbol{\theta}})' \mathbf{I}(\tilde{\boldsymbol{\theta}})(\hat{\boldsymbol{\theta}} - \tilde{\boldsymbol{\theta}}) = (\hat{\boldsymbol{\theta}} - \tilde{\boldsymbol{\theta}})' \tilde{\boldsymbol{\Omega}}^{-1} (\hat{\boldsymbol{\theta}} - \tilde{\boldsymbol{\theta}}),
$$

which is equivalent to the Wald test statistic for the test of the hypothesis that $\boldsymbol{\theta} = \tilde{\boldsymbol{\theta}}$.

When the log-likelihood function is a smooth curve approximated by a quadratic function at the minimum, this allows us to use a quadratic function, such as weighted least squares, to estimate the parameters. Then asymptotically the minimum solution will correspond to the maximum-likelihood solution. And Engle (1984) notes that under this condition the Wald test, the LR test, and the LM test are all equivalent.

Bentler (1993) has incorporated the Wald test and the LM test in his program EQS[©], which is based on minimizing a weighted least-squares criterion $Q = (\mathbf{s} - \boldsymbol{\sigma}(\boldsymbol{\theta}))' \mathbf{W}(\mathbf{s} - \boldsymbol{\sigma}(\boldsymbol{\theta}))$. Karl Jöreskog has used a related index known as the "modification index" (MI) developed by his associate Dag Sörbom (1989) in his program LISREL. This index only shows an "approximate estimate of how much the [likelihood] function will decrease if one adds a parameter θ_1 to the set of free parameters" (Sörbom, 1989, p. 373). The MI is given as

$$
\mathrm{MI} = \frac{1/2 \left(\partial \hat{f}/\partial \theta_1 \right)^2}{\left(\left(\partial^2 \hat{f}/\partial \theta_1^2 \right) - \hat{\mathbf{d}}' \hat{\mathbf{I}}(\theta_2)^{-1} \hat{\mathbf{d}} \right)},
$$

where $\hat{\mathbf{d}} = E \left[\partial^2 \hat{f}/\partial \theta_2 \partial \theta_1 \right]$. The MI can be computed for each fixed parameter. Since the estimated information matrix $\hat{\mathbf{I}}(\theta_2)$ for the free parameters will already have been obtained at convergence of the iterations during estimation of the free parameters, all that remains is to evaluate it at the minimum and compute its inverse. Then one computes the second derivative of the ML fit function f with respect to the parameter θ_1 and the vector $\mathbf{d}$ of second derivatives of the fit function f with respect to the parameters in θ_2 and θ_1 and evaluates these at the minimum. These are then entered in the formula for MI. Sörbom (1989) notes that the MI is then approximately distributed as chi-squared with 1 degree of freedom, which can be a basis for approximate tests of significance of the change in f.

The LM test statistic can also be used to test the effect of freeing a single constrained parameter, for it too will have a distribution as chi-squared with 1 degree of freedom.

Expected Parameter Change

Closely related to the MI is the index of "expected parameter change (EPC)" of Saris, Satorra, and Sörbom (1987). The object of this index is to provide an indication of how much a fixed parameter will change if freed. Let $\theta_{i(0)}$ be the value of the ith parameter under the null hypothesis of the model. Let $\hat{\theta}_i$ be the value of the ith parameter after it has been freed and estimated (along with the other freed parameters in the model). Then the EPC is given by

$$
\mathrm{EPC} = \hat{\theta}_i - \theta_{i(0)} = \frac{\mathrm{MI}}{\left(\partial \hat{f}/\partial \theta_i \right)}.
$$

Saris et al. (1987) suggested using the EPC along with the MI or LM test statistic. If both LM and EPC were large, then that would be a parameter to free. If LM is large but EPC is small, one might not want to free the parameter because it may be an effect of "excess power" to detect small differences that are likely due to artifacts such as minor causal heterogeneity of subjects, the presence of an outlier, minor departure from multivariate normality, and the like. This decision calls for judgment on the part of the researcher. If the LM is small but the EPC is large, this is an ambiguous situation. But it may be just the opposite of the previous situation: low power. In that case one may be inclined not to change the parameter. If both LM and EPC are small, then freeing the parameter is not called for.

Kaplan (2000) notes that Monte Carlo studies of the EPC have shown that it does better than the LM test in finding specification errors. Kaplan also notes that a problem with the EPC is that it is sensitive to the metrics of the observed variables. He notes that he recommended rescaling the EPC in terms of standardized manifest variables. However, for models with latent variables, standardizing both manifest and latent variables to have unit variances will imply a corresponding standardizing of the EPC.

Modifying Models Post Hoc

When a researcher gets an indication that a model fails to fit acceptably, what should he or she do? First, he or she should review again the theory on which his or her model is based in the light of diagnostic information provided by the computer program about where the lack of fit occurs. Large residuals will indicate where the model failed to fit the variables as predicted. What aspects of the model concerned those variables, and how might one revise one's theory and the model in turn to improve the fit? There are also indices and tests that may be performed to discover what constrained parameters are associated with the lack of fit. These are known (in LISREL and some other programs) as "MIs," and in EQS as LM tests. But one must be cautious in their use. Basically, they provide measures of the degree to which lack of fit would be reduced if one freed up a given parameter in question. And so one is inclined to free up the parameter in question and reanalyze the model. But if the lack of fit is not due merely to a misspecified parameter in a correctly specified model structure, this can be misleading. The problem could lie in a misspecified model structure, omission of important latents, and no amount of freeing up parameters in the current model would lead to a correct model, even though the resulting model might fit acceptably. So, in using these indices, one will have to use judgment and control one's temptation to merely free a constrained parameter by considering whether to do so would be consistent with what is plausibly known of the world. And one should not free too many parameters either, if the aim is to salvage most aspects of the current model. Nevertheless, with these warnings, an LM test can be informative.

Kaplan (2000) and Kaplan and Wenger (1993) show that when model parameters covary, as given by the IFIM, simultaneous LM tests of more than one parameter at a time will be influenced by the covariation between the parameter estimates. The magnitude and ordering of multiparameter LM test statistics will vary with the degree of covariation among parameters included in the subset chosen for possible modification. So, choosing which subset of fixed parameters to free could be influenced by the covariances between the parameter estimates. Furthermore, they show that ". . . the covariance matrix of the estimates is determined by the initial specification of the model. After that, each addition (or deletion) of parameters results in a change of the form of the covariance matrix of the estimates, and hence in the ways that specification errors will manifest themselves through the model" (Kaplan, 2000, p. 99). So, the initial ordering of both single parameter LM statistic values for the fixed parameters and individual parameter Wald tests or z statistics for free parameters can and usually will change in later respecified models.

Although these effects on order of these statistics are considered, one must also consider the power of such statistics, since, according to Saris et al. (1987), as summarized by Kaplan and Wenger (1993, p. 468), ". . . the power for detecting such misspecifications of the same magnitude depended on where in the model the misspecification was located." The complexity of interdependencies among parameter estimates thus led Kaplan and Wenger (1993) and Kaplan (2000) to advocate ". . . the more prudent *univariate sequential approach* to model modification whereby restrictions (or inclusions) are made one parameter at a time and careful attention is paid to changes in substantively important parameters . . ." (Kaplan and Wenger, 1993, p. 480). This involves more analyses. The aim should be to free the fixed parameter with the largest LM or MI at each step. After freeing one parameter, one should re-estimate the free parameters of the resulting model, and then compute a new set of LM tests for the remaining fixed parameters. From these the fixed parameter with the now largest LM test statistic is chosen and freed, and so on. However, you should also constrain yourself from freeing parameters for which you can provide no substantive theoretical reason for doing so. And remember, if the model framework is wrongly specified, no freeing and fixing parameters is going to yield a correct model.

Recent Developments

Much of the theory up to now assumes that our variables have multivariate normal distributions. From these assumptions we were able to conclude that the sample "chi-squared statistic" indeed has a chi-square distribution when the hypothesized model is correct, and if not, has a noncentral chi-square distribution, from which we can extract the noncentrality parameter and use a simple sample estimate of it to describe model misfit. At least these assumptions may hold, we believe, asymptotically. But statisticians such as Ke-Hai Yuan (2005) observe that generally our variables do not have a multivariate normal distribution, and this now raises doubts about all those consequences we have deduced from assuming multivariate normality to be the case.

Much of the data in the social sciences consist of Likert scale ratings, which are categorical and not continuous; hence, consequently they cannot possibly have multivariate normal distributions. Furthermore, our samples for study may not be simple random samples, but selected samples to satisfy some criterion, such as a criterion for admission to college or graduate school, the police force, the army, the nursing profession, or medical school, which have minimum requirements on various abilities and education. So, the samples may be truncated at the top or bottom or both. In these cases, not only will we question our chi-square statistic and its nominal cutoff for significance, but even the noncentrality parameter and the GFIs based on it.

Satorra and Bentler (1990) argued that as long as what they called "asymptotic robustness" holds, given the sample "chi-squared" value of $U^* = (N - 1)F_{ML}$, they could find a quantity κ such that

$$T_R = \kappa^{-1}U^* \tag{15.103}$$

has an asymptotic chi-square distribution. κ depends on the fourth-order moments of the distribution of the manifest variables as well as the structure of the model. But the resulting T_R has come to be known as the Satorra–Bentler rescaled chi-square statistic. Several Monte Carlo studies have shown that this statistic behaves well under a number of different conditions.

Yuan (2005) considers cases where the sample U^* is a linear function of the noncentral chi-square statistic:

$$U^* = b\chi^2(df, \lambda) + a, \tag{15.104}$$

which would allow us to transform U^* to a chi-square distributed variate when the noncentrality $\lambda = 0$, that is, the null hypothesis is true:

$$\chi^2(df, \lambda) = b^{-1}(U^* - a). \tag{15.105}$$

Under the null hypothesis that $\lambda = 0$, b adjusts the variance of the chi-square distribution, whereas a adjusts its mean to yield the U^* distribution in question. Since a chi-square variable is the sum of a series of squared unit normal deviates, chi-square represents normalized, squared sampling error. Because the variance of a unit normal deviate is 1, as is the expected value of a squared unit normal deviate, the expected value of the chi-square distribution equals the degrees of freedom, the number of expected squared unit normal deviates.

A statistic T which is distributed as the noncentral chi-square distribution $\chi^2(df, \lambda)$ has for its expected value $E(T) = df + \lambda$. The degrees of freedom, df, represents total average squared sampling error, whereas λ represents total systematic squared error. Thus $\hat{\lambda} = T - df$ is an estimate of systematic error. So, returning to Equation 15.104, given that we have rejected the null hypothesis, all we need to get the noncentrality parameter estimate is to subtract df from the quantity on the left, or $\hat{\lambda} = b^{-1}(U^* - a) - df$. Of course, to get to this point, we would need estimates of b and a. This may be easier said than done. But Yuan (2005) notes that we might develop the values empirically, maybe using bootstrap sampling and regression based on the QQ plots of the quantiles of empirical U^* against quantiles of simulated values derived for the model in question under the assumption that it is true. This makes the probability distribution unique to the application. But in some cases when the elliptical distribution applies, a may not apply and b is estimable.

Yuan (2005) notes that cutoff values such as 0.05 (for RMSEA) or 0.95 (CFI, TLI, and GFI) have "little to do with the mean value of the fit indices or the NCP [noncentrality parameter]" (p. 124). That is true, but these cutoffs are not set probabilistically, but in terms of a putative distance metric of what corresponds to one's view of "close fit." One must not be confused as to what the underlying metaphoric paradigm is for judging fit: Is it closeness in probability or closeness in distance? Both paradigms are used, sometimes combined as in the significance test. But each can be applied ignoring the other, as is the case here for cutoff values that are distance inspired. But Yuan (2005) and Yuan and Chan (2005) note that under different methods of estimation, the resulting so-called chi-squared statistics may differ, as will also any GFI computed from chi-squares under different methods of estimation. Hence a 0.05 cutoff under one method of estimation may not correspond to a 0.05 cutoff under another. Yuan (2005) also suggests that if there is no noncentral chi-square distribution, the noncentrality parameter may be "irrelevant" (p. 124). But this is a bit too strongly stated: All we need for GFIs based on a "noncentrality parameter" is a statistic whose expected value equals the sum of its degrees of freedom and a parameter representing systematic error. This still constrains our possible choices.

Yuan (2005) also reports Monte Carlo studies of a number of approximate chi-square statistics. He concludes that generally chi-square distributions are not obtainable, even when the data are normally distributed. The distribution shapes, he says, will vary substantially when such conditions as sample size, the distribution of the manifest variables, model size, and model misspecification vary. Using unmodified noncentral chi-square distributions to describe properties of fit (especially in the tails) may be inappropriate. In some cases even trying to use a noncentrality parameter estimate will be inappropriate. Whatever robust procedures are then invoked to deal with this situation will be "tailor-made" to the case.

After examining several popular fit indices under the condition that the sample U^* does not follow a chi-square distribution, Yuan (2005) considers possible replacements for the usual chi-square statistic in the formulas for the GFIs. He concludes that two "robust" chi-square measures would function better than the maximum-likelihood-based chi-square, while retaining the property that their expected value equals degrees of freedom:

$$T_{AR} = (N-1)^{-1}[N - 5/3 - (2p+5)/6 - (m-1)/3]T_R. \qquad (15.106)$$

This is the "adjusted rescaled chi-square statistic" derived from the maximum-likelihood chi-square statistic.

The second "robust chi-square measure" is based on ADF estimation. Let $x_i = (x_1, \ldots, x_p)'$ be an observation vector for p random variables from some nonnormal multivariate distribution for the random vector $\mathbf{x}$. We now introduce a new operation known as the vectorization of a symmetric matrix $\mathbf{S}$, which we will denote as $\mathbf{s} = \text{vech}(\mathbf{S})$. This is no more than taking the $p(p+1)/2$ nonduplicated elements in order from $\mathbf{S}$ and arranging them in a column vector $\mathbf{s}$. The conventional way to do this is to select the nonduplicated elements by columns. The nonduplicated elements in the first column of $\mathbf{S}$ are the element in the principal diagonal and all the elements below it. The nonduplicated elements in the second column of $\mathbf{S}$ are the principal diagonal element and all the elements below it. The nonduplicated elements in the third column are the principal diagonal element of that column and all the elements below it. So, in general, we proceed selecting the element in the principal diagonal of $\mathbf{S}$ in the column in question and all the elements below it in that column and packing them in order into the column vector $\mathbf{s}$. Let $\boldsymbol{\sigma}(\boldsymbol{\theta}) = \text{vech}(\boldsymbol{\Sigma}(\boldsymbol{\theta}))$ be a vectorization of a model population variance–covariance matrix derived from its parameters in the vector q. Let $\mathbf{s} = \text{vech}(\mathbf{S})$ be a corresponding vectorization of the sample variance–covariance matrix for the p random variables in $\mathbf{x}$.

From a random sample of N such observation vectors x_i, we may compute the mean vector $\bar{x} = (1/N) \sum_{i=1}^{N} x_i$. For each x_i we may compute a $p \times p$ symmetric matrix $(x_i - \bar{x})(x_i - \bar{x})'$. Now, let us define the $p(p+1)/2$ vector $y_i = \text{vech}[(x_i - \bar{x})(x_i - \bar{x})']$. Next, let $\mathbf{S}_y = (N-1)^{-1}\left[\sum_{i-1}^{N} \bar{y}_i \bar{y}_i' - N \sum_{i-1}^{N} \bar{y}_i \bar{y}_i'\right]$ be the sample variance–covariance matrix for the y_i's. Note that $\mathbf{S}_y$ is a $(p(p+1)/2) \times (p(p+1)/2)$ matrix, whose elements estimate the fourth-order moment matrix Γ with typical elements $\gamma_{ij,kl} = \sigma_{ijkl} - \sigma_{ij}\sigma_{kl}$, where

$$\sigma_{ijkl} = E[(x_i - \mu_i)(x_j - \mu_j)(x_k - \mu_k)(x_l - \mu_l)]$$

and

$$\sigma_{ij} = E[(x_i - \mu_i)(x_j - \mu_j)]$$

are the usual expressions for a covariance. When estimated parameters $\hat{\theta}$ are obtained by GLS estimation, one minimizes

$$F_{\text{GLS}} = (\mathbf{s} - \mathbf{s}(\mathbf{q}))' \mathbf{S}_y^{-1} (\mathbf{s} - \mathbf{s}(\mathbf{q})). \tag{15.107}$$

This is known as the ADF estimator (Browne, 1984). The corresponding chi-square statistic for this estimator is given by Yuan and Bentler (1998) as

$$T_{\text{ADF}} = (N-1)F_{\text{GLS}}(\hat{\boldsymbol{\theta}}). \tag{15.108}$$

Originally Browne (1984) and others hoped that ADF estimation would solve the problem of nonnormality for variables, since it could be applied to most nonnormal distributions of interest. However, Yuan and Bentler (1998) reported that subsequent Monte Carlo studies in the literature revealed that this method of estimation and its test statistic would reject correct models with 68% rejection rates in moderate-to-small samples for large models with many parameters. Furthermore, huge samples in the thousands were needed for the method and its test statistic to behave properly.

Yuan and Bentler (1998) noted that an alternative consistent estimator of Γ would be given by

$$\hat{\Gamma} = (N-1)^{-1} \sum_{i=1}^{N} (y_i - \sigma(\hat{\theta}))(y_i - \sigma(\hat{\theta}))'$$

$$= S_y + \frac{N}{(N-1)}(\bar{y}_i - \sigma(\hat{\theta}))(\bar{y}_i - \sigma(\hat{\theta}))'.$$

By replacing S_y by $\hat{\Gamma}$ in Equation 15.107, we are then in a position to create a new statistic

$$T_{\text{CRADF}}(\hat{\theta}) = T_{\text{ADF}}(\hat{\theta}) / \left(1 + \frac{N}{(N-1)^2} T_{\text{ADF}}(\hat{\theta})\right). \tag{15.109}$$

Because the denominator of Equation 15.109 is necessarily larger than the numerator, the value of T_{CRADF} is relatively smaller than T_{ADF} in samples of smaller N. But as N becomes increasingly larger $N/(N-1)^2$ begins to approach zero, implying then that T_{CRADF} is asymptotically consistent with T_{ADF}. Since in smaller samples T_{ADF} rejected correct models excessively by being biased in being too large, the correction in Equation 15.109 compensates for this bias in smaller samples while performing increasingly the same as T_{ADF} in much larger samples.

For GFIs based on noncentrality parameters, both T_{AR} and T_{CRADF} satisfy the property $E(T) = \text{df}$ when the hypothesized model is correct, implying that when the hypothesized model is incorrect, $T - \text{df}$ provides a relatively unbiased estimate of the noncentrality or systematic lack of fit in nonnormal situations.

Yuan and Chan (2005) studied the behavior of T_{ML}, T_{GLS}, and T_{ADF} under differences in model misspecification, small sample size, and distribution of the sample. They found that the "... difference between T_{ML} and T_{GLS} is due to model misspecification, not the distribution of the sample. The difference between T_{GLS} and T_{ADF} is due to the distribution of the sample and small sample size, not model misspecification. The difference between T_{ML} and T_{ADF} can be due to model misspecification, distribution of the sample, as well as a small sample size" (p. 797). They further noted that systematic differences in the discrepancy functions reflect differences in the manner in which distances are scaled between matrices as well as differences in the matrices compared. Systematic differences both within and between methods of estimation will produce differences for model fit indices, so that cutoff values will not correspond across fit indices, between methods, or even across model size, model discrepancy, and sample size within estimation method. So, one should not use cutoff values uncritically.

Criticisms of Indices of Approximation

Some researchers are opposed to the use of indices of approximation, such as the CFIs, GFIs, TLIs, or RMSEA indices. They argue that it is sufficient to use the chi-square test, and when a model fails to fit by that statistic, one should then pay attention to that fact and proceed to examine diagnostic information given by the computer programs and to produce a more thorough conceptual analysis of the phenomena being modeled to arrive at an understanding of the failure of the model to fit. Other researchers take an almost opposite view: there are no exact truths in science, only relative degrees of approximation. The world is complex, while our concepts of it are gross simplifications and, at best, good approximations. Statistical tests, such as the chi-square test, will increasingly reject almost any model as the sample size increases, because the power to detect miniscule deviations of the model from the data it purports to fit, increases with sample size. But these miniscule deviations most likely concern minor, extraneous influences not anticipated by the researcher at a level of resolution beyond current concepts to handle. It is sufficient to focus on the degree of approximation to learn how well the present model reproduces the data. Good approximations have long been used in engineering, while physical scientists use degree of approximation to gauge the potential of a theory with suitable future modifications to successfully explain the phenomena (Giere, 1988).

The critics of indices of approximation counter that the degree to which a model approximates data has no necessary relation to the degree to which the model is causally misspecified. The "true" model may use quite different latent variables, in a different causal framework, and the approximate fitting model may only be a near-equivalent model to the true model. So, a small misfit may still be a signal of serious causal misspecification.

Advocates of approximate fit indices respond to this argument by noting that it would equally apply to models that fit with nonsignificant chi-squares. Just because the model fits the data does not mean that it is correct. The true model may use quite different latent variables in a different causal framework, and the model that fits with nonsignificant chi-square is only an equivalent model. So, blithely accepting the model with a nonsignificant chi-square may be even more serious, because the researcher has no warning of serious causal misspecification.

Because information about the world is gathered serially in time, it usually is incomplete in representing all facets of a phenomenon at the earlier stages of research. Models in the early stages may be oversimplifications and at best crude approximations. Although exact fit tests reveal the failures of these models, the degrees to which they still approximate the data can be useful information to the researcher in determining provisionally whether he or she is on the right track, but has still to consider refining the model with more features representing the world to obtain better fit.

Approximate fitters note that frequently failure to fit is the result of failure to achieve causal homogeneity in the research subjects studied. Individuals may vary in the degree to which a causal variable has an effect on another variable. The model obtained with a specific causal effect, though with significant chi-square, may yet represent a useful approximation to each individual's effect of the causal variable.

A historic analogue in chemistry (Brady and Humiston, 1982) is the development of the gas laws beginning with Boyle's law relating pressure to volume at a constant temperature, Charles' law relating volume to temperature at a constant pressure, and Avogadro's principle relating volume of a gas under constant temperature and pressure to the number of molecules (moles) of the gas. This gave rise to the ideal gas law $V = nRT/P$, where V is volume, T is temperature, P is pressure, n is the number of moles of the gas, and R is the universal gas constant. But the ideal gas law is only a good approximation for most gases. The Dutch physicist J. D. van der Waals received a Nobel Prize in 1910 for the real gas law, $\left(P + (n^2a/V^2)\right)(V - nb) = nRT$, introducing two parameters a and b which were specific to the gas in question: a represents intermolecular attraction and b an effect due to the size of the gas molecules. The parameters a and b were natural constants that had to be estimated for each real gas, such as helium, oxygen, methane, and so on.

Similarly, one may expect that individuals differ somewhat in the extent to which a unit change in a latent judgment variable will produce a change in a rating on a rating scale, which effect is represented by a factor loading. There may be additional natural constants that reflect individual biases in their ratings, such as leniency and stringency.

My own position is that both chi-square tests of exact fit and indices of approximation provide useful information and should be attended to. However, a failed chi-square test does not necessarily imply that a model is seriously misspecified, nor does a good approximation necessarily imply a partly correct model. It is a matter of scientific judgment which is the case. Because sciences are self-correcting in time as researchers approach the phenomena with different conceptions and perspectives, new experimental designs will yield new data, and errors of judgment will eventually be corrected, either by the researchers themselves or by other researchers.

In structural equation modeling (SEM), the hypothesized models are complex. They have many, sometimes, hundreds of parameters. Ideally a high proportion of these parameters are prespecified by theory. Additionally, the models are studied in certain settings, of which all relevant influences on the outcomes of the observed variables are not always fully known. Some influences may be so small in effect as to have not been previously detected.

We may not be fully aware of causal heterogeneity among experimental subjects, where stimuli have different kinds of effects on different subjects, or words in test items have different idiosyncratic meanings for different subjects. Our models express only linear relations, whereas the causal relations may be nonlinear but monotonic in their effects, so that the linear relations are only approximations. In any experimental or observational setting there are also numerous background conditions that may influence our outcomes, and these may not be modeled (James et al., 1982; Mulaik and James, 1995).

When we try to formulate hypotheses, we try tentatively to identify the principal, obvious causes of their chosen effect variables, along with any other causes. We also try to have indicators of these causal variables. We fix and constrain certain parameters in our models to specify hypotheses about the relations of the causes to the various effect variables. Often we express these as zero coefficients, indicating which variables are *not* causes of other variables. But we can even specify nonzero values, if prior experience or theory is sufficient to provide values. So, the logic of the chi-square test of the hypothesized model goes like this:

If H_1 & $\cdots$ & H_p & B_1 & $\cdots$ & B_m & D_1 & $\cdots$ & D_k are true then $T < \chi^2_{df(0.05)}$.

$H_1, \ldots, H_p$ are hypotheses represented by fixing or constraining certain parameters in the model. Free parameters are not part of the hypothesis, but filler in the model. They are free because the researcher has no knowledge by which to specify their values. They are to be estimated in such a way as to minimize lack of fit of the model conditional on the hypothesized constraints. Thus, if there is any lack of fit, it will be attributed to the fixed and constrained parameters. $B_1, \ldots, B_m$ are background conditions assumed to be the case for the experiment. $D_1, \ldots, D_k$ are probabilistic distributional assumptions made for performing statistical analysis with the data. T is the chi-square test statistic referred to a critical value of the chi-square distribution with df degrees of freedom at the 0.05 level of significance. If all the hypotheses, background assumptions, and distributional assumptions are true, then chi-square should be less than the critical value of chi-square.

Now suppose we observe that it is false that $T < \chi^2_{df(0.05)}$. This then falsifies the joint condition on the left-hand side of the expression. But it does not mean that every one of the assertions on the left-hand side are false. There may be only one condition, say it is a hypothesis about a parameter value, or an assumption about a background condition or a statistical assumption that is false. Then the whole joint expression on the left is negated. Or it could be any number of hypotheses, background assumptions, and statistical assumptions that are false. The chi-square test does not indicate what is false, only that something, somewhere is probably false (Mulaik and James, 1995). So, one must not exaggerate what failure of the chi-square test implies.

The failed chi-square should stimulate an effort to discover the reason for the lack of fit. Most structural equation computer programs have diagnostic information that may be examined for clues to the source of lack of fit. If one believes that there was strong reason to believe the causal framework was correct, then one may consider using LM tests, which are single-degree-of-freedom chi-square tests applied to individual constrained parameters to test whether the constraint contributes to significant lack of fit compared with leaving the parameter free. An LM test should be performed on each constraint, their chi-squares ordered in magnitude. Then the parameter or constraint with the largest significant chi-square value should be freed, and the model reanalyzed. Because the LM tests are not independent, a new set of LM tests is then performed on the remaining constraints, and again the parameter with the largest significant chi-square is freed, and the model reanalyzed again. One should attempt to justify freeing each parameter with theoretical reasons. And one should be judicious in freeing parameters, and not free too many, for the resulting model will have fewer degrees of freedom and represent a less testable model. I do this, hoping to increase the CFI above a value of 0.95 with as few freed parameters as possible. If I can get a nonsignificant chi-square, then all the better.

But residuals sometimes have patterns in them that suggest the presence of other latent variables not included in the model. This can lead to model revisions, but these should have as few freed parameters as possible, because at this point, by consulting the data and freeing parameters, we are not testing hypotheses but exploring. Any parameter that is changed as a result of seeing the data should be treated as a freed parameter unless there is an independent reason not tied to the data for making the change.

Failure of distributional assumptions can be detected by various measures supplied with the computer programs. To test for departures from multivariate normality, one can use Mardia's measure of multivariate kurtosis. Normality is disallowed if the distribution is too light or too heavy in the tails. The raw data itself may be examined for outliers. Of course, if the variables are categorical and not continuous, then departures from multivariate normality are already assured.

But what about the argument that the chi-square test may signal that there is a causal misspecification in the model and accepting an approximation would only seriously obscure this fact? We have already indicated that this same argument would undermine accepting a model with a nonsignificant chi-square, since accepting the model might obscure the misspecification. But there is another reason not to be overly concerned by such arguments. Models should be constructed on the basis of all that is previously known about the phenomenon to be modeled as well as with mechanisms and processes known to exist in other cases in the world. Every effort should be made to rule out alternative models on real-world grounds at this point. We are not simply fitting mathematical models to the numbers in covariance matrices. Thus the mere fact that we can imagine alternative mathematical models that might fit the same data is not sufficient for these to defeat a model that fits exactly or even to a high degree of approximation—if this model has sound, real-world bases for its structure and parameter constraints. The alternative model should have independent real-world foundations for its advocacy before it is taken seriously. The alternative model should also have as many or more degrees of freedom to be a legitimate defeater. Otherwise, alternative imaginary mathematical models with weaker constraints can be used to defeat and undermine any model that fits "exactly" or even approximately. These I call "boogieman models" because they are often invoked by skeptics with an axe to grind against structural equation modeling or the use of approximate fit indices. But like "boogiemen" they have no basis in reality, only in imagination.

Conclusion

Modeling in science is an iterated three-phase process of (1) *abduction* or hypothesis formulation, (2) *deduction*, then (3) *induction* or hypothesis testing (Peirce, 1931–1958). At the outset, in *abduction* one encounters a surprising or little-understood phenomenon in experience, and then after gathering preliminary data about the phenomenon, seeks to consider various possible hypotheses to account for them, evaluating these by how well they conform to the already known or given data. The hypothesis that accounts for the known data in the most parsimonious way is then regarded as one's best "guess" as to the explanation for the phenomenon and the other hypotheses are set aside. But this does not mean that the hypothesis finally chosen is confirmed. It is still only a guess that fits the data that stimulated formulating the hypothesis. One must now take the hypothesis, analyze it, and attempt to deduce possible consequences of it in future experience. This represents the second or *deductive* phase. One then seeks to deduce possible experiments or observations not used in the formulation of the hypothesis that would put the hypothesis to a test. At this point one constructs a model for the anticipated data. Having done that, one then enters the third phase of *induction* or hypothesis testing, where the hypothesized model is compared to data collected by design for the purposes of testing the hypothesis. If the model fits acceptably, this lends provisional support to the hypothesis, but this does not mean it is finally and absolutely confirmed. Further deductions may lead to other models that further test the hypothesis, and other experiments and observations may be performed to test these new deduced consequences. On the other hand, if the initial hypothesis fails to fit acceptably, this leads to a new cycle of abduction, deduction, and induction, where one seeks to refine and revise the model in the light of the initial and subsequent knowledge of the phenomenon. Often the model only achieves an approximation and not a fit to within sampling error. This should still be a sign that something is wrong somewhere in the model. At the same time, one may feel that the principal causes in the model may be appropriate or part of the causes of the data, but one will also need to deduce new consequences formulated into a new, revised model to account for the previous discrepancies as well. The deduction of a revised model will be followed by testing the revised model against new data. This three-phase process goes on and on. But there are other considerations behind this approach.

The philosophy on which my approach to model evaluation is based is the idea that science concerns objective knowledge, that is, knowledge of objects. Objects are invariants in the world that are independent of particular points of view, means of observation, indicators, laboratories, researchers, instruments, personal prejudice, and so on. Objects are perceived by us as having attributes that we order in variables. Causes are functional relations between attributes either within the same object or between objects. Scientists establish causes as invariant functional relations between variables that hold for particular sets of objects under certain conditions. Scientists establish invariants by asserting them as hypotheses or models and testing them against data not used in their formulation. This is essential to establishing the results as independent of the researcher and invariant across whatever prior knowledge of the world he or she possesses in formulating the hypothesis and the data used to test the hypothesis. Invariants also should be found in numerous contexts, and modules of invariants should be translatable from one setting to another and combinable with other modules and retain their invariant properties.

Hypotheses are tested by comparing the model of the data to the data itself. Comparisons may be made in terms of distances between model and data, and/or probabilistically, as to whether the data differ to such a degree in distance and so improbably if the model were true as to strain credibility or not. This requires criteria of when to accept the model provisionally or not. Merely because one model is closer to the data than another is not sufficient to accept it provisionally. A model to be accepted must be within a reasonably close distance of the data and reasonably probable under the hypothesis. So, this approach is not based on generating numerous models and then testing each one and picking the one among them that is merely closest to the data. Closeness must be quite close. And none of the models compared may even fall within the bounds of closeness set by the researcher for provisional acceptance of the model. In large samples closeness may become more important than probability, since sampling error diminishes in magnitude with increases in sample size. Furthermore, probability distributions and models are idealizations and may not capture the fine-grained features of the data visible in high resolution from huge samples. One may not have causally homogeneous subjects so that for some subjects causal variables other than those hypothesized may function, or their structural parameters may vary in value from those predominating in the population. Thus approximations may still be useful and bases for further studies that seek to refine these models further so as to account for the information revealed in high resolution not accounted for by the model.

Now, GFIs are ways of evaluating the distance from the data (say, as measured by a noncentrality parameter) and the probability of the data under the hypothesis. But no model should be accepted merely because it comes within the bounds of provisional acceptance. As Browne and Cudeck (1993) have said, final decisions to accept or reject a model depend on a judgment of many factors and should not be made mechanically on the basis of fit indices.

Specifically, are there causal mechanisms in the world that correspond to the abstract form of the model? Furthermore, fit or closeness alone is not sufficient. The model must be testable and potentially disconfirmable, for this is the mark of a test of an invariant. And the researcher should remain ever clear about what is tested and what is generated as hypothesis for future studies.

To help gauge the disconfirmability of a model, Mulaik (2001) has shown that degrees of freedom are the number of dimensions in which data are free to differ from an incompletely specified model that estimates unspecified and unknown parameters from the data, conditional on the constraints of the hypothesis. So, degrees of freedom can be the basis of an assessment of disconfirmability. But degrees of freedom of a model need to be considered against the maximum possible degrees of freedom, which is given by the number of distinct data points that the model is designed to fit. Each data point presents a potential dimension in which the model may differ from the data, but estimating parameters consumes some of these dimensions and arrogates them to the parameter estimation process. Each parameter estimated takes away a degree of freedom. A PR may then be calculated to determine what portion of the data's dimensions has actually been used to test the model. This involves the ratio of the degrees of freedom of the model to the "degrees of freedom of the data," given by the number of data points to fit.

The PR may be multiplied by a "GFI" that ranges between 0 and 1 to result in an index of model quality. Examples of this are to multiply the PR with the CFI. The result combines a pure measure of fit with a measure of disconfirmability. Lack-of-fit indices such as the estimated noncentrality parameter, or the RMSEA, may be transformed to a GFI ranging from 0 to 1 by the negative exponential function. These too then may be multiplied by the PR to obtain an index of model quality. Model quality criteria should be set high, say, 0.85 or higher. This will mean that a highly disconfirmable model has been shown to have a very high degree of fit to the data. Models with lower PR-adjusted good fit, but high goodness of fit should be evaluated in terms of what was tested and not in terms of the "whole model." Beyond this the PR may be considered separately, for example when evaluating a chi-square statistic, to determine what proportion of the model's parameters was tested by the model constraints.

At the present, the information theoretic indices seem not to be based on the philosophy of establishing objective invariants but on model comparison and selection of the best-fitting model from a group of models, merely on its having the smallest value. These methods have an abductive aspect to them. But *close* may not be *close enough*. Hypothesis testing and hypothesis formulation also become confounded in this framework. Although proponents of information theoretic indices claim to penalize for parameters estimated, these penalties diminish in their impact as sample sizes get larger. They tend in larger samples then to treat estimated parameters as if they are correct parameters because their standard errors become so small that the penalty against them can vanish with increasing sample sizes. Information theoretic approaches do not take into account the possibility of numerous equivalent models in the parts of models with estimated parameters, which undermines the uniqueness of the estimated parameters as objective values. Estimated parameters must be interpreted as hypotheses for future studies, not tested with samples from the same population, but as possible invariants in modules of variables translated to new contexts to establish their invariance. Cross-validation to samples from the same population does not test for objective invariance. What cross-validation measures is sampling stability, and this can usually be best assessed with a single large sample than with two half samples. Cross-validation to random samples will not detect the presence of mixed distributions, of nonhomogeneous populations, especially in larger and larger samples.

Objectivity requires "causal homogeneity" of objects studied in order to establish invariant functional relations across objects and their attributes. This means in factor analysis that subjects should all perform according to the same latent variables and to the same degree of change in observed variables for a unit change of a latent variable. Factor loadings should be invariant across subjects. Otherwise, the factor loadings obtained will be an averaging, so to speak, of the factor loadings of each individual. Such averages may not hold up in modules of variables or sets of individuals taken from one context to another. Establishing "causal homogeneity" is difficult and methodology has not yet been well developed for this task, because it is not a well-recognized problem.

Finally, objectivity should be understood as not a condition in which absolute truth has been established, but as a relative condition that changes with advancing knowledge and more widely applicable models, newer more encompassing theories and wider ranges of data to which they are simultaneously applied. Objectivity also changes with increasing awareness of our roles as subjects in influencing the forms of our data and our conceptions of them, provoking us to modify our models to account for these extraneous influences on our part, for we too are objects in the world interacting with the objects under study. We know too from our personal knowledge of objects that there is always something new, some aspect not before perceived, that may be revealed, changing our view and concept of an object. Much of this philosophy is treated in expanded form in Mulaik (2004).

16

Polychoric Correlation and Polyserial Correlation

Introduction

The theory of factor analysis and structural equation modeling has developed under the assumption that the observed variables have continuous, multivariate normal distributions. But much of the data obtained in the behavioral and social sciences does not involve continuous variables and hence cannot have multivariate normal distributions. For example, responses may be binary and bipolar as

Agree, Disagree

or trinary and bipolar as

Agree, No opinion, Disagree

or ordinal and unipolar as

Never, Rarely, Occasionally, Often, Always

or ordinal and bipolar as

Extremely, Strongly, Somewhat, Neutral, Somewhat, Strongly, Extremely or

Strongly agree	Somewhat agree	Slightly agree	Indifferent	Slightly disagree	Somewhat disagree	Strongly disagree

Ordinal numbers are then assigned to these ordered categories to make quantitative variables.

Conducting analyses with covariances computed from the raw scores on the categorical variables can lead to biased estimates of factor loadings, and in certain cases to improper determinations of the number of factors, especially if the variables are skewed in opposite directions (Olsson, 1979). This will occur if in fact the observed variables are themselves categorizations of underlying latent variables that pairwise have joint bivariate normal distributions. In other words, we should be analyzing the covariances or correlations between these latent variables and not the raw, observed variables. Human judgments in the nervous system may be more nearly continuous in value than the raw categories of binary Agree–Disagree variables or polytomous Likert scale rating variables suggest. And these underlying continuous variables may have joint bivariate normal distributions (which may be a fairly strong substantive assumption). But if these assumptions about underlying latent variables corresponding to our observed variables are reasonable, then we may proceed to first estimate the correlations among these latent variables, and then perform analyses of our models on the resulting correlations.

Polychoric Correlation

Karl Pearson (1901) first proposed polychoric correlation in connection with dichotomous raw variables. In *Biometrika* in 1922 Karl Pearson and his son Egon Pearson generalized the method originally developed by Pearson in 1901 to "polychoric correlation coefficients." But in the days of mechanical calculators the problem was a difficult one to solve with accuracy, and so frequently various approximations were proposed. Olsson (1979) put forth a classic paper on the estimation of the polychoric correlation coefficient by maximum-likelihood estimation. This paper has been a jumping-off point for further developments in the literature. To familiarize ourselves with this method, we will discuss Olsson's paper.

To begin, assume that we observe two ordinal, categorized variables x and y. Variable x has s categories and y has r categories. Their sample joint frequency distribution is given in Table 16.1. Next, assume that corresponding to x and y are two latent variables x and h that have a joint bivariate unit normal distribution. The correspondence between these two sets of variables is given by the manner in which threshold values on the latent variables define

TABLE 16.1

The Form of the Observed Data for the
Polychoric Correlation Problem

	η		b_1	b_1	$\cdots$	b_{r-1}
ξ ╲ y / x		1	2	3	$\cdots$	r
	1	n_{11}	n_{12}	n_{13}	$\cdots$	n_{1r}
a_1	2	n_{21}	n_{22}	$\cdots$	$\cdots$	n_{2r}
a_2	3	n_{31}	$\cdots$	$\ddots$	$\cdots$	n_{3r}
$\vdots$	$\vdots$	$\vdots$	$\vdots$	$\vdots$	$\ddots$	$\vdots$
a_{s-1}	s	n_{s1}	n_{s2}	n_{s3}	$\cdots$	$n_{s,r}$

the category boundaries of the observed variables:

$$x = 1 \quad \text{if } \xi < a_1 \qquad\qquad y = 1 \quad \text{if } \eta < b_1$$
$$x = 2 \quad \text{if } \xi < a_1 \le \xi < a_2 \qquad y = 2 \quad \text{if } b_1 \le \eta < b_2$$
$$x = 3 \quad \text{if } a_2 \le \xi < a_3 \qquad y = 3 \quad \text{if } b_2 \le \eta < b_3$$
$$\vdots \qquad\qquad\qquad\qquad\qquad \vdots$$
$$x = s \quad \text{if } a_{s-1} \le \xi \qquad\qquad y = r \quad \text{if } b_{r-1} \le \eta$$

There is one less threshold value for a latent variable than the number of categories on its corresponding observed variable. The random variable x has category thresholds of $a_1, a_2, a_3, \ldots, a_{s-1}$. The random variable h has category thresholds of $b_1, b_2, b_3, \ldots, b_{r-1}$.

In Table 16.1 we show the format of the raw data for the polychoric correlation problem. The first row and column of the table show the latent variables x and h and their respective threshold values a_i and b_j. The second row and column, excluding the first element in each, indicate the observed polytomous variables x and y and their respective values. Note how the threshold values on the latent variables are positioned. See how, for example, the first threshold value a_1 for the variable x is in the row corresponding to the value $x = 2$. The threshold values correspond to the boundary lines between the values of the observed variables. The value of an observed variable corresponds to the values of its corresponding latent variable that are greater than or equal to the threshold value shown for it, while always less than the value of the threshold shown with the next larger value of the observed variable. The table also shows cross-tabulations of the n's or frequencies in the sample for each joint occurrence of different values of the observed variables. All the n's add up to N, the sample size.

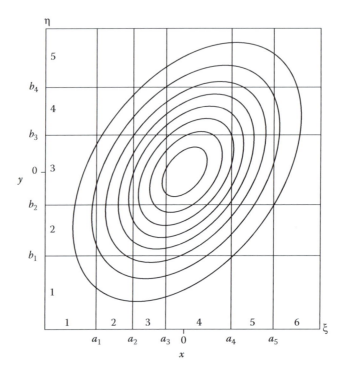

FIGURE 16.1   Contour plot of a bivariate normal distribution for variables x and h with means of zero, variances of 1, $\rho = 0.4$, and threshold lines for determining values of polytomous variables x and y and their values shown.

In Figure 16.1 we show a contour plot of a bivariate normal distribution for two latent standardized continuous variables ξ and η with corresponding polytomous variables x and y. The horizontal and vertical lines in Figure 16.1 represent the thresholds on the continuous variables, which we have labeled with a's and b's. We will assume that $a_0 = \infty$, $b_0 = \infty$, $a_s = \infty$, and $b_r = \infty$. All values of x such that $a_{i-1} \leq \xi < a_i$ equal i, and all values of η such that $b_{j-1} \leq \eta < b_j$ equal j. The values of the observed variables are shown in the left and bottom margins. The threshold lines for latent variable x intersect with threshold lines of latent variable η to form cells (i, j), $i = 1, 2, \ldots, s$; $j = 1, 2, \ldots, r$. The probability that an observation on the observed variables (x, y) will fall within the cell (i, j) is given by π_{ij}.

Overview

The observed relative frequencies $P_{ij} = n_{ij}/N$ of the cases in each cell (i,j) representing the joint values of the observed variables x and y are the raw data. We seek to find a standardized bivariate unit normal distribution for which the relative frequencies P_{ij} jointly have the greatest likelihood across all cells.

Roughly speaking, this corresponds to seeking to fit a bivariate normal distribution to the cells such that the probabilities under the bivariate normal distribution of falling within the respective cells are as much as possible similar to the observed relative frequencies. Since the means and variances of the bivariate unit normal distribution will be fixed to 0 and 1, respectively, to achieve identification, the only parameter governing the variation in the distribution's shape is the correlation ρ between the latent variables. Thus by varying ρ in a systematic manner, we will come to find the value at which the probabilities under the distribution best fit the cells' relative frequencies. In the first method of estimation, we will also have the freedom to adjust the thresholds to vary the cell boundaries. In the second method of estimation, the thresholds will be determined first and will remain fixed throughout variations in ρ.

Derivation

We will assume for the present problem that the means of the latent variables are zero, and the variance of each variable is unity. This establishes an identified solution for ρ. The bivariate normal density function for variables whose means are zero and variances are unity is given by (Biswas, 1991, p. 398):

$$\phi(\xi, \eta; \rho) = \frac{1}{2\pi\sqrt{1 - \rho^2}} \exp\left[-\frac{1}{2(1 - \rho^2)}(\xi^2 - 2\rho\xi\eta + \eta^2) \right]. \qquad (16.1)$$

The cumulative bivariate normal distribution in this case is given by

$$\Phi(\xi, \eta; \rho) = \frac{1}{2\pi\sqrt{1 - \rho^2}} \int_{-\infty}^{\xi} \int_{-\infty}^{\eta} \exp\left[-\frac{1}{2(1 - \rho^2)}(\xi^2 - 2\rho\xi\eta + \eta^2) \right] d\xi \, d\eta.$$

$$(16.2)$$

The derivation of the likelihood equations for estimating ρ is then given as follows: let π_{ij} be the probability that a random observation falls into the cell (i, j). Then the joint likelihood of the sample is given by

$$L = C \cdot \prod_{i=1}^{s} \prod_{j=1}^{r} \pi_{ij}^{n_{ij}}. \qquad (16.3)$$

Note that the reason π_{ij} is raised to the power n_{ij} is because there are n_{ij} cases falling in the cell (i, j), each with likelihood π_{ij}, so, when these are all multiplied together, that results in $\pi_{ij}^{n_{ij}}$. Then these products from each of the cells are in turn multiplied together. C is a constant.

For subsequent differentiation, it is easier to work with the logarithm of L, which is a monotonic transformation that is maximized at the same parameter value(s) as for L:

$$l = \ln L = \ln C + \sum_{i=1}^{s}\sum_{j=1}^{r} n_{ij} \ln \pi_{ij}. \tag{16.4}$$

From the thresholds and the cumulative bivariate normal distribution in Equation 16.2

$$\pi_{ij} = \Phi(a_i, b_j) - \Phi(a_{i-1}, b_j) - \Phi(a_i, b_{j-1}) + \Phi(a_{i-1}, b_{j-1}). \tag{16.5}$$

If one has not worked with cumulative distributions, Equation 16.5 is understood as follows. Suppose we wish to find π_{52}. This is given by

$$\pi_{52} = \Phi(a_5, b_2) - \Phi(a_4, b_2) - \Phi(a_5, b_1) + \Phi(a_4, b_1)$$

π_{52} is the probability of an occurrence of (x, y) in the cell $(5, 2)$. But to obtain this value from the cumulative probability distribution in Equation 16.2, we need to add and subtract the probabilities in certain areas defined by the limits of integration in the formula. The expression $\Phi(5, 2)$ denotes the probability in the rectangular area defined by $(-\infty, -\infty)$ in its lower left corner and $(5, 2)$ in its upper right corner. Of course, that takes in too much area, because we want just the probability in the cell $(5, 2)$. So, next, we subtract the probability $\Phi(4, 2)$. This is the probability of falling within the rectangular area defined by $(-\infty, -\infty)$ in its lower left corner and $(4, 2)$ in its upper right corner. But that still leaves the total probability in cells $(5, 2)$ and $(5, 1)$. So, we next subtract $\Phi(5, 1)$. But this subtracts out not only the probability in the cell $(5, 1)$, but in cells $(1, 1)$, $(2, 1)$, $(3, 1)$, and $(4, 1)$. We need to cancel this extra probability subtracted by adding it back in. The portion to add back in is given by $\Phi(4, 1)$. The result is the probability π_{52} in the cell $(5, 2)$.

There are two methods at this point for how to proceed to obtain maximum-likelihood estimates of the value of ρ. There are several parameters that need to be estimated to make the log likelihood a maximum. These are the various threshold values a_i, $i = 1, \ldots, s - 1$ and b_j, $j = 1, \ldots, r - 1$ and ρ. One method, which we will describe here, is theoretically optimal. It estimates all parameters simultaneously. But this method is also computationally more intensive while being more accurate. The second method first estimates the marginal probabilities of each value of the observed variables and then computes the thresholds from these marginals. This method is less computationally intensive and almost as accurate, except when ρ is large.

The reason why the derivatives are obtained is that we will use them as indications of the slope of the likelihood function in a given parameter's direction of increasing values at a given point in parameter space. The likelihood function is well-behaved in having derivatives that are everywhere continuous. This allows us to seek the extrema or maximum or minimum points of the function where the slope(s) or derivative(s) change sign and pass through zero. In seeking to maximize the likelihood function, we seek the "top of the mountain," so to speak, where in all directions in parameter space the slope is zero or horizontal. When the slope(s) of the (partial) derivative(s) in the parameter direction(s) is (are) zero, we are at a maximum or a minimum. That is why we will obtain the expression(s) for the derivative(s) and set it (them) equal to zero and solve for the parameter values at which the derivative(s) is (are) zero. Because the derivatives themselves may be nonlinear, we may not be able to solve these equations algebraically, but will have to use numerical methods that solve them computationally by iteration. We will also obtain the formulas for the second derivatives of the likelihood function. These formulas tell us how much and in what direction (up or down) the slope is changing at a given point in parameter space. If the second derivative of the function with respect to a parameter is negative at an extreme point, it suggests that the extreme point is at a maximum. At a minimum point, the second derivative is positive. The second derivative(s) are also used in the iterations of the Newton–Raphson method to calculate an optimal correction to the current parameter estimate to produce a new parameter estimate.

Estimating All Parameters Simultaneously

This is a maximization problem in several dimensions. In general, the strategy of the simultaneous method is to iteratively adjust values of ρ, and the a_i's and b_j's until they and the underlying standard bivariate unit normal distribution are positioned in such a way as to maximize the log likelihood in Equation 16.4. Therefore, we need to find initially expressions for the partial derivatives of the log-likelihood function with respect to each of the unknown parameters. Toward this end, we first obtain

$$\frac{\partial l}{\partial a_k} = \sum_{i=1}^{s}\sum_{j=1}^{r} \frac{n_{ij}}{\pi_{ij}}\frac{\partial \pi_{ij}}{\partial a_k}, \quad k = 1, \ s-1 \tag{16.6}$$

$$\frac{\partial l}{\partial b_m} = \sum_{i=1}^{s}\sum_{j=1}^{r} \frac{n_{ij}}{\pi_{ij}}\frac{\partial \pi_{ij}}{\partial b_m}, \quad m = 1, \ r-1 \tag{16.7}$$

$$\frac{\partial l}{\partial \rho} = \sum_{i=1}^{s}\sum_{j=1}^{r} \frac{n_{ij}}{\pi_{ij}}\frac{\partial \pi_{ij}}{\partial \rho}. \tag{16.8}$$

Taking derivatives of Equation 16.5 with respect to a_k, we note that in some cases none of the expressions contain a_k, while in the others, only two do. Any expression that does not contain the parameter with respect to which we take the derivative is zero.

$$\frac{\partial \pi_{ij}}{\partial a_k} = \begin{cases} 0 & \text{if } \pi_{ij} \text{ does not contain } a_k \\[2mm] \dfrac{\partial \Phi(a_k, b_j)}{\partial a_k} - \dfrac{\partial \Phi(a_k, b_{j-1})}{\partial a_k} & \text{if } k = i \\[4mm] \dfrac{\partial \Phi(a_k, b_j)}{\partial a_k} + \dfrac{\partial \Phi(a_k, b_{j-1})}{\partial a_k} & \text{if } k = i - 1 \end{cases} \tag{16.9}$$

Substituting Equation 16.9 into Equation 16.6, and noting that the only cases to consider are those where i goes from k to $k + 1$, we may then rewrite Equation 16.6 as

$$\frac{\partial l}{\partial a_k} = \sum_{j=1}^{r} \left(\frac{n_{kj}}{\pi_{kj}} - \frac{n_{k+1,j}}{\pi_{k+1,j}} \right) \left(\frac{\partial \Phi(a_k, b_j)}{\partial a_k} - \frac{\partial \Phi(a_k, b_{j-1})}{\partial a_k} \right). \tag{16.10}$$

We now take advantage of the rule that $d(\int_a^x f(t)\,dt)/dx = f(x)$. Olsson (1979) cites Tallis (1962, p. 346) for showing that

$$\frac{\partial \Phi(u, v)}{\partial u} = \phi(u) \cdot \Phi \left\{ \frac{(v - \rho u)}{\sqrt{(1 - \rho^2)}} \right\}. \tag{16.11}$$

Here $\phi(u)$ (with one argument) denotes the unit normal density function, while $\Phi(u)$ (again with one argument) denotes the cumulative unit normal density function.

Olsson (1979) then derives

$$\frac{\partial l}{\partial a_k} = \sum_{j=1}^{r} \left(\frac{n_{kj}}{\pi_{kj}} - \frac{n_{k+1,j}}{\pi_{k+1,k}} \right) \cdot \phi(a_k) \cdot \left[\Phi \left\{ \frac{(b_j - \rho a_k)}{\sqrt{1 - \rho^2}} \right\} - \Phi \left\{ \frac{(b_{j-1} - \rho a_k)}{\sqrt{1 - \rho^2}} \right\} \right] \tag{16.12}$$

$$\frac{\partial l}{\partial b_m} = \sum_{j=1}^{r} \left(\frac{n_{im}}{\pi_{kim}} - \frac{n_{i,m+1}}{\pi_{i,m+1}} \right) \cdot \phi(b_m) \cdot \left[\Phi \left\{ \frac{(a_i - \rho b_m)}{\sqrt{1 - \rho^2}} \right\} - \Phi \left\{ \frac{(a_{i-1} - \rho b_m)}{\sqrt{1 - \rho^2}} \right\} \right]. \tag{16.13}$$

Note that

$$\phi(a_k) = \frac{1}{\sqrt{2\pi}} \exp(-a_k^2/2) \tag{16.14}$$

$$\Phi(u) = \frac{1}{\sqrt{2\pi}} \int_{-\infty}^{u} \exp(-t^2/2)\,dt. \tag{16.15}$$

Computing the univariate normal cumulative probability, given u, is usually done today with subroutines in libraries of mathematical computer subroutines.

On the other hand, again following Tallis (1962, p. 344), who showed that

$$\frac{\partial \Phi(u, v)}{\partial \rho} = \phi(u, v) \tag{16.16}$$

where $\phi(u, v)$ (with two arguments) is the bivariate unit normal density function with correlation ρ, Olsson (1979) applied this result to Equation 16.5 to obtain

$$\frac{\partial \pi_{ij}}{\partial \rho} = \phi(a_i, b_j) - \phi(a_{i-1}, b_j) - \phi(a_i, b_{j-1}) + \phi(a_{i-1}, b_{j-1}). \tag{16.17}$$

We can now substitute Equation 16.17 into Equation 16.8 to obtain

$$\frac{\partial l}{\partial \rho} = \sum_{i=1}^{s} \sum_{j=1}^{r} \frac{n_{ij}}{\pi_{ij}} \left[\phi(a_i, b_j) - \phi(a_{i-1}, b_j) - \phi(a_i, b_{j-1}) + \phi(a_{i-1}, b_{j-1}) \right]. \tag{16.18}$$

Equations 16.12, 16.13, and 16.18 thus constitute the first-order partial derivatives of the log-likelihood equation in Equation 16.4. The values of the expressions on the right in Equations 16.17 and 16.18 correspond generally (subscripts omitted) to

$$\phi(a, b) = \frac{1}{2\pi\sqrt{1 - \rho 2}} \exp\left[-\frac{1}{2(1 - \rho^2)}(a^2 - 2\rho ab + b^2) \right] \tag{16.19}$$

computed at the current values of ρ, a, and b, respectively, in the current iteration.

We will postpone for the time being discussing an algorithm for solving for ρ.

Two-Stage, Conditional Maximum-Likelihood Method

Attempting to solve simultaneously for both ρ and the thresholds a_i, $i = 1, \ldots, s - 1$ and b_j, $j = 1, \ldots, r - 1$ was deemed too computationally intensive by early researchers. Pearson and Pearson (1922), Lancaster and Hamdan (1964), and Martinson and Hamdan (1971) presumed that the thresholds were given by the cumulative marginal proportions in the cross-tabulation table. This produces good approximations for ρ and the threshold values. Jöreskog (1994) asserted that this method also has the advantage of consistently using the same thresholds for a variable as it is paired with other variables. Allowing the thresholds of a variable to vary across different other variables with which it is paired assumes interactions between the variables that should otherwise be in the SEM. Furthermore, Jöreskong points out that the thresholds produced by this method are maximum-likelihood estimates.

According to Olsson (1979), the equation system analogous to that in Equations 16.12, 16.13, and 16.15 to be solved after setting the derivative to zero is given by

$$\frac{dl}{d\rho} = \sum_{i=1}^{s}\sum_{j=1}^{r} \frac{n_{ij}}{\pi_{ij}} \left[\phi(a_i, b_j) - \phi(a_{i-1}, b_j) - \phi(a_i, b_{j-1}) + \phi(a_{i-1}, b_{j-1}) \right] = 0$$

$$(16.18a)$$

where the threshold values are given as the values of a unit normal random variable that would correspond to the cumulative marginal proportions:

$$a_i = \Phi^{-1}(P_{i\cdot}), \qquad (16.20)$$

$$b_j = \Phi^{-1}(P_{\cdot j}), \qquad (16.21)$$

where $P_{ij} = n_{ij}/N$ is the proportion in the cell (i, j), and

$$P_{i\cdot} = \sum_{k=1}^{i}\sum_{j=1}^{r} P_{kj} \qquad (16.22)$$

and

$$P_{\cdot j} = \sum_{i=1}^{s}\sum_{k=1}^{j} P_{ik}. \qquad (16.23)$$

Computing Estimates

In the *simultaneous estimation procedure*, there are several numerical algorithms that might be applied to find the estimate of the polychoric correlation ρ and the threshold parameters $a_i, i = 1, \ldots, s - 1$ and $b_j, j = 1, \ldots, r - 1$. These include the Newton–Raphson method, the conjugate gradient method, and the Fletcher–Powell–Davidon method. We have already discussed these methods in Chapter 15. Here we will consider only the Newton–Raphson method. Let $q' = (\rho, a_1, a_2, \ldots, a_{s-1}, b_1, b_2, \ldots, b_{r-1})$ be a $p \times 1$ vector of p parameters. Some suitable initial solution for the parameter vector is denoted as $\hat{q}_{(0)}$.

The Newton–Raphson iterative solution is given by the formula

$$
\begin{bmatrix} \theta_1 \\ \theta_2 \\ \vdots \\ \theta_p \end{bmatrix}_{(j+1)} = \begin{bmatrix} \theta_1 \\ \theta_2 \\ \vdots \\ \theta_p \end{bmatrix}_{(j)} - \begin{bmatrix} \dfrac{\partial^2 l}{\partial \theta_1 \partial \theta_1} & \dfrac{\partial^2 l}{\partial \theta_1 \partial \theta_2} & \cdots & \dfrac{\partial^2 l}{\partial \theta_1 \partial \theta_p} \\[2mm] \dfrac{\partial^2 l}{\partial \theta_2 \partial \theta_1} & \dfrac{\partial^2 l}{\partial \theta_2 \partial \theta_2} & \cdots & \dfrac{\partial^2 l}{\partial \theta_2 \partial \theta_p} \\[2mm] \vdots & \vdots & \ddots & \vdots \\[2mm] \dfrac{\partial^2 l}{\partial \theta_p \partial \theta_1} & \dfrac{\partial^2 l}{\partial \theta_p \partial \theta_2} & \cdots & \dfrac{\partial^2 l}{\partial \theta_p \partial \theta_p} \end{bmatrix}_{(j)}^{-1} \begin{bmatrix} \dfrac{\partial l}{\partial \theta_1} \\[2mm] \dfrac{\partial l}{\partial \theta_2} \\[2mm] \vdots \\[2mm] \dfrac{\partial l}{\partial \theta_p} \end{bmatrix}_{(j)} .
$$

(16.24)

$$
\boldsymbol{\theta}_{(j+1)} = \boldsymbol{\theta}_{(j)} - \mathbf{H}_{(j)}^{-1}\mathbf{g}_{(j)}
$$

or

$$
\boldsymbol{\theta}_{(j+1)} = \boldsymbol{\theta}_{(j)} - \mathbf{H}_{(j)}^{-1}\mathbf{g}_{(j)}.
$$

The vector $\boldsymbol{\theta}_{(j+1)}$ is the new estimated parameter vector produced in iteration (j). The vector $\boldsymbol{\theta}_{(j)}$ is the old estimated parameter vector in iteration (j). The matrix $\mathbf{H}_{(j)}^{-1}$ is the inverse of the matrix of second derivatives of the log-likelihood function with respect to the parameters in q evaluated at $\boldsymbol{\theta}_{(j)}$. $\mathbf{g}_{(j)}$ is the "gradient vector" of first-order partial derivatives of the log-likelihood function with respect to each of the parameters in θ evaluated with values of $\boldsymbol{\theta}_{(j)}$.

Instead of deriving the mathematical expressions for the second derivatives of the log-likelihood function with respect to the parameters, Olsson (1979) used an approximation to $\mathbf{H}$, recommended by Tallis (1962, p. 348), known as the "expected second-order derivatives of l with respect to $\hat{\boldsymbol{\theta}}$":

$$
[\mathbf{I}]_{m,n} = N \sum_{i=1}^{s} \sum_{j=1}^{r} \frac{1}{\pi_{ij}} \left(\frac{\partial \pi_{ij}}{\partial \theta_m} \right) \left(\frac{\partial \pi_{ij}}{\partial \theta_n} \right).
$$

(16.25)

The elements $[\mathbf{I}]_{m,n}$ are computed in each iteration from current values as given by Equations 16.9, 16.11, and 16.17.

In contrast to the simultaneous estimation method, the *two-stage conditional maximum-likelihood method* is much simpler to compute and is the method preferred by Jöreskog (1994). First the maximum-likelihood estimates of the threshold parameters are estimated by Equations 16.20 and 16.21. Programs for estimating the inverse cumulative normal probability values a_i and b_j in Equations 16.20 and 16.21 are usually available in libraries of mathematical computer subroutines. Obtaining these estimates is a one-time procedure, and they remain as constants throughout the rest of the iterations in solving for ρ.

To apply the Newton–Raphson method to this problem to the single unknown ρ, we will need expressions for the first derivative of l with respect of ρ as well as an expression for the second derivative with respect to ρ. The formula corresponding to Equation 16.24 for the Newton–Raphson method using estimated thresholds based on marginals is as follows:

$$\hat{\rho}_{(j+1)} = \hat{\rho}_{(j)} - \left(\frac{\mathrm{d}l^2}{\mathrm{d}^2\rho}\right)^{-1}_{(j)} \left(\frac{\mathrm{d}l}{\mathrm{d}\rho}\right)_{(j)}. \tag{16.26}$$

The first derivative of the log-likelihood function with respect to ρ is given in Equation 16.18. The derivation of the second derivative of the log likelihood with respect to ρ, however, is complex. It is the derivative of the first derivative given in Equation 16.18:

$$\frac{\mathrm{d}^2 l}{\mathrm{d}\rho^2} = \frac{\mathrm{d}}{\partial\,\mathrm{d}\rho}\left\{\sum_{i=1}^{s}\sum_{j=1}^{r}\frac{n_{ij}}{\pi_{ij}}\left[\phi(a_i, b_j) - \phi(a_{i-1}, b_j) - \phi(a_i, b_{j-1}) + \phi(a_{i-1}, b_{j-1})\right]\right\}$$

$$\tag{16.27}$$

Breaking this into parts and substituting simple expressions for more complex ones, we may work out the second derivative.

Let

$$t_{ij} = \frac{n_{ij}}{\pi_{ij}} \tag{16.28}$$

and

$$u_{ij} = [\phi(a_i, b_j) - \phi(a_{i-1}, b_j) - \phi(a_i, b_{j-1}) + \phi(a_{i-1}, b_{j-1})]. \tag{16.29}$$

Then

$$\frac{\mathrm{d}^2 l}{\mathrm{d}\rho^2} = \frac{\mathrm{d}}{\mathrm{d}\rho}\left\{\sum_{i=1}^{s}\sum_{j=1}^{r}t_{ij}u_{ij}\right\} = \left\{\sum_{i=1}^{s}\sum_{j=1}^{r}\left(t_{ij}\frac{\mathrm{d}u_{ij}}{\mathrm{d}\rho} + u_{ij}\frac{\mathrm{d}t_{ij}}{\mathrm{d}\rho}\right)\right\}. \tag{16.30}$$

Then

$$\frac{\mathrm{d}u_{ij}}{\mathrm{d}\rho} = \frac{\mathrm{d}\phi(a_i, b_j)}{\mathrm{d}\rho} - \frac{\mathrm{d}\phi(a_{i-1}, b_j)}{\mathrm{d}\rho} - \frac{\mathrm{d}\phi(a_i, b_{j-1})}{\mathrm{d}\rho} + \frac{\mathrm{d}\phi(a_{i-1}, b_{j-1})}{\mathrm{d}\rho} \tag{16.31}$$

$$\frac{\mathrm{d}t_{ij}}{\mathrm{d}\rho} = \frac{n_{ij}(-\mathrm{d}\pi_{ij}/\mathrm{d}\rho)}{\pi_{ij}^2}$$

$$= \frac{-n_{ij}[\phi(a_i, b_j) - \phi(a_{i-1}, b_j) - \phi(a_i, b_{j-1}) + \phi(a_{i-1}, b_{j-1})]}{\pi_{ij}^2} \tag{16.32}$$

Since $d\phi(a_i, b_j)/d\rho$ is representative of each of the terms on the right-hand side in Equation 16.31, we will substitute Equation 16.19 to obtain this derivative:

$$\frac{d\phi(a_i, b_j)}{d\rho} = \frac{d}{d\rho}\left[\left(\frac{1}{2\pi(1 - \rho^2)^{1/2}}\right) \cdot \exp\left[-\frac{1}{2(1 - \rho^2)}\left(a_i^2 - 2\rho a_i b_j + b_j^2\right)\right]\right].$$

(16.33)

Let us now make the substitutions $w = 1/2\pi(1 - \rho^2)^{1/2}$ and $v = \exp(-gh)$, where $g = 1/2(1 - \rho^2)$ and $h = (a_i^2 - 2\rho a_i b_j + b_j^2)$.

Then

$$\frac{d\phi(a_i, b_j)}{d\rho} = \frac{d}{d\rho}[(w \cdot v)] = w\frac{dv}{d\rho} + v\frac{dw}{d\rho}.$$

(16.34)

Now,

$$\frac{dv}{d\rho} = \frac{d(e^{-gh})}{d\rho} = -e^{-gh}\frac{d(g \cdot h)}{d\rho}$$

(16.35)

and

$$\frac{dw}{d\rho} = \frac{\rho}{2\pi(1 - \rho^2)^{3/2}}.$$

(16.36)

Next,

$$\frac{d(gh)}{d\rho} = \left(g\frac{dh}{d\rho} + h\frac{d(g)}{d\rho}\right)$$

(16.37)

and then

$$\frac{dh}{d\rho} = -2a_i b_j$$

(16.38)

and

$$\frac{d(g)}{d\rho} = \frac{-d\left[2\left(1 - \rho^2\right)\right]/d\rho}{4(1 - \rho^2)^2} = \frac{-2 \cdot (-2\rho)}{4(1 - \rho^2)^2} = \frac{\rho}{(1 - \rho^2)^2}.$$

(16.39)

Consequently

$$\begin{aligned}
\frac{d(gh)}{d\rho} &= \left(\frac{-2a_i b_j}{2(1 - \rho^2)}\right) + \left(a_i^2 - 2\rho a_i b_j + b_j^2\right)\frac{\rho}{(1 - \rho^2)^2} \\
&= \frac{-a_i b_j(1 - \rho^2)}{(1 - \rho^2)^2} + \frac{a_i^2 \rho - 2a_i b_j \rho^2 + b_j^2 \rho}{(1 - \rho^2)^2} \\
&= \frac{-a_i b_j + a_i b_j \rho^2 - 2a_i b_j \rho^2 + a_i^2 \rho + b_j^2 \rho}{(1 - \rho^2)^2} \\
&= \frac{-a_i b_j(1 + \rho^2) + (a_i^2 + b_j^2)\rho}{(1 - \rho^2)^2}.
\end{aligned}$$

(16.40)

Back substituting Equation 16.40 into Equation 16.35, we obtain

$$\frac{dv}{d\rho} = -e^{-gh}\left(\frac{-a_ib_j(1+\rho^2) + (a_i + b_j)\rho}{(1-\rho^2)^2}\right). \tag{16.35a}$$

Then substituting the new Equation 16.35a and Equation 16.36 into Equation 16.34 and using the definitions for w and v in Equation 16.34 as well, we obtain

$$\frac{d\phi(a_i, b_j)}{d\rho} = -\frac{1}{2\pi(1-\rho^2)^{1/2}}e^{-gh}\left(\frac{-a_ib_j(1+\rho^2) + (a_i + b_j)\rho}{(1-\rho^2)^2}\right)$$

$$+ \exp\left(-\frac{1}{2(1-\rho^2)}\left[a_i^2 - 2\rho a_i b_j + b_j^2\right]\right)$$

$$\times \left(\frac{\rho}{2\pi(1-\rho^2)^{3/2}}\right). \tag{16.34a}$$

Since each of the derivative terms in Equation 16.31 are of the same form as in Equation 16.34a except for the coefficients in a and b, we need only to substitute the appropriate coefficient for a_i and b_j in Equation 16.34a to obtain each of these terms in $du_{ij}/d\rho$ in Equation 16.31. Now all we need to obtain is the second derivative of the log-likelihood function with respect to ρ in Equation 16.30. Substituting Equation 16.28 for t_{ij}, Equation 16.29 for u_{ij}, Equation 16.31 for $du_{ij}/d\rho$, and Equation 16.32 for $dt_{ij}/d\rho$ in Equation 16.30, we now have the second derivative needed in Equation 16.26 to carry out the iterations to solve for $\hat{\rho}$.

Polyserial Correlation

When the researcher has both ordinal polytomous and continuous variables, then he or she must consider the case of obtaining the correlations when ordinal polytomous variables are paired with continuous variables, which they will be in obtaining estimates of the correlation matrix among all the variables. Pearson (1909) generalized the concept of an underlying latent continuous variable from the case of tetrachoric correlation to biserial correlation between a dichotomous variable and a continuous measured variable. He went on to (Pearson, 1913) to generalize the polyserial case in a more restricted context wherein he used the mean of the latent variable between the category thresholds as the score of the latent variable, which was also considered by Jaspen (1946). Cox (1974) finally derived the maximum-likelihood estimate for polyserial correlation. Olsson, Drasgow, and Dorans (1982) reviewed these earlier developments and described the derivation of the polyserial correlation.

Derivation

Our derivation closely follows Olsson et al. (1982). Let Y be an observed polytomous variable and X a continuous measured random variable. As in polychoric correlation, we will assume that the polytomous variable Y has associated with it a continuous latent variable η. Variables Y and η are related in the following way: We assume that there exist $r - 1$ threshold values b_j, $j = 1, \ldots, r - 1$, $b_0 = -\infty$, and $b_r = \infty$, defined on η, such that

$$Y = y_j \quad \text{if } b_{j-1} \le \eta < b_j. \tag{16.41}$$

We further assume that $y_{j-1} < y_j$ and $b_{j-1} < b_j$, which indicates the ordinality of the values of the manifest polytomous variable and the associated thresholds.

Figure 16.2 shows a contour plot of a bivariate normal distribution between η and X. We assume that η has a mean of zero and a variance of unity. Variable X has a mean of m and a variance of σ^2. A regression line of η onto X is shown passing from the lower left to the upper right. The regression line determines for a given value x_i the conditional mean of η given X, $\mu_\eta \mid x_i$. The conditional

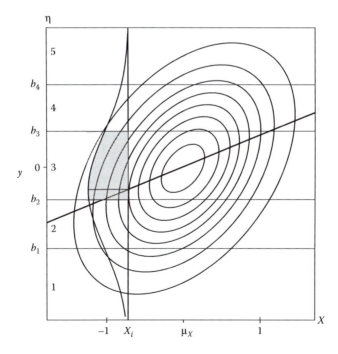

FIGURE 16.2 Contour plot of a bivariate normal distribution between η and X, showing threshold lines and values for discrete Y. Diagonal line is the regression line of η onto X. The conditional distribution of η given x_i is shown. Shaded area equals conditional probability that $Y = 3$ given x_i.

distribution of η given x_i is shown superimposed on the contour plot. The shaded area corresponds to a conditional probability of falling within the thresholds for $y = 3$, given x_i.

We now follow fairly closely the derivation given by Olsson et al. (1982). The joint likelihood of a sample of N observations (x_i, y_i) of X and Y, $i = 1, N$, is given as

$$L = \prod_{i=1}^{N} p(x_i, y_i) = \prod_{i=1}^{N} p(y_i \mid x_i) p(x_i) \qquad (16.42)$$

where $p(x_i, y_i)$ is the likelihood of observing (x_i, y_i) on the ith observation, $p(y_i \mid x_i)$ is the conditional probability of y_i given x_i, and $p(x_i)$ is the unconditional or marginal probability density of x_i (since it is continuous), which is given by

$$p(x_i) = (2\pi\sigma)^{-1/2} \exp\left[-\frac{1}{2}\left(\frac{x_i - \mu}{\sigma} \right)^2 \right]. \qquad (16.43)$$

If we now convert X to a unit normal random variable Z by $Z = (X - \mu)/\sigma$, we can then obtain the conditional probability $p(y_i \mid x_i)$ by considering that the conditional distribution of the latent variable η given x_i has a mean of ρZ and a variance of $(1 - \rho^2)$. Referring now to the cumulative unit normal distribution

$$\Phi(z) = P(Z \le z) = \frac{1}{(2\pi\sigma^2)^{1/2}} \int_{-\infty}^{z} \exp\left[-\frac{1}{2\sigma^2}(t - \mu)^2 \right] dt. \qquad (16.44)$$

Thus, for any $y_i = j$,

$$p(y_i \mid x_i) = \Pr(Y = j \mid x) = \Phi(b_j^*) - \Phi(b_{j-1}^*) \qquad (16.45)$$

where b_j^* and b_{j-1}^* are the thresholds of the conditional distribution of η given x corresponding to the thresholds defining the categories of Y by the equation of transformation

$$b_j^* = \frac{b_j - \rho Z}{(1 - \rho^2)^{1/2}}. \qquad (16.46)$$

In other words, Equation 16.46 transforms the value of the threshold in standard score units on raw η into standard score units of the conditional probability distribution, which has a mean of ρZ and a standard deviation of $(1 - \rho^2)^{1/2}$.

At this point, it is much more convenient to seek to maximize the log joint likelihood

$$l = \log L = \sum_{i=1}^{N} [\log p(y_i \mid x_i) + \log p(x_i)] \qquad (16.47)$$

where "log" refers to the natural logarithms.

We will skip over the case of simultaneously estimating ρ and the threshold parameters, to consider only the two-stage method of conditional maximum likelihood. This method in the first stage, finds estimates of the thresholds directly, computes the estimated mean and variance of the continuous variable X in the usual way, and only estimates ρ iteratively in the second stage while treating the thresholds and mean and variance of X as fixed values.

Let n_k be the number of cases in the sample for which $y_i = k$. Then $P_j = (1/N)\sum_{k=1}^{j} n_k$ is the proportion of cases in the sample where $Y \leq j$. We can now determine values for each b_j:

$$b_j = \Phi^{-1}(P_j), \quad j = 1,\ldots,r-1, \tag{16.48}$$

where $z = \Phi^{-1}(P)$ is the inverse cumulative unit normal distribution function, which should be obtainable by preprogrammed computer subroutines for this function.

To maximize the log joint likelihood function in Equation 16.47, we need to obtain the derivative of this function with respect to ρ, set the derivative to zero, and solve for the value of ρ. The solution is again best obtained using a Newton–Raphson algorithm, which requires both expressions for the first and second derivatives of the log-likelihood function with respect to ρ.

There are two terms in the formula for the log-likelihood function in Equation 16.47, $p(y_i \mid x_i)$ and $p(x_i)$. But only the terms involving $p(y_i \mid x_i)$ are a function of ρ and have nonzero derivatives with respect to ρ. Hence

$$\frac{dl}{d\rho} = \sum_{i=1}^{N} \frac{1}{p(y_i \mid x_i)} \cdot \frac{dp(y_i \mid x_i)}{d\rho}. \tag{16.49}$$

If we let $z_i = (x_i - \hat{\mu})/\hat{\sigma}$, we can now refer to both the cumulative unit normal distribution function and the normal distribution function. Assume that $y_i = j$. Then from Equation 16.45

$$\frac{dp(y_i \mid x_i)}{d\rho} = \frac{d\Phi(b_j^*)}{d\rho} - \frac{d\Phi(b_{j-1}^*)}{d\rho}. \tag{16.50}$$

Then again taking advantage of the fact that $d\left(\int_a^x f(t)\,dt\right)/dx = f(x)$, Olsson et al. (1982) show that

$$\frac{dp(y_i \mid x_i)}{d\rho} = \phi(b_j^*)\frac{db_j^*}{d\rho} - \phi(b_{j-1}^*)\frac{db_{j-1}^*}{d\rho} \tag{16.51}$$

$$= \frac{1}{(1-\rho^2)^{3/2}}\left[\phi(b_j^*)\cdot(b_j\rho - z_i) - \phi(b_{j-1}^*)\cdot(b_{j-1}\rho - z_i)\right], \tag{16.52}$$

where $\phi(b) = (1/(2\pi)^{1/2})\exp(-b^2/2)$, the univariate unit normal distribution density for b. Back substituting now into Equation 16.49, Olsson et al. (1982) obtain

$$\frac{dl}{d\rho} = \sum_{i=1}^{N}\left\{\frac{1}{p(y_i\mid x_i)}\frac{1}{(1-\rho^2)^{3/2}}\left[\phi(b_j^*)\cdot(b_j\rho - z_i) - \phi(b_{j-1}^*)\cdot(b_{j-1}\rho - z_i)\right]\right\}.$$

(16.53)

We are now at a point where we can consider obtaining the second derivative of Equation 16.53 with respect to ρ. Note that we may break down the long expression within the brackets of Equation 16.53 into three terms, each of which is a function of ρ. Then the second derivative takes the form of

$$\frac{d^2l}{d\rho^2} = \frac{d}{d\rho}\sum_{i=1}^{N}\left(\frac{1}{t(\rho)}\frac{1}{u(\rho)}v(\rho)\right)$$

(16.54)

which becomes

$$\frac{d^2l}{d\rho^2} = \sum_{i=1}^{N}\left[\frac{1}{t(\rho)}\frac{1}{u(\rho)}\frac{dv(\rho)}{d\rho} + \frac{1}{t(\rho)}v(\rho)\frac{d}{d\rho}\left(\frac{1}{u(\rho)}\right)\right.$$
$$\left. + \left(\frac{1}{u(\rho)}\right)v(\rho)\frac{d}{d\rho}\left(\frac{1}{t(\rho)}\right)\right].$$

(16.55)

We thus have three derivative expressions that we need to work out:

$$\frac{dv(\rho)}{d\rho} = \frac{d}{d\rho}\left[\phi(b_j^*)\cdot(b_j\rho - z_i) - \phi(b_{j-1}^*)\cdot(b_{j-1}\rho - z_i)\right]$$
$$= \phi(b_j^*)b_j - \phi(b_{j-1}^*)b_{j-1},$$

(16.56)

$$\frac{d}{d\rho}\left(\frac{1}{u(\rho)}\right) = \frac{d}{d\rho}\left(\frac{1}{(1-\rho^2)^{3/2}}\right) = \left(\frac{-3/2(1-\rho^2)^{1/2}(-2\rho)}{(1-\rho^2)^3}\right)$$
$$= \frac{3\rho}{(1-\rho^2)^{5/2}},$$

(16.57)

and

$$\frac{d}{d\rho}\left(\frac{1}{t(\rho)}\right) = \frac{d}{d\rho}\left(\frac{1}{p(y_i\mid x_i)}\right) = \frac{-dp(y_i\mid x_i)/d\rho}{(p(y_i\mid x_i))^2},$$

(16.58)

where $p(y_i \mid x_i)$ is given in Equation 16.45, $dp(y_i \mid x_i)/d\rho$ in Equation 16.52, and

$$t(\rho) = p(y_i \mid x_i), \tag{16.45a}$$

$$u(\rho) = (1 - \rho^2)^{3/2}, \tag{16.59}$$

$$v(\rho) = \left[\phi(b_j^*) \cdot (b_j\rho - z_i) - \phi(b_{j-1}^*) \cdot (b_{j-1}\rho - z_i) \right]. \tag{16.60}$$

So, all that remains is to substitute Equation 16.52, Equations 16.56 through Equation 16.60, and Equation 16.45a into Equation 16.55 to obtain the second derivative.

Computation

We will consider only the Newton–Raphson method, since we have expressions for both the first and the second derivative of the likelihood function for the polyserial correlation, and this method is optimal when feasible.

A close approximation to the final solution is always desired as a starting value for $\rho_{(0)}$. One might use the sample product moment correlation:

$$r_{xy} = \frac{\sum_{i=1}^{N} x_i y_i - (1/N)\sum_{i=1}^{N} x_i \sum_{i=1}^{N} y_i}{\sqrt{\left[\sum_{i=1}^{N} x_i^2 - (1/N)\left(\sum_{i=1}^{N} x_i^2\right)^2 \right]\left[\sum_{i=1}^{N} y_i^2 - (1/N)\left(\sum_{i=1}^{N} y_i^2\right)^2 \right]}}. \tag{16.61}$$

But a better approximation based on the sample product moment correlation is known as the ad hoc *estimator*:

$$\hat{\rho} = \frac{r_{xy} \cdot s_y}{\sum_{j=1}^{r-1} \phi(\hat{b}_j)} \tag{16.62}$$

where

$$s_y = \sqrt{\frac{\sum_{i=1}^{N} y_i^2 - (1/N)\left(\sum_{i=1}^{N} y_i\right)^2}{N}}. \tag{16.63}$$

As a matter of fact, Olsson et al. (1982) report that the ad hoc estimator is itself, surprisingly, a rather accurate estimator. Its RMSE exceeded the RMSE of the maximum-likelihood estimator by less than 0.0058 when $N = 100$ and by less than 0.0040 when $N = 500$ in a Monte Carlo study. The two-stage procedure had a RMSE that never differed from the RMSE of the maximum-likelihood estimator by more than 0.0003 in samples of 500. They note that for all estimators "... a number of general trends are evident ... increasing the sample size decreases the RMSE, increasing the number of categories usually decreases the RMSE, increasing the skew tends to increase RMSE, and increasing ρ decreases the RMSE" (p. 346).

So, let us set $\rho_{(0)}$ equal to the value computed in Equation 16.62. Then the Newton–Raphson algorithm iterates with the following formula:

$$\hat{\rho}_{(j+1)} = \hat{\rho}_{(j)} - \left(\frac{dl^2}{d^2\rho}\right)^{-1}_{(j)} \left(\frac{dl}{d\rho}\right)_{(j)}. \tag{16.64}$$

In each iteration the derivatives are evaluated by substituting in the current value for $\hat{\rho}$. The iterations are halted when $\left|\rho_{(j+1)} - \rho_{(j)}\right| < \varepsilon$, where ε is some very small quantity such as 0.00001.

Evaluation

Computer programs like LISREL compute the thresholds of each of the variables from univariate marginals in a first stage, then compute the matrix of product moment, polychoric and polyserial correlations by maximum-likelihood estimation as a second stage, and then fit the SEM to this matrix in a third stage, using weighted least squares. In programs like EQS the correlation matrix is usually computed automatically when the program detects continuous and categorical variables in the variable mix. The question arises of fit at each of these stages. To what extent is the lack of fit due to the estimation procedure of the correlations, or is it due to the structural model? Maydeu-Olivares (2006) proposes and discusses tests to arrive at answers to these questions. He finds that "... relatively small samples are needed for parameter estimates, standard errors, and structural tests. Larger samples are needed for the distributional and overall tests. Furthermore, parameter estimates, standard errors and structural tests are surprisingly robust to distributional misspecification" (p. 57).

Another problem encountered with biserial, point biserial, polychoric, and polyserial correlations is that the matrices of such coefficients are often not Gramian, meaning they are indefinite, with a few negative eigenvalues. Wothke (1993) discusses the general problems with indefinite covariance matrices and solutions for making them positive definite. He specifically includes a discussion of the correlation matrices involving various correlation coefficients that typically create these problems.

References

Akaiki, H. (1973). Information theory and an extension of the maximum likelihood principle. In B. N. Petrov and F. Csaki (Eds), *Second International Symposium on Information Theory*, pp. 26–281. Budapest: Akademiai Kiado.

Akaiki, H. (1987). Factor analysis and AIC. *Psychometrika, 52*, 317–332.

Anderson, J. C. & Gerbing, D. W. (1988). Structural equation modeling in practice: A review and recommended two-step approach. *Psychological Bulletin, 103*, 411–423.

Anderson, T. W. (1955). Some recent results in latent structure analysis. In *Proceedings of the 1954 invitational conference on testing problems*, pp. 49–53. Princeton, NJ: Educational Testing Service.

Anderson, T. W. (1959). Some scaling models and estimation procedures in the latent class model. In O. Grenander (Ed.), *Probability and Statistics, the Harold Cramer Volume*, pp. 9–38. New York: Wiley.

Arbuckle, J. L. & Wothke, W. (1999). *Amos 4.0 User's Guide*. Chicago, IL: SmallWaters Corporation.

Baars, B. J. (1997). *In the Theater of Consciousness: The Workspace of the Mind*. New York: Oxford University Press.

Bandura, A. (1989). Human agency in social cognitive theory. *American Psychologist, 44*, 1175–1184.

Baron, R. M. & Kenny, D. A. (1986). The moderator–mediator variable distinction in social psychological research: Conceptual, strategic, and statistical considerations. *Journal of Personality and Social Psychology, 51*, 1173–1182.

Bast, J. & Reitsma, P. (1997). Matthew effects in reading: A comparison of latent growth curve models and simplex models with structured means. *Multivariate Behavioral Research, 32*, 135–167.

Bazaraa, M. S. & Shetty, C. M. (1979). *Nonlinear programming: Theory and algorithms*. New York: Wiley.

Bentler, P. M. (1989). *EQS Structural Equations Program Manual*. Los Angeles, CA: BMDP Statistical Software.

415

Bentler, P. M. (1990). Comparative fit indexes in structural models. *Psychological Bulletin, 107*, 238–246.

Bentler, P. M. (1992). *EQS: Structural equations program manual.* Los Angeles: BMDP Statistical Software.

Bentler, P. M. (1993). *EQS structural equations program manual.* Los Angeles, CA: BMDP Statistical Software.

Bentler, P. M. (1994). On the quality of test statistics in covariance structure analysis: Caveat emptor. In C. R. Reynolds (Ed.), *Cognitive Assessment, a Multidisciplinary Perspective*, pp. 237–260. New York: Plenum Press.

Bentler, P. M. (1995). *EQS structural equations program manual.* Encino, CA: Multivariate Software.

Bentler, P. M. & Bonett, D. G. (1980). Significance tests and goodness of fit in the analysis of covariances structures. *Psychological Bulletin, 88*, 588–606.

Bentler, P. M. & Chu, E. J. C. (1995). *EQS for Macintosh User's Guide.* Encino, CA: Multivariate Software.

Bentler, P. M. & Weeks, D. G. (1982). Linear structural equations with latent variables. *Psychometrika, 45*, 289–308.

Bigham, C. C. (1932). *A Study of Error.* New York: College Entrance Examination Board.

Biswas, S. (1991). *Topics in Statistical Methodology.* New York: Wiley.

Blalock, H. M., Jr. (1964). *Causal Inferences in Nonexperimental Research.* Chapel Hill, NC: University of North Carolina Press.

Blalock, H. M., Jr. (1969). Multiple indicators and the causal approach to measurement error. *American Journal of Sociology, 75*, 264–272.

Blumenthal, A. L. (1977). *The Process of Cognition.* Englewood Cliffs, NJ: Prentice Hall.

Bock, R. D. (1989). Measurement of human variation: A two-stage model. In R. D. Bock (Ed.), *Multilevel Analysis of Educational Data*, pp. 319–342. San Diego: Academic Press.

Bock, R. D. & Bargmann, R. E. (1966). Analysis of covariance structures. *Psychometrika, 31*, 507–534.

Bollen, K. A. (1989). *Structural Equations with Latent Variables.* New York: Wiley.

Bollen, K. A. (1990). Overall fit in covariance structure models: Two types of sample size effects. *Psychological Bulletin, 107*, 256–259.

Born, M. (1951). *The Restless Universe.* New York: Dover.

Bozdogan, H. (1987). Model selection and Akaike's information criteria (AIC). The general theory and its analytical extensions. *Psychometrika, 52*, 345–370.

Bozdogan, H. (1990). Theory and applications of information-based measure of complexity ICOMP in Bayesian confirmatory factor analysis. Invited paper presented at the Symposium on Goodness of Fit Procedures at the American Educational Research Association (AERA), Boston, MA, April 16–20.

Bozdogan, H. (2000). Akaike's information criterion and recent developments in information complexity. *Journal of Mathematical Psychology, 44*, 62–91.

Brady, J. E. & Humiston, G. E. (1982). *General Chemistry*, 3rd edition. New York: Wiley.

Breusch, T. & Pagan, A. (1980). The Lagrange-multiplier test and its applications to model specification in econometrics. *Review of Economic Studies, 47*, 239–253.

Brown, T. L. (2003). *Making Truth: Metaphor in Science.* Urbana, IL: University of Illinois Press.

Browne, M. W. (1974). Generalized least-squares estimators in the analysis of covariance structures. *South African Statistical Journal, 8*, 1–24.

Browne, M. W. (1977). Generalized least-squares estimators in the analysis of covariance structures. In D. J. Aigner & A. S. Goldberger (Eds), *Latent Variables in Socio-Economic Models*, pp. 205–226. Amsterdam: North-Holland.

Browne, M. W. (1982). Covariance structures. In D. M. Hawkins (Ed.), *Topics in Multivariate Analyses*, pp. 72–141. Cambridge: Cambridge University Press.

Browne, M. W. (1984). Asymptotically distribution-free methods for the analysis of covariance structures. *British Journal of Mathematical and Statistical Psychology*, 37, 62–83.

Browne, M. W. & Cudeck, R. (1989). Single sample cross-validation indices for covariance structures. *Multivariate Behavioral Research*, 24, 445–455.

Browne, M. W. & Cudeck, R. (1992). Alternative ways of assessing model?. *Sociological Methods & Research*, 21, 230–258.

Browne, M. W. & Cudeck, R. (1993). Alternative ways of assessing model fit. In K. H. Bollen & J. S. Long (Eds), *Testing Structural Equation Models*. Newburg Park, CA: Sage.

Bunge, M. (1959). *Causality: The Place of the Causal Principle in Modern Science*. Cambridge: Harvard University Press.

Byrne, B. M. (1994). *Structural Equation Modeling with EQS and EQS/Windows*. Thousand Oaks, CA: Sage.

Byrne, B. M. & Goffin, R. D. (1993). Modeling MTMM data from additive and multiplicative covariance structures: An audit of construct validity concordance. *Multivariate Behavioral Research*, 28, 67–96.

Campbell, D. T. & Fiske, D. W. (1959). Convergent and discriminant validation by the multitrait–multimethod matrix. *Psychological Bulletin*, 56, 81–105.

Carlson, M. & Mulaik, S. A. (1993). Trait ratings from descriptions of behavior as mediated by components of meaning. *Multivariate Behavioral Research*, 28, 111–159.

Chen, H.-T. (1983). Flowgraph analysis for effect decomposition. *Sociological Methods and Research*, 12, 3–29.

Costner, H. L. (1969). Theory, deduction and rules of correspondence. *American Journal of Sociology*, 75, 245–263.

Cox, N. R. (1974). Estimation of the correlation between a continuous and a discrete variable. *Biometrics*, 30, 171–178.

Cudeck, R. (1988). Multiplicative models and MTMM matrices. *Journal of Educational Statistics*, 13, 131–147.

Cudeck, R. & Browne, M. W. (1983). Cross-validation of covariance structures. *Multivariate Behavioral Research*, 18, 147–167.

Cudeck, R. & Henly, S. J. (1991). Model selection in covariance structures analysis and the "problem" of sample size: A clarification. *Psychological Bulletin*, 109, 512–519.

Copi, I. M. (1978). *Introduction to Logic*. New York: Macmillan.

Dowe, P. (1999). The conserved quantity theory of causation and chance raising. *Philosophy of Science*, 66(Proceedings), S486–S501.

Dowe, P. (2000). *Physical causation*. New York: Cambridge University Press.

Dowe, P. & Noordhof, P. (Eds) (2004). *Causality in an Indeterminate World*. New York: Routledge.

Duncan, O. D. (1966). Path analysis: Sociological examples. *American Journal of Sociology*, 72, 1–16.

Duncan, O. D. (1975). *Introduction to Structural Equation Models*. New York: Academic Press.

Efron, B. & Hinkley, D. V. (1978). Assessing the accuracy of the maximum likelihood estimator: Observed versus expected Fisher information. *Biometrika, 65*, 457–487.

Engle, R. F. (1984). Wald, likelihood ratio, and Lagrange multiplier tests in econometrics. In Z. Griliches & M. D. Intriligator (Eds), *Handbook of Econometrics*, vol. II, pp. 775–826. New York: Elsevier Science Publishers.

Fan, X. & Sivo, S. A. (2005). Sensitivity of fit indexes to misspecified structural or measurement model components: Rationale of two-index strategy revisited. *Structural Equation Models, 12*, 343–367.

Fisher, F. M. (1966). *The Identification Problem in Econometrics.* New York: McGraw-Hill.

Fisher, R. A. (1925). *Statistical Methods for Research Workers.* London: Oliver and Boyd.

Fleishman, E. A. & Hempel, W. W., Jr. (1955). The relation between abilities and improvement with practice in a visual discrimination reaction task. *Journal of Experimental Psychology, 49*, 301–312.

Fletcher, R. (1981). *Practical Methods of Optimization*, Wiley-Interscience, John Wiley & Sons.

Fletcher, R. & Powell, M. J. D. (1963). A rapidly convergent descent method for minimization. *The Computer Journal, 2*, 163–168.

Forster, P. (2001). Scientific inquiry as a self-correcting process. *Digital Encyclopedia of Charles S. Peirce.* http://www.digitalpeirce.fee.unicamp.br/home.htm.

Gibson, J. J. (1950). *The Perception of the Visual World.* Boston: Houghton Mifflin.

Gibson, J. J. (1966). *The Senses Considered as Perceptual Systems.* London: Allen & Unwin.

Giere, R. N. (1988). *Explaining Science: A Cognitive Approach.* Chicago: University of Chicago Press.

Glymour, C., Scheines, R., Spirtes, P., & Kelly, K. (1987). *Discovering causal structure.* San Diego, CA: Academic Press.

Goldberger, A. S. (1964). *Econometric Theory.* New York: Wiley.

Goldfarb, D. (1976). Factorized variable metric methods for unconstrained optimization. *Mathematics of Computation, 30*, 796–811.

Graham, J. W. & Collins, N. L. (1991). Controlling correlational bias via confirmatory factor analysis of MTMM data. *Multivariate Behavioral Research, 26*, 607–629.

Graybill, F. A. (1969). *Introduction to Matrices with Applications in Statistics.* Belmont, CA: Wadsworth Publishing Co.

Guilford, J. P. (1967). *The Nature of Human Intelligence.* New York: McGraw-Hill.

Gustafsson, J. & Balke, G. (1993). General and specific abilities as predictors of school achievement. *Multivariate Behavioral Research, 28*, 407–434.

Guttman, L. (1952). Multiple group methods for common-factor analysis: Their basis, computation, and interpretation. *Psychometrika, 17*, 209–222.

Guttman, L. (1954). A new approach to factor analysis: The radex. In P. F. Lazarsfeld (Ed.), *Mathematical Thinking in the Social Sciences.* Glencoe, IL: The Free Press.

Guttman, L. (1959). A structural theory for intergroup beliefs and attitudes. *American Sociological Review, 24*, 318–328.

Guttman, L. (1965). A faceted definition of intelligence. In R. Eiferman (Ed.), *Studies in psychology, Scripta Hierosolymitana*, vol. 14. Jerusalem, Israel: The Hebrew University.

Hart, B. & Spearman, C. (1913). General ability, its existence and nature. *British Journal of Psychology, 5*, 51–84.

Haughton, D. M. A., Oud, J. J. H. L., & Jansen, A. R. G. (1997). Information and other criteria in structural equation model selection. In *Communications in*

Statistics: Simulation and Computation 1996, Part B. Simulation and computation, pp. 1477–1516. New York: Marcel Dekker.

Hayduk, L. A. (1996). *LISREL©, Issues, Debates, and Strategies.* Baltimore: Johns Hopkins University Press.

Hayduk, L. A. & Glaser, D. N. (2000). Jiving the four-step, waltzing around factor analysis, and other serious fun. *Structural Equation Modeling, 7,* 1–35.

Heck, R. H. & Thomas, S. L. (2000). *An Introduction to Multilevel Modeling Techniques.* Mahwah, NJ: Lawrence Erlbaum Associates.

Heise, D. R. (1969). Problems in path analysis and causal inference. In E. F. Borgatta (Ed.), *Sociological Methodology 1969*, Chapter 2, pp. 38–73. San Francisco, CA: Jossey-Bass.

Heise, D. R. (1975). *Causal Analysis.* New York: Wiley.

Hempel, C. G. (1965). Aspects of scientific explanation. In C. G. Hempel (Ed.), *Aspects of Scientific Explanation and Other Essays in Philosophy of Science*, pp. 331–496. New York: Macmillan.

Herting, J. R. & Costner, H. L. (2000). Another perspective on "The proper number of factors" and the appropriate number of steps. *Structural Equation Modeling, 7,* 92–110.

Hertz, H. (1894). *Die Prinzipien der Mechanik in neuem Zusammenhange dargestellt* (P. Lenard, Ed., Leipzig: J. A. Barth, 1894; D. E. Jones & J. T. Walley, Trans). [*The Principles of Mechanics Presented in a New Form.*] London: Macmillan and Co. 1899; reprinted New York: Dover Publications, 1956.

Hogg, R. V. & Craig, A. T. (1965). *Introduction to Mathematical Statistics*, 2nd edition. New York: Macmillan.

Holzinger, K. J. (1944). A simple method of factor-analysis. *Psychometrika, 9,* 257–262.

Holzinger, K. J. & Swineford, F. (1937). The bifactor method. *Psychometrika, 2,* 41–54.

Hox, J. (2002). *Multilevel analysis.* Mahwah, NJ: Lawrence Erlbaum Associates.

Hu, L. & Bentler, P. M. (1995). Evaluating model fit. In R. H. Hoyle (Ed.), *Structural Equation Modeling: Concepts, Issues and Applications.* Thousand Oaks, CA: Sage.

Hu, L. & Bentler, P. M. (1999). Cutoff criteria for fit indexes in covariance structure analysis: Conventional criteria versus new alternatives. *Structural Equation Modeling, 6,* 1–55.

Hume, D. (1739/1969). *A Treatise of Human Nature.* Baltimore, MD: Penguin. (Original work published in 1739 and 1740.)

Hume, D. (1777/1975). *Enquiries Concerning Human Understanding and Concerning the Principles of Morals.* Oxford: Clarendon. (Original work published in 1777.)

Hunter, J. E. & Gerbing, D. W. (1982). Unidimensional measurement, second order factor analysis and causal models. In B. M. Staw & L. L. Cummings (Eds), *Research in organizational behavior*, vol. IV, pp. 267–320. Greenwich, CT: Jai Press.

Ingmire, A. E., Taylor, C. W., Darragh, R., Geitgey, D., Gross, Y., Orwig, B. I., Popiel, E., Smith, F. L., Smith, H. H., Van Sant, G. E., & Mulaik, S. A. (1967). *The effectiveness of a leadership program in nursing.* Boulder, CO: Western Interstate Commission for Higher Education.

James, L. R. & Brett, J. M. (1984). Mediators, moderators, and test for mediation. *Journal of Applied Psychology, 69*(2), 307–321.

James, L. R. & Jones, A. P. (1980). Perceived job change and job satisfaction: An examination of reciprocal causation. *Personnel Psychology, 33,* 97–135.

James, L. R., Mulaik, S. A., & Brett, J. M. (1982). *Causal analysis: Assumptions, models and data.* Beverly Hills, CA: Sage.

Janik, A. & Toulmin, S. (1973). *Wittgenstein's Vienna.* New York: Simon & Schuster.

Jaspen, N. (1946). Serial correlation. *Psychometrika, 11,* 23–30.

Jaynes, J. (1990/1976). *The Origins of Consciousness in the Breakdown of the Bicameral Mind.* Boston: Houghton-Mifflin & Co. Second Edition with an Afterward. First published in 1976.

Jones, M. B. (1959). *Simplex theory.* Monograph series no. 3. Pensacola, FL: U.S. Naval School of Aviation Medicine.

Jones, M. B. (1960). *Molar correlational analysis.* Monograph series no. 4. Pensacola, FL: U.S. Naval School of Aviation Medicine.

Jones, W. T. (1952). *A History of Western Philosophy.* New York: Harcourt, Brace and Co.

Jöreskog, K. G. (1967). Some contributions to maximum likelihood factor analysis. *Psychometrika, 32,* 443–482.

Jöreskog, K. G. (1969). A general approach to confirmatory maximum likelihood factor analysis. *Psychometrika, 34,* 183–202.

Jöreskog, K. G. (1970). A general model for analysis of covariance structures. *Biometrika, 57,* 239–251.

Jöreskog, K. G. (1971). Simultaneous factor analysis in several populations. *Psychometrika, 36,* 409–426.

Jöreskog, K. G. (1973). A general method for estimating a linear structural equation system. In A. S. Goldberger & O. D. Duncan (Eds), *Structural equation models in the social sciences,* pp. 85–112. New York: Seminar Press.

Jöreskog, K. G. (1974). Analyzing psychological data by structural analysis of covariance matrices. In D. Krantz, R. C. Atkinson, R. D. Luce, & P. Suppes (Eds), *Measurement, Psychophysics and Neural Information Processing,* Vol. II. San Francisco, CA: W. H Freeman & Co., pp. 1–56.

Jöreskog, K. G. (1979). Structural equation models in the social sciences: Specification, estimation and testing. In J. Magidson (Ed.), *Advances in Factor Analysis and Structural Equation Models.* Cambridge, MA: Abt Books.

Jöreskog, K. G. (1994). On the estimation of polychoric correlations and their asymptotic covariance matrix. *Psychometrika, 59,* 381–389.

Jöreskog, K. G. & Sörbom, D. (1979). J. Magidson (Ed.), *Advances in Factor Analysis and Structural Equation Models.* Cambridge, MA: Abt Books.

Jöreskog, K. G. & Sörbom, D. (1981). *LISREL V: Analysis of Linear Structural Relationships by the Method of Maximum Likelihood.* Chicago: National Educational Resources.

Jöreskog, K. G. & Sörbom, D. (1989). *LISREL 7 A Guide to the Program and Applications 2nd Edition.* Chicago: SPSS Inc.

Kant, I. (1790/1987). *Critique of Judgment.* (W. S. Pluhar, Trans., with Forward by Mary J. Gregor). Indianapolis, IN: Hackett Publishing Co.

Kant, I. (1787/1996). *Critique of Pure Reason,* revised edition (W. S. Pluhar, Trans). Indianapolis, IN: Hackett Publishing Co. (First edition published in 1781.)

Kaplan, D. (2000). *Structural Equation Modeling.* Thousand Oaks, CA: Sage.

Kaplan, D. & Wenger, R. N. (1993). Asymptotic independence and separability in covariance structure models: Implications for specification error, power and model modification. *Multivariate Behavioral Research, 28,* 483–498.

Kelly, E. L. & Fiske, D. W. (1951). *The Prediction of Performance in Clinical Psychology.* Ann Arbor, MI: University of Michigan Press.

Kendall, M. G. & Stuart, A. (1969). *The Advanced Theory of Statistics,* vol. 1, 3rd edition. London: Griffin.

Keynes, J. M. (1936). *The General Theory of Employment, Interest and Money*. Cambridge: Cambridge University Press.

Kieseppä, I. A. (2003). AIC and large samples. In S. D. Mitchell (Ed.), *Proceedings of the 2002 biennial meeting of the Philosophy of Science Association*, pp. 1265–1276, Milwaukee, WI, November 7–9, 2002.

Klein, L. R. (1953). *A textbook of econometrics*. Evanston, IL: Row, Peterson.

Klein, L. R. (1969). Estimation of interdependent systems in macroeconometrics. *Econometrica*, 37, 171–192.

Koopmans, T. C. & Hood, W. C. (1953). The estimation of simultaneous linear economic relationships. In W. C. Hood & T. C. Koopmans (Eds), *Studies in econometric method*, Chapter 6. New York: Wiley.

Lakoff, G. (1987). *Women, Fire and Dangerous Things: What Categories Reveal About the Mind*. Chicago: University of Chicago Press.

Lakoff, G. (1993). The contemporary theory of metaphor. In A. Ortony (Ed.), *Metaphor and Thought*, 2nd edition, pp. 202–251. Cambridge: Cambridge University Press.

Lakoff, G. & Johnson, M. (1980). *Metaphors We Live By*. Chicago: University of Chicago Press.

Lakoff, G. & Johnson, M. (1999). *Philosophy in the Flesh*. New York: Basic Books.

Lakoff, G. & Nuñez (2000). *Where Mathematics Comes From. How the Embodied Mind Brings Mathematics into Being*. New York: Basic Books.

Lakoff, G. & Turner, M. (1989). *More Than Cool Reason: A Field Guide to Poetic Metaphor*. Chicago: University of Chicago Press.

Lancaster, H. O. & Hamdan, M. A. (1964). Estimation of the correlation coefficient in contingency tables with possibly nonmetrical characters. *Psychometrika*, 29, 383–391.

Lance, C. E. (1991). Evaluation of a structural model relating job satisfaction, organizational commitment, and precursors to voluntary turnover. *Multivariate Behavioral Research*, 26, 137–162.

Land, K. C. (1969). Principles of path analysis. In E. F. Borgatta (Ed.), *Sociological Methodology 1969*, Chapter 1, pp. 3–37. San Francisco, CA: Jossey-Bass.

Laudan, L. (1980). Why was the logic of discovery abandoned? In T. Nickles (Ed.), *Scientific Discovery, Logic, and Rationality. Boston studies in the Philosophy of Science*, vol. 56. Dordrecht: D. Reidel.

Lazersfeld, P. F. and Henry, N. W. (1968). *Latent Structure Analysis*. Boston: Houghton Mifflin.

Lee, S. & Hershsberger, S. (1990). A simple rule for generating equivalent models in covariance structure modeling. *Multivariate Behavioral Research*, 25, 313–334.

Lewis, D. (1973). Causality. *The Journal of Philosophy*, 70, 556–567.

Lewis, D. (1979). Counterfactual dependence and Time's Arrow. *Noûs*, 13, 445–476.

Lewis, D. (2000). Causation as influence. *The Journal of Philosophy*, 97, 182–197.

Little, T. D., Schanabel, K. U., & Baumert, J. (Eds.) (2000). *Modeling Longitudinal and Multilevel Data*. Mahwah, NJ: Lawrence Erlbaum Associates.

Locke, J. (1694/1905/1962). *Locke's Essay Concerning Human Understanding, 2nd edition, Books II and IV* (M. W. Calkins, Ed.). La Salle, IL: Open Court. (Originally published in 1905 and reprinted in 1962.)

Lohmöller, J. (1989). *Latent Variable Path Modeling with Partial Least Squares*. Heidelberg: Physica-Verlag.

Lord, F. M. & Novick, M. R. (1968). *Statistical Theories of Mental Test Scores*. Reading, MA: Addison-Wesley.

Losee, J. (1980). *A Historical Introduction to the Philosophy of Science*. Oxford: Oxford University Press.

MacCallum, R. C., Wegener, D. T., Uchino, B. N., & Fabrigar, L. R. (1993). The problem of equivalent models in applications of covariance structure analysis. *Psychological Bulletin, 114*, 185–199.

Magnus, J. R. & Neudecker, H. (1988). *Matrix Differential Calculus with Applications in Statistics and Econometrics*. New York: Wiley.

Marsh, H. W. (1988). Multitrait multimethod analysis. In J. P. Keeves (Ed.), *Educational Research Methodology, Measurement and Evaluation: An International Handbook*. Oxford: Pergamon Press.

Marsh, H. W. (1989). Confirmatory factor analyses of multitrait-multimethod data: Many problems and a few solutions. *Applied Psychological Measurement, 13*, 335–361.

Marsh, H. W., Balla, J. R., & McDonald, R. P. (1988). Goodness-of-fit indices in confirmatory factor analysis: The effect of sample size. *Psychological Bulletin, 103*, 391–411.

Marsh, H., Byrne, B. M., & Craven, R. (1992). Overcoming problems in confirmatory factor analyses of MtMM data: The correlated uniqueness model and factorial invariance. *Multivariate Behavioural Research, 27*, 489–507.

Martinson, E. O. & Hamdan, M. A. (1971). Maximum likelihood and some other asymptotically efficient estimators of correlation in two way contingency tables. *Journal of Statistical Computation and Simulation, 1*, 45–54.

Mason, S. J. (1953). Feedback theory: Some properties of signal flow graphs. *Proceedings of IRE, 41*, 1144–1156.

Mason, S. J. (1956). Feedback theory: Further properties of signal flow graphs. *Proceedings of IRE, 44*, 920–926.

Mawson, C. O. S. (Ed.) (1963). *Roget's Pocket Thesaurus*. New York: Pocket Books.

Maydeu-Olivares, A. (2006). Limited information estimation and testing of discretized multivariate normal structural models. *Psychometrika, 71*, 57–77.

McArdle, J. J. (1979). The development of general multivariate software. In J. Hirschbuhl (Ed.), *Proceedings of the Association for the Development of Computer Based Instruction Systems*. Akron, OH: University of Akron Press.

McArdle, J. J. (1980). Causal modeling applied to psychonomic systems simulation. *Behavioral Research Methods and Instrumentation, 12*, 193–209.

McArdle, J. J. & Boker, S. M. (1990). *RAMpath: A Computer Program for Automatic Path Diagrams*. Hillsdale, NJ: Lawrence Erlbaum Publishers.

McArdle, J. J. & McDonald, R. P. (1984). Some algebraic properties of the Reticular Action Model for moment structures. *British Journal of Mathematical and Statistical Psychology, 37*, 234–251.

McDonald, R. P. (1979). The structural analysis of multivariate data: A sketch of a general theory. *Multivariate Behavioral Research, 14*, 221–228.

McDonald, R. P. (1980). A simple comprehensive model for the analysis of covariance structures: Some remarks on application. *British Journal of Mathematical and Statistical Psychology, 33*, 161–183.

McDonald, R. P. (1989). An index of goodness-of-fit based on non-centrality. *Journal of Classification, 6*, 97–103.

McDonald, R. P. & Hartmann W. M. (1992). A procedure for obtaining initial values of parameters in the RAM model. *Multivariate Behavioral Research, 27*, 57–76.

McDonald, R. P. & Marsh, H. W. (1990). Choosing a multivariate model: Non-centrality and goodness-of-fit. *Psychological Bulletin, 107*, 247–255.

McDonald, R. P. & Swaminathan, H. (1973). A simple matrix calculus with applications to multivariate analysis. *General Systems, 18*, 37–54.

Mellenbergh, G. J., Kelderman, H., Stijlen, J. G., & Zondag, E. (1979). Linear models for the analysis and construction of instruments in a facet design. *Psychological Bulletin, 86*, 766–776.

Meredith, W. (1964). Notes on factorial invariance. *Psychometrika, 29*, 177–185.

Meredith, W. (1993). Measurement invariance, factor analysis, and factorial invariance. *Psychometrika, 58*, 525–543.

Meredith, W. (1964a). Notes on factorial invariance. *Psychometrika, 29*, 177–185.

Meredith, W. (1964b). Rotation to achieve factorial invariance. *Psychometrika, 29*, 187–206.

Meredith, W. & Tisak, J. (1990). Latent curve analysis. *Psychometrika, 55*, 107–122.

Michotte, A. (1946/1963). *The perception of causality* (T. Miles & G. Miles, Trans). Louvain: Institute superior de Pholosophie, 1956 edition. New York: Basic Books.

Mill, J. S. (1874). *A system of logic*, 8th edition. New York: Harper.

Monahan, J. F. (2001). *Numerical Methods of Statistics.* Cambridge: Cambridge University Press.

Muirhead, R. J. (1982). *Aspects of Multivariate Statistical Theory.* New York: Wiley.

Mulaik, S. A. (1971). A note on some equations of confirmatory factor analysis. *Psychometrika, 36*, 63–70.

Mulaik, S. A. (1975). Confirmatory factor analysis. In D. J. Amick and H. J. Walberg (Eds), *Introductory Multivariate Analysis.* Berkeley, CA: McCutchan, pp. 170–207.

Mulaik, S. A. (1985). Exploratory statistics and empiricism. *Philosophy of Science, 52*, 410–430.

Mulaik, S. A. (1986). Toward a synthesis of deterministic and probabilistic formulations of causal relations by the functional relation concept. *Philosophy of Science, 53*, 313–332.

Mulaik, S. A. (1987). A brief history of the philosophical foundations of exploratory factor analysis. *Multivariate Behavioral Research, 22*, 267–305.

Mulaik, S. A. (1988). Confirmatory factor analysis. In J. R. Nesselroade & R. B. Cattell (Eds), *Handbook of Multivariate Experimental Psychology*, 2nd edition, pp. 259–288. New York: Plenum Press.

Mulaik, S. A. (1990). An analysis of the conditions under which the estimation of parameters inflates goodness of fit indices as measures of model validity. Paper presented at the Annual Meeting, Psychometric Society, Princeton, NJ, June, 28–30.

Mulaik, S. A. (1993). Objectivity and multivariate statistics. *Multivariate Behavioral Research, 28*, 171–203.

Mulaik, S. A. (1994). Kant, Wittgenstein, objectivity, and structural equations modeling. In Cecil R. Reynolds (Ed.) *Cognitive Assessment: A Multidisciplinary Perspective.* New York: Plenum Press, pp. 209–236.

Mulaik, S. A. (2001). The curve-fitting problem: An objectivist view. *Philosophy of Science, 68*, 218–241.

Mulaik, S. A. (2004). Objectivity in science and structural equation modeling. In D. Kaplan (Ed.), *The Sage Handbook of Quantitative Methodology for the Social Sciences*, pp. 425–446. Thousand Oaks, CA: Sage.

Mulaik, S. A. & James, L. R. (1995). Objectivity and reasoning in science and structural equation modeling. In R. H. Hoyle (Ed.), *Structural Equation Modeling: Concepts, Issues and Applications*, pp. 118–137. Thousand Oaks, CA: Sage.

Mulaik, S. A. & Millsap, R. E. (2000). Doing the four-step right. *Structural Equation Modeling, 7*, 36–73.

Mulaik, S. A. & Quartetti, D. G. (1997). First-order or higher-order general factor? *Structural Equation Modeling, 4*, 193–211.

Mulaik, S. A., James, L. R., Van Alstine, J., Bennett, N., Lind, S., & Stillwell, C. D. (1989). An evaluation of goodness-of-fit indices for structural equation models. *Psychological Bulletin, 105*, 430–445.

Muthén, B. (1989). Latent variable modeling in heterogenous populations. *Psychometrika, 54*, 557–585.

Muthén, B. (1990). *Means and Covariance Structure Analysis of Hierarchical Data.* Los Angeles: UCLA Statistics series, #62.

Niles, H. E. (1922). Correlation, causation and Wright's theory of "path coefficients." *Genetics, 7*, 258–273.

Olsson, O. H., Foss, T., & Breivik, E. (2004). Two equivalent discrepancy functions for maximum likelihood estimation: Do their test statistics follow a non-central chi-square distribution under model misspecification? *Sociological Methods and Research, 32*, 453–500.

Olsson, U. (1979). Maximum likelihood estimation of the polychoric correlation coefficient. *Psychometrika, 44*, 443–460.

Olsson, U., Drasgow, F., & Dorans, N. J. (1982). The polyserial correlation coefficient. *Psychometrika, 47*, 337–347.

Osgood, C. E., Suci, G. J. & Tannenbaum, P. H. (1957). *The Measurement of Meaning.* Urbana-Champaign, IL: University of Illinois Press.

Pearl, J. (2000). *Causality.* Cambridge: Cambridge University Press.

Pearl, J. & Verma, T. (1991). A theory of inferred causation. In J. A. Allen, R. Fikes, & E. Sandewall (Eds), *Principles of Knowledge Representation and Reasoning: Proceedings of the Second International Conference*, pp. 441–452. San Mateo, CA: Morgan Kaufmann.

Pearson, K. (1892). *The Grammar of Science.* Part I: Physical. London: Adam & Charles Black.

Pearson, K. (1895). Contributions to the mathematical theory of evolution, II. *Philosophical Transactions of the Royal Society of London, 186*, 343.

Pearson, K. (1901). Mathematical contributions to the theory of evolution, VII: On the correlation of characters not quantitatively measurable. *Philosophical Transactions of the Royal Society of London, Series A, 195*, 1–47.

Pearson, K. (1909). On a new method for determining the correlation between a measured character A and a character B. *Biometrika, 7*, 96.

Pearson, K. (1911). *The Grammar of Science: Part I. Physical.* London: Adam & Charles Black. (Original work published 1892.)

Pearson, K. (1913). On the measurement of the influence of "broad categories" on correlation. *Biometrika, 9*, 116–139.

Pearson, K. & Pearson, E. (1922). On polychoric coefficients of correlation. *Biometrika, 14*, 127–156.

Peirce, C. S. (1931–1958). *Collected Papers of Charles Sanders Peirce* (Vols 1–8, C. Hartshorne & P. Weiss, Eds; Vols 1–6, A. W. Burks, Ed.; Vols 7–8). Cambridge, MA: Harvard University Press.

Pike, R. W. (1986). *Optimization for Engineering Systems.* New York: Van Nostrand Reinhold.

Popper, K. R. (1961). *The Logic of Scientific Discovery* (translated and revised by the author). New York: Science Editions. (Original work published in 1934.)

Press, W. H., Teukolsky, S. A., Vetterling, W. T., & Flannery, B. P. (1992). *Numerical Recipes,* 2nd edition. Cambridge, MA: Cambridge University Press.

Raftery, A. E. (1986). A note on Bayes factors for log-linear contingency table models with vague prior information. *Journal of the Royal Statistical Society, Series B, 48,* 249–250.

Raftery, A. E. (1993). Bayesian model selection in structural equation models. In K. A. Bollen & J. S. Long (Eds), *Testing Structural Equation Models,* pp. 163–180. Newbury Park, CA: Sage.

Rao, C. R. (1958). Some statistical methods for comparison of growth curves. *Biometrics, 14,* 1–17.

Rao, C. R. (1973). *Linear Statistical Inference and its Applications,* 2nd edition. New York: Wiley.

Reise, S. P. & Duan, N. (Eds) (2003). *Multilevel modeling: Methodological Advances, Issues, and Applications.* Mahwah, NJ: Lawrence Erlbaum Associates.

Rencher, A. C. (2002). *Methods of Multivariate Analysis,* 2nd edition.New York: Wiley.

Rindskopf, D. (1983). Parameterizing inequality constraints on unique variances in linear structural models. *Psychometrika, 48,* 73–83.

Rindskopf, D. (1984). Using phantom and imaginary latent variables to parameterize constraints in linear structural models. *Psychometrika, 49,* 37–47.

Rogers, G. S. (1980). *Matrix Derivatives.* New York: Marcel Dekker.

Russell, B. (1918, 1919). The philosophy of logical atomism. *Monist, 28,* 495–527; *29,* 32–63, 190–222, 345–380.

Sakamoto, Y., Ishiguro, M., & Kitagawa, G. (1986). *Akaike Information Criterion Statistics.* Dordrecht, The Netherlands: Reidel.

Saris, W. E., Satorra, A., & Sörbom, D. (1987). The detection and correction of specification errors in structural equation models. In C. C. Clogg (Ed.), *Sociological Methodology,* pp. 105–129. San Francisco: Jossey-Bass.

Satorra, A. & Bentler, P. M. (1988a). Scaling corrections for chi-square statistics in covariance structure analysis. *Proceedings of the business and economics sections,* pp. 308–313. Alexandria, VA: American Statistical Association.

Satorra, A. & Bentler, P. M. (1988b). *Scaling corrections for statistics in covariance structure analysis* (UCLA statistics series 2). Los Angeles: University of California, Department of Psychology.

Satorra, A. & Bentler, P. M. (1990). Model conditions for asymptotic robustness in the analysis of linear relations. *Computational Statistics and Data Analysis, 10,* 235–249.

Satorra, A. & Bentler, P. M. (1994). Corrections to test statistic and standard errors in covariance structure analysis. In A. Von Eye & C. C. Clogg (Eds), *Analysis of Latent Variables in Developmental Research,* pp. 399–419. Newbury Park, CA: Sage.

Scheines, R., Cooper, G., Yoo, C. W., & Chu, T. J. (2001). Piece-wise linear instrumental variable estimation of causal influence. In *Proceedings of eighth international workshop on artificial intelligence and statistics.* Morgan Kauffman.

Scher, A., Young, A. C., & Meredith, W. M. (1960). Factor analysis of the electrocardiograph. *Circulation Research, 8,* 519–526.

Schlick, M. (1932/1959). Causality in everyday life and in recent science. In E. Sprague & P. W. Taylor (Eds), *Knowledge and Value*, pp. 193–210. New York: Harcourt, Brace.

Schmid, J. & Leiman, J. M. (1957). The development of hierarchical factor solutions. *Psychometrika*, 22, 53–61.

Schmitt, N., Coyle, B. W., & Saari, B. B. (1977). A review and critique of analyses of multitrait–multimethod matrices. *Multivariate Behavioral Research*, 12, 447–478.

Schumacker, R. E. & Marcoulides, G. A. (1998). *Interaction and Nonlinear Effects in Structural Equation Modeling*. Mahwah, NJ: Lawrence Erlbaum Associates.

Schwartz, G. (1978). Estimating the dimension of a model. *Annals of Statistics*, 6, 461–464.

Sen, P. K. (1989). The mean–median–mode inequality and non-central chi square distributions. *Sankhya: The Indian Journal of Statistics*, 51, 106–114.

Shapiro, J. M. & Whitney, D. R. (1967). *Elementary analysis and statistics*. Columbus, OH: Charles E. Merrill Publishing Co.

Simon, H. A. (1952). On the definition of the causal relation. *Journal of Philosophy*, 49, 517–528.

Simon, H. A. (1953). Causal ordering and identifiability. In W. C. Hood & T. C. Koopmans (Eds), *Studies in Econometric Methods*. New York: Wiley.

Simon, H. A. (1977). *Models of discovery*. Dordrecht, Holland: R. Reidel.

Simon, H. A. & Rescher, N. (1966). Cause and counterfactual. *Philosophy of Science*, 33, 323–340.

Singh, J. (1959). *Great Ideas of Modern Mathematics: Their Nature and Use*. New York: Dover Publications.

Sörbom, D. (1989). Model modification. *Psychometrika*, 54, 371–384.

Spirtes, P., Glymour, C., & Scheines, R. (1993). *Causation, Prediction and Search*. New York: Springer.

Spirtes, P., Glymour, C., & Scheines, R. (2000). *Causation, Prediction and Search*, 2nd edition. Cambridge, MA: The MIT Press.

Steiger, J. H. (1989). *EZPATH: A Supplementary Module for SYSTAT and SYGRAPH*. Evanston, IL: SYSTAT.

Steiger, J. H. (1994). SEPATH—A statistica for Windows structural equations modeling program. In F. Faulbaum (Ed.) *Softstat '93: Advances in Statistical Software 4*. Stuttgart: Gustav Fischer Verlag.

Steiger, J. H. (1995). Technical aspects of SEPATH. *Statistica*, pp. 3651–3684. Tulsa, OK: StatSoft.

Steiger, J. H. & Lind, J. C. (1980). Statistically-based tests for the number of common factors. Paper presented at the Annual Meeting of the Psychometric Society, Iowa City, IO.

Steiger, J. H., Shapiro, A., & Browne, M. W. (1985). On the multivariate asymptotic distribution of sequential chi-square statistics. *Psychometrika*, 50, 253–263.

Stelzl, I. (1986). Changing a causal hypothesis without changing the fit: Some rules for generating equivalent path models. *Multivariate Behavioral Research*, 21, 309–331.

Stone-Romero, E. F. & Rosopa, P. J. (2004). Inference problems with hierarchical multiple regression-based tests of mediating effects. In J. J. Martocchio (Ed.), *Research in Personnel and Human Resources Management*, vol. 23, pp. 249–290. San Diego, CA: Elsevier.

Tallis, G. M. (1962). The maximum likelihood estimation of correlation from contingency tables. *Biometrics, 18,* 342–353.

Tanaka, J. S. & Huba, G. J. (1985). A fit index for covariance structure models under arbitrary GLS estimation. *British Journal of Mathematical and Statistical Psychology, 38,* 197–201.

Thurstone, L. L. (1947). *Multiple factor analysis.* Chicago: University of Chicago Press.

Thurstone, L. L. (1949). Note about the multiple group method. *Psychometrika, 14,* 43–45.

Thurstone, L. L. (1951). *An Analysis of Mechanical Aptitude.* Psychometric Laboratory Report No. 62. Chicago: University of Chicago.

Topper, D. R. (1983). Art in the realist ontology of J. J. Gibson. *Synthese, 54,* 71–83. Retrieved July 16, 2007 from http://www.gambrich.co.uk/showdis.php?id=17.

Tryon, R. C. (1939). *Cluster analysis.* Ann Arbor, Mich.: Edwards.

Tucker, L. R. (1958). Determination of parameters of a functional relation by factor analysis. *Psychometrika, 23,* 19–23.

Tucker, L. R. & Lewis, C. (1973). The reliability coefficient for maximum likelihood factor analysis. *Psychometrika, 38,* 197–201.

Turner, M. E. & Stephens, C. D. (1959). The regression analysis of causal paths. *Biometrics, 15,* 236–258.

Twardy, C. R. & Bingham, G. P. (2002). *Perception and Psychophysics, 64,* 956–968.

Van Emden, M. H. (1971). *An Analysis of Complexity.* Amsterdam: Mathematical Centre Tracts 35.

Watkins, E. (2005). *Kant and the Metaphysics of Causality.* Cambridge: Cambridge University Press.

West, S. G., Finch, J. F., & Curran, P. J. (1995). Structural equation models with nonnormal variables: Problems and remedies. In R. H. Hoyle (Ed.), *Structural Equation Modeling: Concepts, Issues and Applications,* pp. 56–75, Thousand Oaks, CA: Sage Publications.

Widaman, K. F. (1985). Hierarchically tested covariance structure models for multitrait–multimethod data. *Applied Psychological Measurement, 9,* 1–26.

Williamson, R. E. & Trotter, H. F. (1979). *Multivariable Mathematics.* Englewood Cliffs, NJ: Prentice-Hall.

Wishart, J. (1928). The generalized product moment distribution in samples from a normal multivariate population. *Biometrika, A20,* 32–52.

Wittgenstein, L. (1922/1978). *Tractatus Logico-Philosophicus.* London: Routledge & Kegan Paul.

Wittgenstein, L. (1953). *Philosophical Investigations.* New York: Macmillan.

Wittgenstein, L. (1975). *Philosophical Remarks* (R. Rhees, Ed.; R. Hargreaves & R. White, Trans). Oxford: Blackwell.

Wold, H. O. A. (1975). Soft modeling by latent variables: The non-linear iterative partial least squares approach. In J. Gani (Ed.), *Perspectives in probability and statistics, Papers in Honour of M. S. Bartlett,* pp. 520–540. London: Academic Press.

Wold, H. O. A. & Jureen, L. (1953). *Demand analysis.* New York: Wiley.

Wolff, P. (2007). Representing causation. *Journal of Experimental Psychology, 136,* 82–111.

Woodward, J. (2001). Causation and manipulability. In E. N. Zalta (Ed.), *The Stanford encyclopedia of philosophy,* Fall 2001 edition. Available at http://plato.stanford.edu/archives/fall2001/entries/causation-mani/.

Woodward, J. (2007). *Making things happen: A theory of causal explanation.* Oxford: Oxford University Press.

Wothke, W. (1984). *The estimation of trait and method components in multitrait–multimethod measurement.* Unpublished doctoral dissertation, Department of Behavioral Science, University of Chicago, Chicago.

Wothke, W. (1993). Nonpositive definite matrices in structural modeling. In K. A. Bollen & J. S. Long (Eds), *Testing structural equation models,* pp. 256–293. Newbury Park, CA: Sage.

Wothke, W. (1996). Models for multitrait-multimethod matrix analysis. In G. A. Marcoulides & R. E. Schumacker (Eds), *Advanced Structural Equation Modelling: Issues and Techniques.* Mahwah, NJ: Lawrence Erlbaum Associates, pp. 7–56.

Wright, S. (1921). Correlation and causation, *Journal of Agricultural Research, 20,* 557–585.

Wright, S. (1923). The theory of path coefficients: A reply to Niles' criticism. *Genetics, 8,* 239–255.

Wright, S. (1931). Statistical methods in biology. *Journal of the American Statistical Association, 26,* 155–163.

Wright, S. (1934). The method of path coefficients. *Annals of Mathematical Statistics, 5,* 161–215.

Wyatt, G. (2004). *Macroeconomic Models in a Causal Framework.* Edinburgh: Harmony House.

Yuan, K. (2008). Noncentral chi-square versus normal distributions in describing the likelihood ratio statistic: The univariate case and its multivariate implication. *Multivariate Behavioral Research, 43,* 109–136.

Yuan, K. H. (2005). Fit indices versus test statistics. *Multivariate Behavioral Research, 40,* 115–148.

Yuan, K. H. & Bentler P. M. (1998). Normal theory based test statistics in structural modeling. *British Journal of Mathematical and Statistical Psychology, 51,* 289–309.

Yuan, K.-H. & Chan, W. (2005). On nonequivalence of several procedures of structural equation modeling. *Psychometrika, 70,* 791–798.

Index